T0200744

Self-Organization in Sensor and Actor Networks

WILEY SERIES IN COMMUNICATIONS NETWORKING & DISTRIBUTED SYSTEMS

Series Editor: David Hutchison, *Lancaster University, UK*
Series Advisers: Serge Fdida, *Université Pierre and Marie Curie, Paris, France*
                   Joe Sventek, *University of Glasgow, Glasgow, UK*

The 'Wiley Series in Communications Networking & Distributed Systems' is a series of expert-level, technically detailed books covering cutting-edge research, and brand new developments as well as tutorial-style treatments in networking, middleware and software technologies for communications and distributed systems. The books will provide timely and reliable information about the state-of-the-art to researchers, advanced students and development engineers in the Telecommunications and the Computing sectors.

Other titles in the series:

Wright: *Voice over Packet Networks* 0-471-49516-6 (February 2001)
Jepsen: *Java for Telecommunications* 0-471-49826-2 (July 2001)
Sutton: *Secure Communications* 0-471-49904-8 (December 2001)
Stajano: *Security for Ubiquitous Computing* 0-470-84493-0 (February 2002)
Martin-Flatin: *Web-Based Management of IP Networks and Systems*, 0-471-48702-3 (September 2002)
Berman, Fox, Hey: *Grid Computing. Making the Global Infrastructure a Reality,* 0-470-85319-0 (March 2003)
Turner, Magill, Marples: *Service Provision. Technologies for Next Generation Communications* 0-470-85066-3 (April 2004)
Welzl: *Network Congestion Control: Managing Internet Traffic* 0-470-02528-X (July 2005)
Raz, Juhola, Serrat-Fernandez, Galis : *Fast and Efficient Context-Aware Services* 0-470-01668-X (April 2006)
Heckmann: *The Competitive Internet Service Provider* 0-470-01293-5 (April 2006)

Self-Organization in Sensor and Actor Networks

Falko Dressler

University of Erlangen, Germany

John Wiley & Sons, Ltd

Other Wiley Editorial Offices

John Wiley & Sons Inc., 111 River Street, Hoboken, NJ 07030, USA

Jossey-Bass, 989 Market Street, San Francisco, CA 94103-1741, USA

Wiley-VCH Verlag GmbH, Boschstr. 12, D-69469 Weinheim, Germany

John Wiley & Sons Australia Ltd, 42 McDougall Street, Milton, Queensland 4064, Australia

John Wiley & Sons (Asia) Pte Ltd, 2 Clementi Loop #02-01, Jin Xing Distripark, Singapore 129809

John Wiley & Sons Canada Ltd, 6045 Freemont Blvd, Mississauga, Ontario, L5R 4J3, Canada

Wiley also publishes its books in a variety of electronic formats. Some content that appears
in print may not be available in electronic books.

Library of Congress Cataloging-in-Publication Data

Dressler, Falko.
    Self-Organization in sensor and actor networks / Falko Dressler.
       p. cm.
    Includes bibliographical references and index.
    ISBN 978-0-470-02820-9 (cloth)
  1. Sensor networks. 2. Self-organizing systems. 3. Biologically-inspired
computing   4. Computer networks–Management–Data processing.   I. Title.
  TK7872.D48D74 2007
  681'2 – dc22

                                             2007028888

British Library Cataloguing in Publication Data

A catalogue record for this book is available from the British Library

ISBN   978-0-470-02820-9 (HB)

Typeset in 10/12pt Times by Laserwords Private Limited, Chennai, India
Printed and bound in Great Britain by Antony Rowe Ltd, Chippenham, Wiltshire
This book is printed on acid-free paper responsibly manufactured from sustainable forestry
in which at least two trees are planted for each one used for paper production.

To Bettina

Contents

V Bio-inspired Networking

Foreword

It seems likely that in the future there will be a strong degree of self-organization in many deployed networked systems. This is an intuitively appealing notion, and one that has fired up the imaginations of researchers who are now actively pursuing research in this area.

Self-organization is still very much an evolving subject – one of the reasons why a new book in the area is most welcome – but there are aspects that have been known for many years, and that derive from the study of Systems in the 1950s, 60s and 70s. General Systems Theory held out the promise of understanding any system, whether natural or man-made, and this was a strong theme in some Universities (indeed, in Systems Departments such as the one at Lancaster University). Much of this work later became embedded in Management Science and in Engineering Departments where 'soft' and 'hard' variants of Systems work have been respectively studied.

One can easily see that ad hoc wireless networks need to self-organize just to get communications under way; they also need to self-reorganize whenever changes occur to the infrastructure. Less clear is the extent to which fixed-line networks including the Internet will in future need to become more self-organizing than they are today. However, the Internet increasingly supports enterprises of many kinds that demand (or at least expect) a very high level of availability, while at the same time experiencing more and more attacks on its infrastructure and services. Therefore, there is a need to find approaches to providing resilience without human intervention and where the system, guided by appropriate policies, learns how to improve its strategies and mechanisms for detection and remediation of the many challenges.

In this book, Falko Dressler sets out to give a general introduction to self-organization and to specialize in sensor and actor networks, where a detailed description is given of the challenges and possibilities offered by self-organization in practice. His text, in five major parts, covers theory through network technologies to the application of self-organization in various types of network. Although it is an advanced text, I believe it is accessible to others who have a general interest in the subject.

This book provides a comprehensive coverage of the principles and practices of network self-organization. It is an excellent addition to the Wiley CNDS Series.

Additional material including lecture slides are available at http://selforg.org

David Hutchison
Lancaster University

Preface

Self-organization is a rather fascinating concept that enables systems consisting of huge numbers of autonomously acting subsystems to perform a collective task. Moreover, self-organizing systems show an overall behavior that cannot easily be predicted or even preprogrammed in a scalable way. It was in the early 1960s that people like Ashby and Eigen investigated self-organization properties in (natural) systems. Since then, a great number of (technical) solutions have been developed, which, either on purpose or unintentionally, inherently formed the basic concepts of self-organization.

The aim of this book is to investigate the concepts of self-organization in the context of autonomous sensor and actor networks. The primary objective is to categorize the basic self-organization methods and to survey techniques for communication and coordination in massively distributed systems according to the developed classification scheme. Basically, two possible approaches can be thought of for organizing this book. First, we could start analyzing sensor and actor network technology and figure out what basic mechanisms are employed and how these relate to self-organization. A second approach would be to introduce self-organization as a methodology, apparently used everywhere in our life (in nature and in technical systems), and afterwards to continue with technical issues in sensor and actor networks, searching for previously learned self-organization methods. I decided to follow the second approach in order to keep the focus on self-organization while studying the term in the world of sensor and actor networks. The term 'self-organization' is still often misunderstood and misinterpreted. Therefore, this textbook is intended to be a basis for a better understanding of the concepts of self-organization, especially in the domain of sensor and actor networks. It provides a stepwise introduction of definitions, methodologies and corresponding techniques relevant in the context of self-organization.

Recent advances in miniaturization and wireless communication enabled the development of low-cost sensor nodes. Additionally, new application domains of sensor and actor networks emerged that demand huge numbers of interacting devices. Thus, the relevance of self-organization methods is rapidly increasing, as it is considered the primary control paradigm for distributed and massively distributed systems. The reader will see that self-organization has a number of advantages compared with other control paradigms. So, it becomes possible to operate huge numbers of collaborating subsystems, even in cases of limited resources, unreliable communication and massive failures of single systems. Unfortunately, these advantages are accompanied by some rather annoying side effects, such as the increasing complexity and a nondeterministic behavior. By using optimal combinations of the basic methods of self-organization, these disadvantages can be minimized to some extent.

According to the objective of this textbook – to study sensor and actor networks – the most relevant domains of communication and coordination are deeply investigated based on well known algorithms and mechanisms and a number of case studies. This includes networking aspects of medium-access control, ad hoc routing, data-centric communication and clustering techniques. Additionally, control mechanisms for cooperation, task and resource allocation, and collaborative actuation are investigated. The book is concluded by a brief introduction to the domain of bio-inspired algorithms. This study is included for two reasons – first, to demystify the term bio-inspired networking, and, secondly, to show the capabilities of such bio-inspired approaches.

What is unique about this textbook?

This book represents the first comprehensive overview of self-organization techniques in the context of wireless sensor and actor networks. It also provides a detailed classification of the basic mechanisms of self-organization. There are many reasons to study self-organization, such as the fascinating effects, which, if correctly understood and employed, provide the possibility of envisioning new kinds of previous system limitations. Additionally, this book is the first comprehensive study of technical solutions focusing particularly on sensor and actor networks.

Audience

This textbook is intended for graduate students, researchers and practitioners who are interested in the broad field of self-organization techniques as well as in application domains in sensor and actor networks. The book is structured to accompany a graduate course in computer science. Thus, some basic knowledge of networking, communication protocols and distributed systems is required. As this textbook provides a global view of algorithms and protocols developed for building self-organizing networking architectures, it can also serve as a reference resource for researchers, engineers and developers working in the field of sensor and actor networks.

Structure and organization

The book is organized into five parts. We start with an introduction to self-organization as a control paradigm for massively distributed systems. Thus, Part I introduces the main ideas and concepts of self-organization. It can be seen as a reference for general studies in the field of self-organization. All relevant terms are introduced, based on examples of natural and technical self-organization. The primary intention is to become familiar with concepts of self-organization and to understand its opportunities and limitations.

Networking aspects relevant to ad hoc and sensor networks are investigated in Part II. This part can be regarded as a broad introduction to algorithms and protocols needed to develop and to maintain wireless sensor networks. Therefore, aspects of protocols for wireless communication, ad hoc routing, data-centric communication and clustering are discussed. The bridge to self-organization methods introduced in Part I is built in the form of permanent discussions of self-organizing aspects of investigated algorithms.

Coordination and control aspects are studied in Part III. This part starts with an introduction to the concepts, challenges and opportunities that emerged with the development of sensor and actor networks. Besides the networking issues discussed in Part II, communication and coordination are relevant for sensor–actor control. Additionally, concepts of task and resource allocation are investigated that allow collaborative executions of complex tasks by autonomously self-organizing systems. Again, the basic self-organization methods are outlined in all analyzed techniques.

Part IV is intended as a conclusion and summary of the investigations in the previous parts of the book. The basic self-organization methods are revisited in the context of the algorithms, techniques and protocols investigated in the context of sensor and actor networks. Additionally, evaluation criteria are discussed that are relevant for estimating the quality and performance of self-organization techniques in sensor and actor networks.

Finally, Part V introduces a very special field that is strongly related to self-organization, namely bio-inspired networking. After a brief introduction to this research domain, the principles and concepts of three selected areas of bio-inspired research are investigated. In all three domains, case studies are depicted that provide solutions for the efficient operation of sensor and actor networks.

Acknowledgments

Since I began working on this textbook on self-organization in sensor and actor networks, many people have given me invaluable help and have been influential in shaping my thoughts on how to organize and teach a course on this topic. I want to thank all these people, including my students, colleagues, faculty members and helpful friends from around the world. Namely, I would like to mention Ian F. Akyildiz and Adam Wolisz, who encouraged me to stay with this topic, Imrich Chlamtac, who shared my visions in the bio-inspired networking domain, Özgür B. Akan, who helped me in various ways and with whom I co-edited a special issue on bio-inspired computing and communication for *Ad Hoc Networks*, Reinhard German, who encouraged me to initiate my studies on integrated robot-sensor networks, and to Bettina Krüger, Isabel Dietrich and Christoph Sommer for proofreading the manuscript. Last, but not least, I wish to thank Birgit Gruber from Wiley, who contacted me at an Infocom conference and convinced me of the idea to write a book on this topic, and all the staff at Wiley (Sarah Hinton, Wendy Hunter, Richard Davies and Joanna Tootill) for their assistance during the writing and the production phase of this book.

Falko Dressler
Erlangen, Germany

About the Author

Falko Dressler is an assistant professor leading the Autonomic Networking Group at the Department of Computer Sciences, University of Erlangen-Nuremberg. He teaches on self-organizing sensor and actor networks, network security and communication systems. Dr Dressler received his M.Sc. and Ph.D. degree from the Dept of Computer Sciences, University of Erlangen in 1998 and 2003, respectively. From 1998 to 2003, he worked at the Regional Computing Center at the University of Erlangen as a research assistant. In 2003, he joined the Computer Networks and Internet group at the Wilhelm-Schickard-Institute for Computer Science, University of Tuebingen. Since 2004, he has been with the Computer Networks and Communication Systems group at the Department of Computer Sciences, University of Erlangen-Nuremberg.

Dr Dressler is an Editor for the *ACM/Springer Wireless Networks* (WINET) journal and Editor for the *Journal of Autonomic and Trusted Computing* (JoATC). He is a guest editor of a special issue on bio-inspired computing and communication in wireless ad hoc and sensor networks for the *Elsevier Ad Hoc Networks* journal. Dr Dressler is General Chair of the 2nd ICST/IEEE/ACM International Conference on Bio-Inspired Models of Network, Information, and Computing Systems (Bionetics 2007). He was co-chair and TPC member of a number of international conferences and workshops organized and sponsored by ACM, ICST, IEEE and IFIP.

Dr Dressler is a member of ACM, IEEE, IEEE Communications Society and IEEE Computer Society. He is actively participating in several working groups of the IETF. His research activities are focused on (but not limited to) Autonomic Networking addressing issues in Wireless Ad Hoc and Sensor Networks, Self-Organization, Bio-inspired Mechanisms, Network Security, Network Monitoring and Measurements and Robotics.

List of Abbreviations

ACK   Acknowledgment

ACO   Ant Colony Optimization

AI   Artificial Intelligence

AIS   Artificial Immune System

ANN   Artificial Neural Network

AODV   Ad Hoc on Demand Distance Vector

ASCENT   Adaptive Self-Configuring Sensor Network Topologies

BEB   Binary Exponential Back-off

CADR   Constrained Anisotropic Diffusion Routing

CBR   Constant Bit Rate

CSMA   Carrier Sense Multiple Access

CTS   Clear To Send

DAA   Dynamic Address Allocation

DAAP   Dynamic Address Allocation Protocol

DACP   Dynamic Address Configuration Protocol

DAD   Duplicate Address Detection

DBTMA   Dual Busy Tone Multiple Access

DEPR   Distributed Event-driven Partitioning and Routing

DHCP   Dynamic Host Configuration Protocol

DINTA   Distributed In-network Task Allocation

DNA   Deoxyribonucleic Acid

DSDV   Destination Sequenced Distance Vector

DSR   Dynamic Source Routing

DYMO   Dynamic MANET on Demand

EA   Evolutionary Algorithm

FCM   Fuzzy C-Means

FEC   Forward Error Correction

FORP   Flow Oriented Routing Protocol

GA   Genetic Algorithm

GEAR   Geographical end Energy-Aware Routing

GPS   Global Positioning System

GPSR   Greedy Perimeter Stateless Routing

HEED   Hybrid Energy-Efficient Distributed Clustering Approach

HSR   Hierarchical State Routing

IDSQ   Information Driven Sensor Querying

IETF   Internet Engineering Task Force

ILP   Integer Linear Programming

IVC   Inter-Vehicle Communication

LEACH   Low-Energy Adaptive Clustering Hierarchy

MAC   Medium Access Control

MACA   Multiple Access with Collision Avoidance

MACA-BI MACA By Invitation

MACAW   Multiple Access with Collision Avoidance for Wireless

MANET   Mobile Ad Hoc Network

MINLP   Mixed Non-Linear Programming

MILD   Multiplicative Increase and Linear Decrease

MRTA   Multi-Robot Task Allocation

NAV   Network Allocation Vector

NHDP   Neighborhood Discovery Protocol

NTP   Network Time Protocol

OAA   Open Agent Architecture

OAP   Optimal Assignment Problem

OLSR   Optimized Link State Routing

PACMAN   Passive Auto-Configuration for Mobile Ad Hoc Networks

PCM   Power-Control MAC

PDAD   Passive Duplicate Address Detection

PEGASIS   Power-Efficient Data-Gathering Protocol for Sensor Information Systems

PLL   Phase-Locked Loop

QoS   Quality of Service

RABR   Route-Lifetime Assessment Based Routing

RBS   Reference Broadcast Synchronization

RERR   Route Error

RFID   Radio Frequency ID

RIP   Routing Information Protocol

RON   Resilient Overlay Network

RREQ   Route Request

RREP   Route Reply

RPF   Reverse Path Forwarding

RSN   Rule-based Sensor Network

RTOS   Real-Time Operating System

RTS   Request To Send

RWP   Random Way Point

S-MAC   Sensor MAC

SANET   Sensor and Actor Network

SI   Swarm Intelligence

SOTIS   Self-Organizing Traffic Information System

TBRPF   Topology Broadcast based on Reverse Path Forwarding

TDM   Time Division Multiplexing

TDMA   Time Division Multiple Access

ToF   Time of Flight

TPSN   Timing-sync Protocol for Sensor Networks

TTL   Time To Live

VANET   Vehicular Ad Hoc Network

WLAN   Wireless LAN

WMN   Wireless Mesh Network

WPDD   Weighted Probabilistic Data Dissemination

WSN   Wireless Sensor Network

ZRP   Zone Routing Protocol

Part I

Self-Organization

Self-organization is one of the most fascinating properties of many natural systems. Often, huge numbers of individual subsystems participate on a common objective. This collaborative behavior can be observed in the form of visible patterns emerging on a higher level. The fundamentals of self-organization are simple algorithms executed by autonomously acting systems, interactions among these systems, and probabilistic decision processes. The objective of this part is to introduce the basic concepts of self-organization and to provide a better understanding of the employed paradigms. Additionally, the basis methods are identified and analyzed. During these discussions, the main focus lies on technical systems with respect to networking, ad hoc networking, wireless sensor networks, and sensor and actor networks.

Outline

- Introduction to self-organization

  Characteristics of self-organizing systems are outlined to enable a common understanding of self-organization. Application scenarios show the broad range of applicability.

- System management and control – a historical overview

  A historical overview of typical system architectures in communications and networking demonstrates the demand for supporting massively distributed systems. The control paradigms changed from centralized to distributed and finally to self-organizing management and control.

- Natural self-organization

  An overview of natural self-organization that should deepen the understanding of self-organization as a control theory. Examples are provided from biology, geology, economy and chemistry. Additionally, the differences between self-organization and bio-inspired solutions are discussed to demystify both terms.

- Self-organization in technical systems

  The general applicability of self-organization in technical systems is discussed, focusing on the following application domains: autonomous systems, multi-robot systems, autonomic networking, mobile ad hoc networks, and sensor and actor networks.

- Methods and techniques

  The basic methods for self-organization, i.e. feedback, interactions and probabilistic behavior, are summarized. This finally leads to a number of design paradigms for self-organizing systems. Additionally, developing and modeling aspects of nature-inspired self-organizing systems are outlined.

- Self-organization – further reading

1

Introduction to Self-Organization

Self-organization is not an invention; nor was it developed by an engineer. The principles of self-organization evolved in nature before we finally managed to study and apply these ideas to technical systems. The first articles on self-organization date back to the early 1960s. Ashby (1962) and von Foerster (1960) analyzed self-organizing mechanisms and Eigen and Schuster (1979) finally made the term 'self-organization' popular in natural and engineering sciences. Meanwhile, various application scenarios for self-organization methods have been identified. Nevertheless, the term 'self-organization' and its context and meaning are still often misused or misunderstood.

One common ambiguity is the essentially different meanings of 'self-organization' and 'bio-inspired'. We will discuss this difference in more detail in Section 4.3. In short, self-organization is a general paradigm for operation and control of massively distributed systems. We can observe many examples of self-organization in nature and many proposed solutions in this domain have their roots in biological mechanisms, and thus can be named bio-inspired. Nevertheless, self-organization is only one example of bio-inspired algorithms. Additionally, not all self-organization techniques are related to bio-inspired research. An introduction to bio-inspired methods in the context of self-organization is provided in Part V.

In the domain of networked computers, different control concepts have been developed. Starting with monolithic, centralized controlled systems, the demand for improved scalability and simplified deployment strategies was growing. The research domain of distributed systems is working on such solutions. Novel approaches lead to control and collaboration paradigms that show the same behavior as that described for self-organizing systems (Gerhenson and Heylighen 2003). Their common objective is to reduce global state information by achieving the needed effects based on local information or probabilistic approaches only. Most of these solutions are using, whether explicitly or implicitly, methodologies similar to biological systems. Even though bio-inspired algorithms represent a novel research domain, first reviews and surveys of such approaches are already available (Dressler 2006b; Prehofer and Bettstetter 2005).

The main goal of this chapter is to analyze the concepts behind self-organization, its applicability and its limitations. Additionally, we develop a comprehensive definition of

Self-Organization in Sensor and Actor Networks   Falko Dressler
© 2007 John Wiley & Sons, Ltd

'self-organization' out of various former approaches. This clarification, together with a summary of properties and capabilities of self-organizing systems, will hopefully lead to a deeper understanding of self-organization in general and its applicability in the context of Sensor and Actor Networks (SANETs) in particular.

1.1   Understanding self-organization

The term 'self-organization' refers to a specific control paradigm for complex systems. Eigen and Schuster (1979) investigated the main concepts behind self-organization in the context of natural – mostly biological – systems and tried to transfer methods and techniques to control mechanisms for technical systems. Basically, he discovered that systems consisting of huge numbers of subsystems need some kind of controlled autonomy that enables a proper functioning in a highly scalable system. Similarly, self-organization mechanisms are required if systems need to face changing environmental conditions or have to adapt to previously unknown application scenarios. If a system is able to fulfill the described requirements, i.e. if it is scalable and adaptive (we will later discuss the meaning of scalability and adaptivity in detail) and if it does not rely on external control, the system is said to be self-organizing.

One of the first definitions of self-organization by Yates *et al.* (1987) characterizes self-organization as follows: 'Technological systems become organized by commands from outside, as when human intentions lead to the building of structures and machines. But many natural systems become structured by their own internal processes: these are the self-organizing systems, and the emergence of order within them is a complex phenomenon that intrigues scientists from all disciplines.' Thus, he already discovered some of the hidden properties of self-organization: completely decentralized control, emerging structures and a high complexity of the overall system.

A particular feature of self-organizing systems was discovered by many researchers in this field: the operation principles always lead to a globally visible effect – the creation of patterns. Depending on the application scenario (we will discuss some basic scenarios in the next subsection), such patterns might be directly visible based on observations of the environment. In other cases, such patterns appear if the system and the environment that it operates in are studied more closely in order to see effects such as the – pattern-like – change of system parameters or environmental conditions.

There are many examples visible in our everyday life. The most cited examples are perhaps the oscillating reactions of the Belousov-Zhabotinskiy reaction (Winfree 1972). The oscillation occurs due to two simultaneously conducted processes: reaction and diffusion. These processes cause a system in which the concentrations of reactants and products oscillate temporally and spatially. These oscillations lead to the creation of spectacular patterns, such as those shown in Figure 1.1.[1]

Based on the described observations, Camazine *et al.* (2003) developed the following definition for self-organization:'Self-organization is a process in which pattern at the global level of a system emerges solely from numerous interactions among the lower-level components of a system. Moreover, the rules specifying interactions among the systems'

---

[1]http://www.swisseduc.ch/chemie/orderchaos/

Figure 1.1 Pattern formation in the Belousov-Zhabotinskiy reaction [Reproduced by permission of Juraj Lipscher]

Table 1.1   Properties of self-organizing systems

Property	Description
No central control	There is no global control system or global information available. Each subsystem must perform completely autonomously.
Emerging structures	The global behavior or functioning of the system emerges in the form of observable patterns or structures.
Resulting complexity	Even if the individual subsystems can be simple and perform basic rules, the resulting overall system becomes complex and often unpredictable.
High scalability	There is no performance degradation if more subsystems are added to the system. The system should perform as requested, regardless of the number of subsystems.

components are executed using only local information, without reference to the global pattern.' This definition takes into consideration that the effect of self-organization, i.e. the result of the interactions between the systems' components, can be described by an observable pattern. Additionally, it explicitly denies global state information that would allow each system to analyze the current (global) pattern and to modify its internal parameterization according to an overall goal.

So far, self-organization was discussed in a fairly sketchy way. In order to understand the meanings of self-organization and how to develop self-organizing systems, a more detailed discussion on principles, mechanisms and operation methods is necessary. Nevertheless, some outstanding properties of self-organizing systems can already be summarized as shown in Table 1.1.

1.2   Application scenarios for self-organization

Self-organization can be found in every stretch of our life. Essentially, self-organization as the key paradigm applies to all aspects of system control. Nevertheless, there are many scenarios in which self-organization is not the optimal solution. Before discussing the limitations of self-organization in a further section, we will try to become more familiar with self-organization and its usual application scenarios.

Basically, it can be said that self-organization can be found in all kinds of system control whenever properties described in Table 1.1 must be considered. Generally, self-organization is always employed for the operation and control of complex systems that consist of numerous subsystems. Such systems can be individuals building a loose group or subsystems that cannot be separated without harming the global system.

Most often cited examples of natural self-organization are swarms of animals (Bonabeau *et al.* 1999; Kennedy and Eberhart 2001), the cellular signaling pathways studied in molecular biology (Alberts *et al.* 1994) and the mammalian immune system (Janeway *et al.* 2001). These natural application scenarios fit very well in the context of self-organization as the key paradigm for operation and control. Swarms of animals such as ants, bees or fishes represent loosely coupled systems of a huge number of subsystems. A control mechanism is required in order to enable the swarm to fulfill global tasks such as foraging. In this context, each individual works on its own intention but collaborates on a global task.

Differently, cells in an organism are tidily coupled. In this context, self-organization can be found in two system aspects. First, the development of the organism is said to be self-organizing because there is no central control that determines which cell has to be created at which time. Secondly, the communication between the cells, i.e. the so-called signaling pathways, is strongly self-organizing. Without a global communication plan such as well known paths, signals are efficiently transmitted to appropriate destinations. This enables the cellular system to perform common tasks without reference to each other.

The mammalian immune system also represents a massively distributed detection device that searches for known and unknown anomalies and attacks against the body. After a successful detection, countermeasures are initiated. All these activities are done without direct reference to a global goal and without expensive coordination. Nevertheless, biology is not the only domain in which natural self-organization can be found. Examples can be found in physics or chemistry as well (Winfree 1972).

As this book aims to support and improve the understanding of self-organization in more technical domains, the application of self-organization should be extended to technical systems. Independently from the specific application scenario, self-organization helps if huge numbers of subsystems are to be managed and controlled and if there is no possible way to provide global state information, e.g. due to limited communication pathways. A well known example that is obviously self-organizing is the Internet. The increasing number of end systems, which are efficiently participating on a loosely coupled compound of autonomous networks, is supported. With yet higher numbers of inter-networked systems and less reliant communication media, the demand for self-organization techniques is increasing. Ad hoc networks, wireless sensor networks, and sensor and actor networks represent such systems. The common objective is to reduce global state information by achieving the needed effects based on local information or probabilistic approaches only. Thus, the inter-networking between huge numbers of devices with possibly limited communication resources increases the demand for well understood self-organization techniques.

Similarly, the robotics community searches for adequate methods to control the behavior of the robot systems. In addition to the requirements for the communication between the individual machines, task and resource allocation algorithms are needed to operate a distributed system consisting of several robots.

2

System Management and Control – A Historical Overview

Technical systems are becoming more complex every year. When focusing on communication and computer systems, usually Moore's Law is cited to characterize the rapidly changing world of technology (Moore 1965). As depicted in Figure 2.1, not only the speed of computer systems strongly increases every year, but also the complexity in the form of subsystems, e.g. transistors, that must be controlled in order to make efficient use of the available resources.

Other examples are the forthcoming achievements in the field of parallel computing. Different domains in computer science and electrical engineering collaborate to enable the usage of tens up to thousands of processors, embedded systems and other computer systems in parallel. This includes novel programming paradigms, changing communication infrastructures and new logics for handling the huge number of resources by introducing abstraction layers such as middleware architectures.

In this section, the need for self-organization as a new paradigm for management and control of modern communication and computer systems is outlined based on a brief historical review. More technical issues such as the application in particular scenarios of SANETs will be discussed in Parts II and III. Before we study the basic architecture of computer systems and the corresponding methods for management and control in the following, a brief introduction to the terms management and control is provided. In the context of this book, we refer to management as the process of administrating systems, performing parameter updates, and verifying the system state. Thus, it is about operating a system or a group of systems. In contrast, control is about feedback and runtime control of the system according to dynamic changes of arbitrary environmental or system-inherent conditions. Together, management and control define the overall process of maintaining, operating, and adapting systems.

Self-Organization in Sensor and Actor Networks   Falko Dressler
© 2007 John Wiley & Sons, Ltd

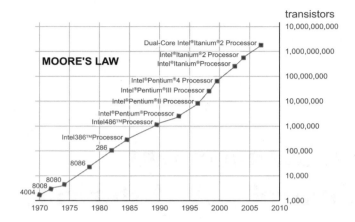

Figure 2.1 Moore's Law depicting the rapid growth of size and complexity of computer systems [Reproduced by permission of Intel]

2.1 System architecture

The architecture of computer systems has changed dramatically since the early beginnings in the middle of the 20th century. The major development stages are shown in Figures 2.2 and 2.3. These figures also depict the corresponding usage profiles, e.g. which systems and resources are used by a user or a cooperating subsystem, respectively.

The first computers have been so large and expensive that people, i.e. users, were required to share such systems. Resources, e.g. processing power and storage space, were limited and careful handling and management were required to enable an efficient and economic use. Only a few individuals were able to access such computer systems and even they had time-limited access.

$n:1$ $1:1$

Figure 2.2 Changing system architecture and usage profiles: mainframes and mini-computers were used by many people simultaneously (left); PCs are usually associated to a single user (right)

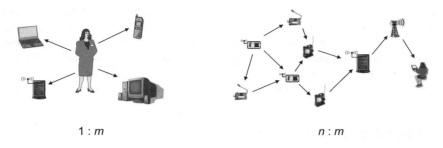

1 : *m*   *n* : *m*

Figure 2.3 Changing system architecture and usage profiles: pervasive computing involves many devices controlled by a single user (left); massively distributed systems involve the interaction between huge numbers of systems (right)

The age of the personal computer changed both the architecture and the usage profiles. Basically, everyone is able to buy and operate his own PC. Starting in the early 1980s, the available resources, such as processing speed, memory and storage, have grown exponentially. Nevertheless, the usage profile stayed the same. A single computer system is operated to perform tasks on behalf of a single person.

Another paradigm change was motivated by the introduction of pervasive computing (Hansmann *et al.* 2003). The development of PDAs, mobile phones and other devices that could be interconnected for several reasons made it necessary to update the architecture of the overall system and the corresponding usage profiles. From now on, not only a single computer system was used by an individual user, but a number of smaller, less powerful devices that form a loosely coupled system performing tasks on behalf of a single person. The pervasive computing domain includes scenarios from wearable devices and home entertainment to more sophisticated security systems. The corresponding control scheme also changed in order to support simplified interconnection of the different systems informally summarized as plug-and-play.

Currently, we are at the border to yet another change towards even higher dynamics and increased numbers of cooperating subsystems. Ubiquitous computing, grid computing and Wireless Sensor Networks (WSNs) are the key players in this development. All these domains share the same usage principles and corresponding architectural features. Ubiquitous computing includes necessary features to identify and to use resources that are hidden in the environment for, usually, specific use in a given scenario (Weiser 1993a,b). On the other hand, grid computing is about to locate available computational resources in a large network consisting of thousands of participating nodes (Foster 2001; Foster and Kesselman 2004), usually PCs or other more powerful systems. Similarly, WSNs provide distributed resources but focus on much smaller subsystems, sensor nodes, with strongly limited resources (Akyildiz *et al.* 2002b). The usage profile mainly changes in terms of communication relationships. While previously the user, i.e. the human, stands in the middle of the control system, now, resources are primarily used by other electronic devices and computer systems. This process can no longer be organized by a human operator but should be performed somehow automatically.

In summary, it can be said that systems are becoming more complex and consist of more subsystems. This includes an increasing heterogeneity of systems as well as of communication pathways. Therefore, the changing system architecture demands updated management and control schemes, focusing on the changing usage profiles. This process finally leads to the development of the principle of massively distributed systems that will be the major system paradigm in the next decade.

2.2 Management and control

Besides the changing system architecture, there is a paradigm change in the management and control of systems. As shown in Figure 2.4, one can see a process of changing the control methodology from monolithic/centralized systems over distributed control to massively distributed/self-organized systems. The motivation for this paradigm change is discussed in the following subsections. Overall, this leads to a strong motivation and demand for further studies and developments in the context of self-organizing technical systems.

2.2.1 Centralized control

'Monolithic' and 'centralized' refer to the first generation of computer systems and the corresponding control method. In the literature, both terms were often used in the same context. This needs to be clarified in the context of management and control. 'Monolithic' means systems consisting of a single computer, its peripherals and perhaps some remote terminals, whereas 'centralized' refers to a single point of control for a group of systems. Therefore, we use the term 'centralized' in the following to describe the corresponding control paradigm.

Figure 2.5 shows the basic principle of centralized control. A well defined control process is responsible for managing all subsystems. This process is pre-configured with the properties, e.g. addresses and capabilities, of all controlled subsystems. Changes in the

- Monolithic / centralized systems
  - Monolithic: systems consisting of a single computer, its peripherals, and perhaps some remote terminals.
  - Centralized: systems with a well defined centralized control process.
- Distributed systems
  - A collection of independent subsystems that appears to the application as a single coherent system.
- Self-organizing autonomous systems
  - Autonomously acting individual systems performing local programs and acting on local data, but participating on a global task, i.e. showing an emergent behavior.

Figure 2.4 Changing paradigms for system management and control

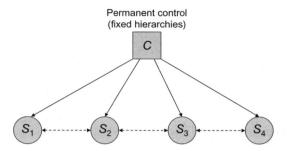

Figure 2.5 Centralized control principle (C depicts the control process and S_i the controlled systems, respectively)

configuration and topology are possible, but involve a complete (manual) reconfiguration of the central control process.

Examples of centralized control can be found in all domains of computer and communication systems. Some examples showing the importance of centralized control are:

- centralized services – e.g. a single server for all users;

- centralized data – e.g. a single on-line telephone book;

- centralized algorithms – e.g. a routing algorithm using global topology information.

Even though other control paradigms have been developed, centralized control is still the preferred solution in some application scenarios. The main advantages are its simplicity and effectiveness. If only a few well known systems have to be managed, then there is no need for high-cost (either time or memory) distributed algorithms or possibly less deterministic self-organization methods.

On the other hand, at least two problems must be discussed that finally demand other control paradigms: transparency and scalability. Transparency refers to more flexible resource specification and scalability to the number of systems under control. This includes also the possibility of dynamic changes to the configuration at runtime. The resulting paradigm for management and control is called 'distributed systems'.

2.2.2 Distributed systems

During the last several decades of distributed systems research, various definitions of distributed systems have been given in the literature. In this book, we will follow the characterization of distributed systems by Tannenbaum and van Steen (2002):

Definition 2.2.1 Distributed system – A distributed system is a collection of independent computers that appear to its users as a single coherent system.

This definition depicts all necessary characteristics of distributed systems. Their internal structure and the differences between the various subsystems and the ways in which they communicate are hidden from the application. The users and applications can interact with

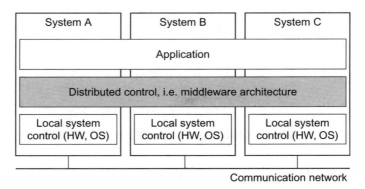

Figure 2.6 System architecture of a distributed system

a distributed system in a consistent and uniform way. A distributed system is assumed to be easily expandable and scalable. In order to support heterogeneous subsystems, distributed systems are usually organized in the form of a 'middleware' (the middleware layer extends over multiple machines), i.e. a software that is logically placed between the application and the operating systems at each participating machine (Tannenbaum and van Steen 2002). This system architecture is shown in Figure 2.6.

As already stated, the centralized control paradigm must not be changed without reason. There are several goals that demonstrate the need for system control and management using the distributed systems principle:

- Resource access – All available resources should be easily accessible from the application as well as from every subsystem. This includes mechanisms to locate and to access resources such as data or processing units. At the same time, interoperability issues should be solved by abstraction layers and open interfaces.

- Transparency – Possibly the most important goal is transparency. A distributed system that is able to present itself to users and applications similar to a single computer system or a centrally controlled system, respectively, is said to be transparent. There are several aspects to be considered when discussing the degree of transparency, as depicted in Table 2.1.

  The quality of a distributed system is described by the degree of transparency. It should be noted that there is always a trade-off between the degree of transparency and system performance.

- Scalability – According to Neuman (1994), the scalability of a distributed system can be measured along at least three different dimensions. First, the size of a system with respect to the number of involved subsystems is one important measure. This includes the possibility to easily add systems and resources to the distributed system without degrading the performance. Secondly, scalability includes the geographically size, i.e. the distance between individual subsystems. For example, the message delay, which is directly proportional to the distance, must be dealt with. Thirdly, and most

Table 2.1   Different forms of transparency in a distributed system (ISO 1995)

Transparency	Description
Access	Hide differences in data representation and how a resource is accessed
Location	Hide where a resource is located
Migration	Hide that a resource may move to another location
Relocation	Hide that a resource may be moved to another location while in use
Replication	Hide that a resource is replicated
Concurrency	Hide that a resource may be shared by several competitive users
Failure	Hide the failure and recovery of a resource
Persistence	Hide whether a (software) resource is in memory or on disk

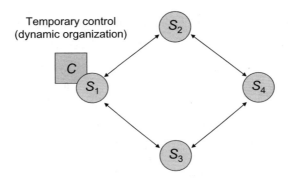

Figure 2.7 Distributed control principle (C depicts the dynamically placed control process and S_i the controlled systems, respectively)

important in our discussion of system control, the manageability must be considered. This includes administrative boundaries as well as the number of interconnected systems and resources.

Figure 2.7 shows the control paradigm as used in distributed systems. A specific control process is dynamically allocated to one of the participating nodes. This process maintains control for at least a set of resources that is needed by the distributed application. In contrast to the centralized control principle, the following three advantages can be achieved:

1. Adaptability – Depending on the resource handling, transparency mechanisms allow the incorporation of new systems and resources that might have been unknown at design time. A well known example is JINI (Keith 2000; Waldo 2000)–JAVA-based, from middleware platform Sun Microsystems. JINI allows each device to bring not only its own interfaces, but also the corresponding Code for accessing them. This opens a great spectrum for developments of distributed applications.

2. Fault tolerance – Two kinds of faults must be dealt with in distributed systems. First, single resource failures can be solved by redundancy – there is basically no difference from centralized control. Secondly, the control process itself can be failing.

Previously, it was stated that the control process is dynamically placed within the network. This dynamic allocation is a perfect source for fault tolerance. Every time the control process fails, it is re-allocated to another network node. Usually, election algorithms are used for this procedure (Garcia-Molina 1982; Singh and Kurose 1994).

3. Scalability – Usually, distributed systems consist of several interconnected computers that are controlled by a middleware. The main differences in the proposed solutions include the (logical) network topology, e.g. mesh networks, rings or hierarchies, and the employed synchronization method, e.g. global synchronization, barriers or task-oriented mechanisms. In this context, scalability is achieved by following a couple of guidelines (Tannenbaum and van Steen 2002):

   • No machine has complete information about the (overall) system state.

   • Machines make decisions based only on local information.

   • Failure of one machine does not ruin the algorithm.

   • There is no implicit assumption that a global clock exists.

Techniques involved in distributed systems include, for example, asynchronous communication, distribution, replication and caching. These mechanisms indeed solve the mentioned problems of centralized control, i.e. transparency and scalability, but may lead to other issues such as consistency problems.

In summary, distributed systems provide a control paradigm that focuses on systems consisting of many subsystems. Based on modern middleware architectures, they allow efficient resource usage with a high grade of fault tolerance. If too many systems have to be controlled or if the communication network does not provide sufficient resources, yet another control paradigm is needed to solve the following problems: impossible synchronization, overhead for resource management and (still) scalability. These issues also illustrate the limitations of current middleware solutions.

2.2.3   Self-organizing systems

As already motivated, a new control theory is required for managing massively distributed systems. Self-organization is discussed as the ultimate solution to the mentioned problems. While pure self-organization comes along with inherent limitation that we will discuss later, complex systems become manageable and the scalability is increased even more. The basic principles of self-organization have been investigated by several researchers for more than 40 years (Dressler 2006b; Heylighen 1999; Prehofer and Bettstetter 2005; von Foerster and Zopf 1962). We will revisit the meanings and mechanisms in more detail in the following section.

Management and control in self-organizing systems are completely distributed, i.e. each participating subsystem has its own control process, as shown in Figure 2.8. Basically, the control paradigm of such systems can be compared to the centralized one. Each system operates on its own, without the need for perfect coordination with all other participating systems. Nevertheless, there must be an additional component that finally ensures the cooperation of all subsystems on a common objective. Several mechanisms for self-organization

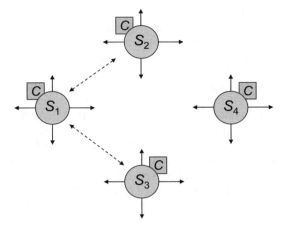

Figure 2.8 Management and control in self-organizing systems (C depicts the local control process and S_i the controlled systems, respectively)

can be distinguished that will be outlined in the following sections. This includes the following features that enable self-organization with respect to the final objective – management and control of complex systems:

- autonomous behavior control;

- loose coupling of subsystems;

- no global state maintenance;

- no (global) synchronization;

- strong dependence on the environment;

- possibly cluster-based collaboration.

Heylighen (1999) summarized the basic properties the enable self-organization as follows (we will later see that this list already considers the most important self-organization techniques and mechanisms):

- Global order from local interactions focuses on the need for collaborations between the individual subsystems in order to achieve a global goal. The observable system behavior represents an impression of all system interactions at a lower level. Correlation can be used as a measure to describe the global system state. In this context, 'local' means interactions between neighboring systems.

- Distributed control refers to the complete absence of any external or internal agent that is responsible for guiding, directing or controlling the system. In contrast to classical distributed systems, not even dynamically chosen control instances are present. All systems contribute evenly to the resulting arrangement. In order to achieve scalability, any centralized control instance must be replaced by some kind of distributed control.

- Robustness and resilience are system characteristics that are important in most scenarios. Additionally, they represent general characteristics of self-organizing systems that are meant to be insensitive to perturbations or errors. The primary reason for this fault tolerance is the redundant, distributed organization. Together with the operation principles of self-organizing systems as depicted in Chapter 6, the level of fault tolerance is inherently increased.

- Non-linearity and feedback can be seen as effect and cause. Most effects in self-organizing systems cannot be described by simple causal relations. This non-linearity can be understood from the relation of feedback that holds between the system's components. Feedback loops not only affect a single destination, but also influence multiple components that again contribute to the feedback. In this context, positive and negative feedback must be distinguished. In short, positive feedback amplifies effects while negative feedback suppresses or counteracts particular effects.

- Organizational closure, hierarchy, and emergence are discussed controversially in the academic community. Organizational closure refers to the closure of the system against influences from the outside. Depending on the viewpoint, the environment has a special function that enables a self-organizing system to behave in a particular way, i.e. it adapts to the environmental conditions. Nonetheless, the environment must not rule the system, i.e. represent a central control. Complex systems tend to have a hierarchical architecture, i.e. 'boxes within boxes'. In this case, the term 'self-organizing system' can easily be misused. Single hierarchies might be self-organizing but relate to other levels that employ other control paradigms. Finally, emergence means that some properties of the self-organizing system cannot be reduced to the properties of its elements. Thus, the overall system can perform more powerful tasks than can be estimated by summing up the actions of the system's components.

3

Self-Organization – Context and Capabilities

The main purpose of this section is to provide adequate definitions of self-organization and related terms, such as emergence and the well known self-X capabilities. Hereby, we concentrate on WSNs and SANETs. Additionally, the term 'self-organization' should be distinguished from manual or external organization and control. We will also discuss the capability to generate globally visible patterns, i.e. the (desired) global goal.

3.1 Complex systems

Self-organizing systems are meant to be dynamic and structures emerge through interactions of the system's components, e.g. by obeying nonlinear kinetics. Such structures are called dissipative. The complex systems theory looks exactly for systems showing such a nonlinear behavior. Therefore, we start our discussion with a short review of complex systems research.

The complexity of the world is contrasted with the simplicity of the basic laws of physics. In recent years, considerable study has been devoted to systems that exhibit complex outcomes. This experience has not given us any new laws of physics, but has instead given us a set of lessons about appropriate ways of approaching complex systems (Goldenfeld and Kadanoff 1999). To us, complexity means that we have structure with variations.

Natural complex systems are modeled using the mathematical techniques of dynamical systems, which include differential equations, difference equations and maps. Because they are nonlinear, complex systems are more than the sum of their parts. Most biological systems are complex systems in the sense outlined above, while, traditionally, most humanly engineered systems are not. A proper definition of complex systems can be deduced from these observations:

Definition 3.1.1 Complex system – The term complex system formally refers to a system of many parts which are coupled in a nonlinear fashion. A linear system is subject to the principle of superposition, and hence is literally the sum of its parts, while a nonlinear system is not. When there are many nonlinearities in a system (many components), its behavior can be as unpredictable as it is interesting.

The literature on nonlinear systems often mentions self-organization, emergence and complexity (Kauffman 1993). All these terms are relative and must be discussed in the context of this book (the terms 'self-organization' and 'emergence' will be defined in the following section). Camazine *et al.* (2003) depicted the meaning of complexity in a way that generates the need for system control, even in the case of complex systems. Individual systems may use relatively simple behavioral rules to generate structures and patterns at a global level that are more complex than the components and the processes from which they emerge. According to Camazine *et al.* (2003), systems are complex, not because they involve many behavioral rules and large numbers of different systems, but because of the nature of the system's global response.

Complexity and complex systems generally refer to a system of interacting units that display global properties not present at lower levels. These systems may show diverse responses that are often sensitively dependent on both the initial state of the system and the nonlinear interactions among its components. Since these nonlinear interactions involve amplifications or cooperativity, complex behavior may emerge, even though the system components may be similar and follow simple rules. Complexity in a system does not require complicated components or numerous complicated rules of interaction.

In the direct context of self-organization, complexity theory is often discussed as follows:[1] Critically interacting components self-organize to form potentially evolving structures exhibiting a hierarchy of emergent system properties. The elements of this definition are summarized in Table 3.1.

We will see that the research on complex systems notably overlaps with self-organization. Thus, we need management and control mechanisms for dynamic, highly scalable and adaptive complex systems. Self-organization is seen as the paradigm for this challenge.

Table 3.1   Elements and properties related to complexity theory

Element	Description
Critical interaction	The system is information-rich, neither static nor chaotic
Components	Modularity and autonomy of part behavior implied
Self-organization	The attractor structure is generated by local contextual interactions
Potential evolving	Environmental variation selects and mutates attractors
Hierarchy	Multiple levels of structure and responses appear (hyper-structure)
Emergent system properties	New features are evident which require a new vocabulary

[1] Self-Organizing Systems (SOS) FAQ – http://www.calresco.org/sos/sosfaq.htm

3.2 Self-organization and emergence

The characteristics of self-organizing systems include their capability to exhibit emergent properties. Emergence refers to a process by which a system of interacting subunits acquires qualitatively new properties that cannot be understood as the simple addition of their individual contributions. Since these system-level properties arise unexpectedly from nonlinear interactions among a system's components; the term 'emergent property' may suggest to some a mysterious property that materializes magically.

There are two primary terms that characterize and form the shape of this book: 'self-organization' and 'emergence'. Unfortunately, there is no single, commonly agreed definition of these terms available. Therefore, we will elaborate our own definition that – in the case of self-organization – will also lead to some classification of the different meanings of self-organization.

Definition 3.2.1 Self-organization – Self-organization is a process in which structure and functionality (pattern) at the higher level of a system emerge solely from numerous interactions among the lower-level components of a system without any external or centralized control. The system's components interact in a local context either by means of direct communication or environmental observations, and, usually without reference to the global pattern.

Figure 3.1 depicts the most important properties of a self-organizing system. A number of individual systems are working in a given environment. Each component operates on its local information that represents the neighboring systems and the environment. Based on a set of simple rules, the system participates in a global goal. There is no particular component that can directly refer to the global objective or pattern.

This definition of 'self-organization' focuses on the emergence of patterns. Similar definitions can be found in the literature concerning well studied methodologies in biological

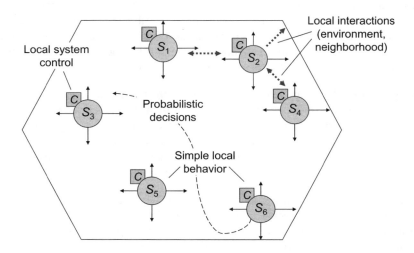

Figure 3.1 Properties of self-organizing systems

systems (Ashby 1962; Camazine *et al.* 2003) or in technical systems such as computer networks (Dressler 2006b; Prehofer and Bettstetter 2005). The interaction of single components finally defines the behavior of the global system. Applied to computer networks, self-organization can be seen as the interactions between nodes in the network leading to globally visible effects, e.g. the transport of messages from a source node to a sink node.

Since we speak of the emergence of pattern or system behavior, the term emergence needs to be defined as well. In short, emergence refers to the appearance of a property of feature not previously observed as a functional characteristic of the system or a system's component. Generally, higher level properties are regarded as emergent. Nevertheless, this does not include aggregates of lower level functionalities and emergent properties are not epiphenomenal, i.e. neither illusions nor descriptive simplifications only. This means that the higher-level properties should have causal effects on the lower-level ones. Based on these characteristics of emergent behavior, the following definition concludes the term emergence.

Definition 3.2.2 Emergence – Emergent behavior of a system is provided by the apparently meaningful collaboration of components (individuals) in order to show capabilities of the overall system (far) beyond the capabilities of the single components.

One striking property of self-organization – corresponding to the provided definition – is the missing determinism of the involved algorithms. Each solution can be only evaluated by means of particular environmental conditions that influence the systems' behavior. Often, probabilistic methods are employed for building the solutions that make up the local decisions together with direct or indirect communications.

Before we discuss the mechanisms behind self-organization and emergence and the corresponding design paradigms, the best known properties and features need to be summarized and explained:

- Absence of external control – As already depicted, one of the most important characteristics of self-organization is the completely distributed control. Each participating system acts on local decisions, i.e. we speak of autonomous systems collaborating on a global objective. Even if this objective might be known by each component, it is not possible to review the current global state and act accordingly.

- Adaptation to changing conditions – An inherent feature of self-organizing systems is their capability to adapt to changing environmental conditions. This is a direct result of the distributed working principle. Self-organization incorporates local interactions with the environment and the system's behavior is regulated according to the observations. If the environment changes too rapidly or if the modifications are out of a tolerance range, fluctuations and instability might occur to the overall system.

- Global order and local interactions – The emergence of a global order or pattern from local interactions is another feature that makes a control process self-organizing. The local interactions between the system's components replace coordination paradigms such as global state information or precise synchronization. The differentiating factor (sometimes called the degree of self-organization) is the definition of the term 'local' in the considered system.

- Complexity – Complexity is not only an inherent property of self-organizing systems as discussed before. Additionally, complexity arises out of the global demands and goals. Usually, multiple concurrent values or objectives need to be considered simultaneously. We also speak of 'multi-objective optimization', which is a research domain on its own. Interestingly, the search for optimal results in a self-organizing system might become easier to achieve due to the system-inherent fluctuations and instabilities.

- Control hierarchies – Self-organizing systems are often arranged in multiple nested self-organized levels. Such control hierarchies allow the more efficient specification or description of self-organization methods and techniques. In general, there is no requirement to perform management and control on all hierarchies based on self-organization paradigms. Nevertheless, if other mechanisms are employed – let's assume a centralized control in the highest system level – then only the lower levels can be described with the features discussed in the context of self-organization.

- Dynamic operation – Self-organization is a dynamic process. It evolves over a period of time and, usually, does not hold similarly to other algorithms with an optimal solution. Instead, self-organization processes can be seen as search algorithms looking for the best system state according to the current environmental conditions. As these conditions are subject to change over time, e.g. through the system's interactions themselves, the observable dynamics are increased even more.

- Fluctuations and instability – The mentioned dynamics of self-organizing systems occur, for example, through fluctuations and instabilities. Fluctuations refer to noise and randomized selections of behavioral options. Analogically, instabilities occur through nonlinearities or self-reinforcing choices.

- Dissipation – It seems to be obvious that self-organization rarely leads to optimal solutions. We will discuss this issue in detail later. For the moment, we assume that in most cases, suboptimal results for the selected system behavior can be found. Thus, this will always lead to dissipation compared with other algorithms. Examples are energy usage and operations far-from-equilibrium.

- Multiple equilibria and local optima – Multiple equilibria refer to the presence of many possible attractors that affect the local decision processes in the system's components. Additionally, every multi-objective optimization problem can lead to local optima. Such cases are hard to detect and may represent rather bad solutions. Based on randomization and the self-organization inherent features of fluctuation and instability, such local optima can be abandoned.

- Redundancy – We have already discussed one of the main features of self-organizing systems – the inherent tolerance to faults or the system's redundancy. If system's components become unavailable, e.g. due to failures and damage, other components take over the work. This refers to some degree of insensitivity to damage. This degree finally illustrates the ability of a system to tolerate faults and to the resulting quality. Generally, the degree of redundancy is one of the most important quality of service measures, e.g. in communication networks, the reliability of a network connection or the ability of the application to tolerate packet loss.

- Self-maintenance – Self-maintenance is often used in direct relation to, and sometimes even in the same sense as, redundancy. Nevertheless, redundancy means the ability to tolerate faulting systems, whereas self-maintenance refers to the ability to repair or to reproduce these systems.

3.3   Systems lacking self-organization

The given definition of self-organization allows the determination of which systems can be called self-organizing and which cannot. Certainly, not all patterns arise through self-organization and even those primarily based on self-organization may involve other mechanisms as well. In this section, we will elaborate two of the best known examples that are often mixed up with self-organization: external organization and blueprints or templates.

3.3.1   External control

External control or organization is generally referred to as the presence of a well informed supervisory leader. This leader directs the activities of the system's components by providing each system with detailed instructions about what to contribute to the overall goal. By contrast, a group of systems that builds a pattern by means of self-organization needs no supervision by a leader.

Camazine *et al.* (2003) suggest the following procedure to verify whether a system is self-organizing or not. One should always look for key individuals that may play a leadership role in the organization of the system, i.e. that influence the system's behavior according to the final goal. The discovery of such leaders provides some evidence that the overall system is not self-organized. Likewise, the discovery of interactions between the system's components supports the assumption that the system might be self-organized. These assumptions can be evaluated in an experiment. One has to manipulate the system by removing the potential leader or by blocking interactions between the components in order to test whether the pattern can be generated under these altered conditions.

The open question is, how can a leader or any other kind of external control influence a self-organizing system? This refers to the control paradigms for self-organized systems. In short, external control can operate a fully self-organized system by manipulating the environmental conditions. Usually, self-organized systems will directly interact with the surrounding environment. Thus, appropriate changes somewhat control the system behavior. Such control mechanisms are called 'external control'.

3.3.2   Blueprints and templates

Blueprints or recipes predefine directives for all individual systems participating at the global goal. Basically, such blueprints are compact representations of the spatial or temporal characteristics of the global pattern. Systems using blueprints or recipes need to integrate mechanisms for synchronization (either between groups of systems or even between all systems) and for adjustments. Synchronization is required if multiple systems collaborate on a particular task and will be reassembled into other groups after it was finished. Adjustments are necessary if tolerances in the work of individual systems might happen and must be accepted.

The main difference between blueprints and recipes is that blueprints specify the goal, i.e. what has to be created, while recipes provide step-by-step instructions on how the system has to be built. In conclusion, we refer to directives for a desired system behavior based on blueprints or recipes.

Similarly, preexisting patterns or templates restrict the system to certain levels of freedom and flexibility. A template usually specifies the global goal with some degree of uncertainty. Compared with blueprints and recipes, a template can be seen as an incomplete blueprint, describing only parts of the global objective. Therefore, templates define a gray area between pure self-organizing systems and systems organized by means of external or internal control.

3.4 Self-X capabilities

Often, the term 'self-organization' is used in conjunction with other mechanisms – the so-called 'self-X capabilities'. Terms such as self-management, self-protection, self-repair and self-optimization actually haunt through several academic communities, providing buzzwords to relate latest research results to. It is not our intension to evaluate the meaning or the quality of all these mechanisms. Nonetheless, a short summary should be provided in order to integrate the features into the context of self-organization or to distinguish self-organization from them, respectively. Table 3.2 summarizes the best known self-X capabilities.

Some selected features qualify for a more detailed discussion due to their importance in the context of self-organization. 'Self-configuration' can be seen as the process of local system management based on observations of the environment as well as on interactions

Table 3.2   Selected well known self-X capabilities

Feature	Description
Self-configuration	Methods for (re-)generating adequate configurations, depending on the current situation in terms of environmental circumstances
Self-management	Capability to maintain systems and devices, depending on the current system parameters
Adaptability	Ability of the system's components to adapt to changing environmental conditions
Self-diagnosis	Mechanisms to perform system autonomous checks and to compare the results with reference values
Self-protection	Capability to protect the system and its components against unwanted or even aggressive environmental influences
Self-healing	Methods for changing configurations and operational parameters of the overall system to compensate failures
Self-repair	Techniques similar to self-healing but focusing on actual repair mechanisms for failing system parts
Self-optimization	Ability of the system to optimize the local operation parameters according to global objectives

with other systems. Environmental conditions include, for example, the connectivity between neighboring systems and other quality-of-service parameters. Additionally, direct interactions with the environment, such as measurements or actuation results, can result in the demand for configuration updates. 'Self-management' is often used in the meaning of self-configuration. In general, self-management can be seen not as a technique, but as an inherent feature or a characteristic of self-organizing systems. Each component must be able to maintain and control itself according to the current local system state. In this context, 'adaptability' refers to the capability of the overall system to adapt to changing environmental conditions, e.g. the operation parameters or the changing number of neighboring nodes. Together, self-management (local system state maintenance), self-configuration (adaptation of the system's components) and adaptability (adaptation of the overall system) enable the autonomous behavior of several systems participating together on a global goal.

So far, management and control in normal conditions have been targeted. Two additional mechanisms are needed in order to make efficient use of massively distributed systems. 'Self-healing' stands for the ability of the system to tolerate faults. Robustness and resilience are strongly required in systems lacking central control. Basically, we consider mechanisms that allow the detectection, localization, and repair of failures automatically. Such mechanisms can be primarily distinguished by the cause of the failure, e.g. breakdown, overload or malfunction, and the preferred solution, e.g. reconfiguration, replacement or repair. Finally, 'self-optimization' capabilities are needed to optimize the operational parameters of the system's components according to the overall objective by choosing optimal methods and their parameters. The term 'optimal' always reflects the quality of the solution as requested by the application scenario, i.e. the collective efforts of all participating systems.

While there are definitely aspects covered by particular self-X capabilities that are not addressed by self-organization, e.g. antivirus software is self-protecting a PC or modern hard discs employ self-diagnosis techniques – both examples do not refer to self-organization. Nevertheless, we will summarize all the self-X terms in the context of this book under the term 'self-organization'.

3.5   Consequences of emergent properties

Emergence and the corresponding emergent properties can have important consequences for self-organizing systems. As previously discussed, emergence reflects nonlinearity in behavioral reactions of self-organizing systems. Amplification effects and sensibility to noise finally lead to the primary consequence of emergent properties: a small change in a system parameter can result in a large change in the overall behavior of the system. The advantage of this behavior is the introduced adaptability and flexibility of self-organizing systems.

These kinds of emergent properties obviously leads to the question of whether adaptability to environmental changes can be handled by simple rules. If small changes may have such strong effects, how can a system find a stable state? The role of environmental factors reflects the behavioral patterns caused by some specific initial conditions. Positive feedback as used by most self-organization methods results in great sensitivity to these conditions.

The solution lies in the use of suppressing factors that counteract uncontrolled amplification leading to chaotic behavior. Negative feedback is used for this purpose. Feedback mechanisms, both positive and negative, are explained in detail in Chapter 6.

As an example, we examine the growth rate of a certain population of non-overlapping generations such as studies for many species of insect. Often, the logistic difference equation or quadratic recurrence equation is used to model the growth rate:

$$x_{n+1} = rx_n(1 - x_n). \tag{3.1}$$

This equation, where r (sometimes also denoted μ) is a positive constant known as the 'biotic potential' gives the so-called logistic map. This quadratic map is capable of very complicated behavior. While John von Neumann had suggested using the logistic map $x_{n+1} = 4x_n(1 - x_n)$ as a random number generator in the late 1940s, it was not until work by W. Ricker in 1954 and detailed analytic studies of logistic maps beginning in the 1950s with Paul Stein and Stanislaw Ulam that the complicated properties of this type of map beyond simple oscillatory behavior were widely noted (Wolfram 2002).

Camazine *et al.* (2003) discuss the parameters of Equation 3.1 in detail. Here, only the applicability and resulting effects (especially the relevance of the parameter r) are explained. x_n denotes the population size of the n-th generation. The population size is scaled to vary between 0 (no individuals) and 1 (the maximum number of individuals). r corresponds to the productive rate so that rx_n refers to the next-generation population size in absence of any limiting factors. Negative feedback is introduced in the form of the term $(1 - x_n)$. It also makes the system nonlinear. In conclusion, $(1 - x_n)$ limits the population growth as it nears its carrying capacity, because as x_n approaches 1, the factor $(1 - x_n)$ approaches 0.

Figure 3.2 shows the system behavior for varying r. As explained in Wolfram (2002), the population will extinct for $0 \leq r < 1$, independently of the initial population size x_0. A constant population size will be reached after a number of generations for $1 \leq r < 3$.

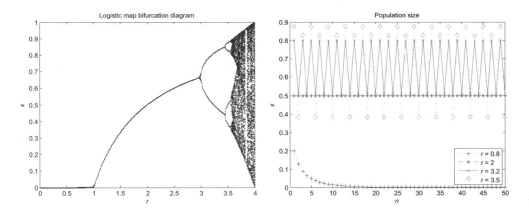

Figure 3.2 Population growth model illustrating the influence of small parameter changes on the resulting pattern. The bifurcation diagram depicts the dependence on the parameter r (left), while the other plot shows the evolution of the population size over the time (right)

Values greater than $r > 3$ will lead to multi stability, i.e. x_n will oscillate between two or four population sizes. Beyond $r > 3.57$, deterministic chaos prevents a stable population size.

Due to emergence and emergent properties, self-organization always leads to the evolution of patterns and structure. Intuitively, we can speak about the generation of adaptive structures and patterns by tuning system parameters in self-organized systems rather than by developing new mechanisms for each new structure. Nonetheless, the presented example has shown that this 'parameter tuning' is a complex and sometimes unpredictable process. In conclusion, it can be said that the concept of self-organization offers the possibility that strikingly different patterns result from the same mechanisms operating in a different parameter range. Thus, emergence is consistent with simple rules and resulting complex patterns – the solution to a paradox.

3.6   Operating self-organizing systems

Self-organization and emergence enable a new kind of operation and control in completely autonomous or massively distributed systems, respectively. So far, we discussed the main idea of self-organization as well as the occurring nonlinearities in the pattern formation. Thus, we have to deal with non-deterministic behavior of sell-organizing systems. So the question on controllability arises. If we speak about control, we do not mean a centralized supervisory control, as depicted in Section 3.3. So the precise question is 'Can we make control systems self-organizing?' or 'Can we make self-organization controllable?' The preliminary answer is 'Yes, but . . . '. If we give up on deterministic behavior, i.e. using self-organization as the control paradigm, we'd have to insert the equivalent of Asimov's Laws of Robotics, i.e. specifically disallow certain harmful characteristics or behaviors, or to be satisfied with a certain reaction/output quality.

3.6.1   Asimov's Laws of Robotics

Asimov saw a promising technological innovation – the development of autonomous robots – to be exploited and managed. Asimov's intent was to devise a set of rules that would provide reliable control over semi-autonomous machines. In order to control them, he created the following set of rules, known as the Laws of Robotics (Asimov 1968):

- *First Law:* A robot may not injure a human being, or, through inaction, allow a human being to come to harm.

- *Second Law:* A robot must obey orders given it by human beings, except where such orders would conflict with the First Law.

- *Third Law:* A robot must protect its own existence as long as such protection does not conflict with the First or Second Law.

Clarke (1993) discussed the process of creation and development of Asimov's Laws from an engineering point of view. His goal was to determine whether such an achievement is likely or even possible in the real world. In the process, he focused on practical, legal and ethical matters that may have short- or medium-term implications for practicing

Table 3.3   Key features exhibited by robot systems

Feature	Description
Programmability	Implying computational or symbol-manipulative capabilities that a designer can combine as desired (a robot is a computer)
Mechanical capability	Enabling it to act on its environment rather than merely function as a data-processing or computational device (a robot is a machine)
Flexibility	Allowing it to operate using a range of programs and manipulate and transport materials in a variety of ways

information technologists. The basis of his investigations was the following definition of robots (Chandor 1985). As can be seen, its main characteristics are closely related to self-organizing systems. The three key elements exhibited by robot systems are summarized in Table 3.3.

Asimov's early stories discussed several problems that relate to the applicability of the three laws. Clarke (1993) summarized and explained these problems by evaluating real world robot systems:

- The ambiguity and cultural dependence of terms – A robot must be able to distinguish robots from humans. This includes another critical question: how does a robot interpret the term 'human'. Obviously, this term is ambiguous and depends on cultural variations. What if robots become more human-like? How can a subject prove to be human or android?

- The role of judgment in decision making – The assumption that there is a literal meaning for any given series of signals is currently considered naive. Similarly, conflicting orders may have to be prioritized, such as when two humans give inconsistent instructions. Whether the conflict is overt, unintentional or even unwitting, it nonetheless requires a resolution. Even in the absence of conflicting orders, a robot may need to recognize foolish or illegal orders and decline to implement them, or at least question them.

- The sheer complexity – The strategies as well as the environmental variables involve complexity. This widens the scope for dilemma and deadlock. Again, what about conflicting orders?

- Audit of robot compliance – Safety issues demand an audit process, i.e. a check whether the laws are imposed in precisely the manner intended. That is currently not yet solved for any given software product. Additionally, could the laws be overridden or modified? Obviously, we are faced with impending security issues as well.

- Robot autonomy – Sometimes humans may delegate control to a robot and find themselves unable to regain it, at least in a particular context. One reason is that to avoid deadlock, a robot must be capable of making arbitrary decisions. An example is the requested execution of two conflicting orders.

Based on these observations, the robot laws were reformulated in order to achieve unambiguity and to increase security and safety. Clarke (1994) discussed an extended set of the Asimov's Laws of Robotics as follows:

- *The Meta-Law:* A robot may not act unless its actions are subject to the Laws of Robotics.

- *Law Zero:* A robot may not injure humanity, or, through inaction, allow humanity to come to harm.

- *Law One:* A robot may not injure a human being, or, through inaction, allow a human being to come to harm, unless this would violate a higher-order Law.

- *Law Two:* (a) A robot must obey orders given it by human beings, except where such orders would conflict with a higher-order Law. (b) A robot must obey orders given it by superordinate robots, except where such orders would conflict with a higher-order Law.

- *Law Three:* (a) A robot must protect the existence of a superordinate robot as long as such protection does not conflict with a higher-order Law. (b) A robot must protect its own existence as long as such protection does not conflict with a higher-order Law.

- *Law Four:* A robot must perform the duties for which it has been programmed, except where that would conflict with a higher-order law.

- *The Procreation Law:* A robot may not take any part in the design or manufacture of a robot unless the new robot's actions are subject to the Laws of Robotics.

3.6.2 Attractors

To control the behavior of self-organizing systems, we must be able to model the evolution of such a system. Heylighen (1999) describes this procedure as follows. Basically, we need rules that tell us how the system moves from one state into another in the course of time t. This might be expressed by a function $f_T : S \rightarrow S : s(t) \rightarrow s(t + T)$, which is usually the solution of a differential or difference equation. Considering self-organizing or complex systems, this evolution is often nondeterministic and irreversible. Most models still assume the dynamics to be deterministic; thus, for a given initial state s, there will in general be only one possible later state $f(s)$. In practice, the lack of information about the precise state will make the evolution unpredictable. The resulting stochastic process can, for example, be modeled as a Markov chain. Following the used notations, for each state s_i, the Markov chain describes the probabilities for a state change $P(s_j|s_i) = \mathbf{M}_{ij} \in [0, 1]$ in the form of a transition matrix $\mathbf{M}$. Thus, the probability distribution of any state at $t + 1$ can be calculated, as the corresponding distributions are known for t:

$$P(s_j, t + 1) = \sum_i \mathbf{M}_{ij} P(s_i, t). \tag{3.2}$$

Based on this behavioral model, Heylighen (1999) defines an attractor as follows. We assume that in self-organizing systems, there will always be state transitions $s_i \rightarrow s_j$ such that the inverse transition $s_j \rightarrow s_i$ is either impossible or less probable (nonlinearity of the system behavior). This means that when the system is allowed to evolve, it will tend to leave certain states, e.g. s_i, and to enter and stay in other states, e.g. s_j. Such a final state is called an 'attractor'. Formally, an attractor state means that $\forall s_i \in A, \forall s_j \notin A, \forall n, T : f_T(s_i) \in A$ or $\mathbf{M}_i^n j = 0$. Basically, an attractor is a mathematical model of causal closure. Inside an attractor, the process can no longer reach out.

In summary, an attractor is a preferred position for the system, such that if the system is started from another state, it will evolve until it arrives at the attractor, and will then stay there in the absence of other factors. An attractor can have different shapes, sizes and dimensions. All specify a restricted volume of state space (a compression). The larger area of state space that leads to an attractor is called its basin of attraction and comprises all the pre-images of the attractor state. The ratio of the volume of the basin to the volume of the attractor can be used as a measure of the degree of self-organization.

Figure 3.3 illustrates a system that, depending on its configuration, has one or two attractors. The system is modeled as a Markov chain consisting of three states: 1, 2 and 3. Each state has an associated state probability π_i. Only a single parameter controls the system behavior – the transition probability p. If p is very large, i.e. if it converges to 1, then π_1 and π_3 become attractors, i.e. the system tends to stay either in 1 or in 3. On the other hand, if p is very small, i.e. if it converges to 0, then π_2 becomes an attractor.

When a system enters an attractor, it thereby loses its freedom to reach states outside the attractor, and thus decreases its statistical entropy. During the state transitions of the system towards the attractor, the entropy will gradually decrease, reaching a minimum in the attractor itself. Since the reaching of an attractor is an automatic process, entailed by the system's dynamics, it can be viewed as a general model of self-organization: the spontaneous reduction of statistical entropy.

A complex system can have many attractors and these can alter with changes to the environmental conditions of the system. Therefore, studying self-organization is equivalent to investigating the attractors of the system, their form and dynamics. The attractors in complex systems vary in their persistence; some have long durations, so can appear as fixed 'objects', some are of very short duration (transient attractors) and many are intermediate.

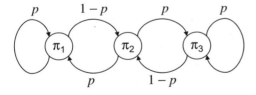 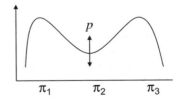

Figure 3.3 Example of an attractor system. If p is very large, i.e. converges to 1, then π_1 and π_3 become attractors, i.e. the system tends to stay either in 1 or in 3. If p is very small, then π_2 becomes an attractor

3.7   Limitations of self-organization

In this chapter, we discussed the meanings, challenges and advantages of self-organization as an operation and control paradigm in massively distributed systems. This section, however, focuses on the weak points and the limitations of self-organization. Even though self-organization definitely solves many problems that accompany the development of entities that are acting autonomously and which are entitled to collaboratively work on a global goal, some problems are still unsolved or might even appear by introducing self-organization. In the following, we discuss some examples of still open issues. These can be seen as starting points to conduct further research on self-organization and distributed control.

- Controllability – As shown in Figure 3.4, the predictability of the behavior of a self-organizing system is rapidly decreasing while increasing its scalability. This problem is directly related to the controllability of the system. For example, network management solutions as known so far cannot be successfully employed in self-organizing ad hoc networks. Therefore, one can believe that the network is operating well but it is hard to prove the availability. Even harder is the guarantee of quality of service parameters.

- Cross-mechanism interference – The composition of multiple self-organizing mechanisms can lead to unforeseen effects. For example, two different energy-aware methods implemented at the MAC and the network layer in ad hoc networks might interfere and lead either to reduced throughput or reliability or to much higher energy consumption compared with the non-optimized behavior. Cross-layer design and cross-method validation techniques are needed to identify such interferences and to eliminate them.

- Software development – The development of self-organizing systems and appropriate applications demands new software engineering approaches. Multiple questions have to be considered simultaneously: where to run which (part of the) application and how to distribute data and activity over the network. Recent examples that have to be elaborated on more clearly are policy-based approaches and profile-matching techniques.

- System test – The test of the system, its components and the installed software becomes a complex task. It is not possible to create a lab environment showing exactly the properties of the desired deployment scenario. The same holds for field tests because it is not possible to predict future conditions influencing the system. System test approaches must be changed to incorporate the unpredictable environment.

Figure 3.4 depicts the main problem of self-organizing systems – the reduced determinism. The more scalable a system becomes, e.g. by using the described self-organization techniques, the less control in the collaboration of all participating entities is possible. The primary conclusion is that the predictability of the system behavior must be reduced for such a self-organizing system. Admittedly, some research groups try to elaborate new control mechanisms that focus on the manipulation of the environment. Nevertheless, even this

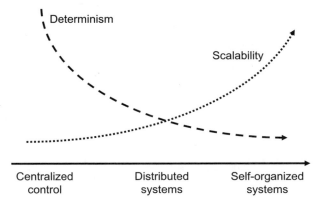

Figure 3.4 Scalability vs determinism in centralized controlled and self-organizing systems

approach will not be able to guarantee a particular behavior of the system – it only allows increasing the probability to achieve the designated goal, i.e. show the demanded behavior.

Similarly, one may argue that localized algorithms often may not result in a global optimum with respect to a certain property but only provide solutions close to the optimum. This sub-optimality, however, is not a real disadvantage in a dynamic system. Here, optimum configurations are subject to change frequently anyway, and fast convergence to a stable configuration is important.

These observations directly lead to the question of how to evaluate self-organizing systems. Actually, most self-organizing systems under study will be evaluated by statistical measures. These techniques are also available for evaluating self-organization methods under development. Simulation models can be created to study the complex system behavior. Again, statistical measures can be used to obtain performance and functional characteristics of the entire system (Law and Kelton 2000). In summary, it can be said that the behavior of self-organizing systems can be predicted to a certain degree in the form of probabilities or statistical evaluation criteria such as mean or median values with corresponding variance information.

4

Natural Self-Organization

The following two sections provide examples for self-organizing systems and processes. We will start with examples observed in nature, for two reasons. First, these are more comprehensible and easier to be categorized as 'self-organized'. Secondly, we want to develop a kind of understanding that easily arises out of these mechanisms. In a later step, we show how to adapt mechanisms that have been observed as self-organization phenomena in nature to technical systems.

4.1 Development of understandings

The turn to nature for solutions to technological questions has bought us many great unforeseen concepts. This encouraging course seems to hold for many aspects of technology. First studies on biological self-organization and its possible adaptation to technical solutions date back to the 1960s. von Foerster (1960) and Eigen and Schuster (1979) proposed employing self-organization methods known from many areas of biology. They saw the primary application in engineering in general. Nevertheless, it has been shown that communications can benefit from biologically inspired mechanisms as well.

In contrast to manual or automated centralized or decentralized control, self-organization is intended to work without any global state (Gerhenson and Heylighen 2003). Local information and its adaptation based on the observation of the environment and interactions with neighboring entities are used for all decision processes. The modification of local state is done using feedback loops. Basically, there exist two kinds of such control loops: positive feedback used as an amplifier for observed effects, and negative feedback to prevent overreaction due to amplification effects. The behavior of complex systems can be controlled using such feedback loops as depicted in Figure 4.1, leading to an emergent behavior of the overall system. Figure 4.1 shows two activities that are triggered by observations of the environment. On the right-hand side, 'activation' in terms of positive interaction is depicted. This positive feedback leads to upregulation of system parameters, usually in the form of nonlinear amplification. In order to control this amplification, negative feedback,

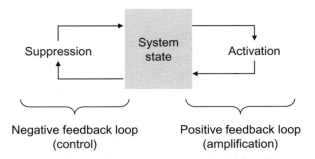

Figure 4.1 Positive and negative feedback used for behavior control of autonomous systems

as shown on the left-hand side, is employed. Measurements or observations of ongoing activations lead to the enabling of suppression mechanisms that directly counteract the activations.

Using this simple model of behavior control, a better understanding of the interaction of components building a complex pattern can be achieved: each entity is acquiring and acting upon information gathered from the environment. Therefore, self-organization arises entirely from multiple interactions among the single components. Two kinds of interactions have to be distinguished: interactions with the environment, which are also known as stigmergy (Theraulaz and Bonbeau 1999), and information transfer between individuals.

Understanding self-organization enables the power of prediction. Whatever the final verdict on chaos in nature may turn out to be, the success of nonlinear dynamics won't stand or fall with it (Zimmer 1999). Especially, bio-inspired approaches show many possible techniques that might influence the development of technical systems. Basically, self-organization effects can be observed in all domains of natural sciences. In the following, we discuss selected examples as known from biology and chemistry. Additionally, typical examples from economy are included in our discussion.

4.2   Examples in natural sciences

Selected examples of natural self-organization should conclude the general discussion of self-organization. Afterwards, we will continue with its applicability in technical systems and the methods and techniques behind making systems self-organizing.

Natural self-organization can be observed and studied in multiple domains. We will study selected examples from biology and chemistry. This list is definitely incomplete. Nevertheless, this book is on self-organization in technical systems or, more precisely, in Sensor and Actor Networks. Other examples can be found in the literature, e.g. as provided by Eigen and Schuster (1979) and Camazine *et al.* (2003). We will emphasize biological self-organization because of its close relationship to the general behavior of computer networks. Nevertheless, self-organization must not be used as a synonym for bio-inspired techniques.

4.2.1 Biology

The following list is an incomplete collection of the diverse phenomena in biology, which have been described as 'self-organizing' in the literature:

- spontaneous folding of proteins and other biomacromolecules;
- formation of lipid bilayer membranes;
- homeostasis (the self-maintaining nature of systems from cell to the whole organism);
- morphogenesis, or how the living organism develops and grows;
- the coordination of human movement;
- the creation of structures by social animals, such as social insects (bees, ants, termites) and many mammals;
- grouping behavior (such as the formation of flocks by birds, schools of fish, etc.).

In general, this list of phenomena must be extended to include the origin of life itself from self-organizing chemical systems, in the theories of hypercycles and autocatalytic networks (Eigen and Schuster 1979).

A typical example of biological self-organization is the system-inherent nature of the structuring of growing cells. An example is shown in Figure 4.2. This figure shows a cell culture of immortalized epithelial cells. From left to right, four stages in the development from single separated cells to a monolayer with tight junctions between the cells are shown. A detailed description of this process is, for example, provided by Freshney and Freshney (2002):

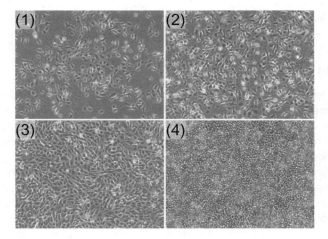

Figure 4.2 Cell culture of immortalized epithelial cells; over a time period of 9 days, the cells are proliferating and forming a tight monolayer [Reproduced by permission of Bettina Krüger]

1. One day after seeding cells on a plastic cell culture dish, the cells are attaching to the bottom and spreading. Thin connections to other cells (filopodia) are formed to contact other cells. Growth factors are secreted by the cells which promote to growth the other cells in the culture.

2. Cell culture of epithelial cells after 2 days. The cells are proliferating. Cells which loose contact to the bottom and other cells, mostly die. Closer connections between the cells are formed.

3. Culture of epithelial cells after 4 days. The cells start to form a monolayer.

4. Culture of epithelial cells after 9 days. The cells form a tight monolayer (coble stone pattern). Polarization of the cell surfaces in an apical (top, 'medium-site') and a basolateral site (bottom) and tight junctions between the cells are characteristic for such a monolayer.

Weng *et al.* (1999) studied the complexity in biological signaling systems. Biological signaling pathways interact with one another to form complex networks. Complexity arises from the large number of components, many with isoforms that have partially overlapping functions, from the connections among the components, and from the spatial relationship between components. Signaling in biological systems occurs at multiple levels. Basically, the term 'signaling' can be used to describe interactions between single molecules to interactions between species in ecological systems. There is simply too much essential detail in biological signaling for the unaided human mind to organize and understand. It appears that a paradigm shift from the qualitative to the quantitative is taking place in biology – we are moving from a descriptive to a predictive science. Understanding complex signaling networks may also provide a clear molecular view of the interactions of individuals with their environment, i.e. the self-organizing processes in the biological environment.

A similar example that was discussed in the early 1950s is the diffusion–reaction theory of morphogenesis (Turing 1952). Observed patterns of morphogen systems led to the hypothesis that nonlinear models are needed to describe the overall system behavior. We will discuss diffusion–reaction mechanisms in more detail by studying chemical reactions.

4.2.2   Chemistry

Complexity is also a subject of growing importance in chemistry (Whitesides and Ismagilov 1999). Oscillating chemical reactions raised the interest of the academic society early on. Nonetheless, the description and explanation of such processes became just possible in the early 1970s. At that time, researchers started to interpret the observations by means of reaction–diffusion systems.

The oscillating reactions of the Belousov–Zhabotinskiy type are perhaps the best known example (Winfree 1972). This class of reactions has the characteristics that the simultaneous operation of two processes–reaction and diffusion–results in a system in which the concentrations of reactants and products oscillate temporally and spatially and in which this oscillation can result in ordered patterns. The resulting process leads to fascinating patterns, as depicted in Figure 1.1.

Belousov observed that a solution of citric acid, acidified bromate (BrO_3^-) and a ceric salt oscillated periodically between yellow and clear. An example experiment is based on the mixture of potassium bromate ($KBrO_3^-$), malonic acid ($CH_2(COOH)_2$) and manganese sulfate ($MnSO_4$). This mixture needs to be prepared in a heated solution of sulfuric acid (H_2SO_4). The reaction produced changing colors as the manganese oscillates between two oxidation states.

These reactions can be described mathematically by systems of nonlinear equations. Equations 4.1 and 4.2 represent a minimum set of two reaction–diffusion equations (Whitesides and Ismagilov 1999; Zykov *et al.* 1998):

$$\partial u/\partial t = D_u \nabla^2 u + F(u, v) \tag{4.1}$$

$$\partial v/\partial t = D_v \nabla^2 v + \epsilon G(u, v). \tag{4.2}$$

In these equations, u is the concentration of a species that catalyzes reaction; v is the concentration of a species that inhibits reaction. Therefore, $\partial u/\partial t$ and $\partial v/\partial t$ describe changes in concentrations of u and v, respectively, with time. The terms $F(u, v)$ and $\epsilon G(u, v)$ characterize reactions between u and v. The most interesting parameters are D_u and D_v the diffusion coefficients of u and v, respectively. These parameters indicate whether activators are diffused faster than inhibitors and vice versa.

Equation 4.1 depicts the reaction part – a positive feedback loop that amplifies the reaction. On the other hand, Equation 4.2 shows the diffusion part – a negative feedback loop controlling the system and preventing system over-reactions.

4.3 Differentiation self-organization and bio-inspired

The purpose of this section is manifold. First, some kind of demystification of the buzzword 'bio-inspired' should be achieved. Secondly, a brief overview of bio-inspired techniques is provided. Finally, the differences between bio-inspired mechanisms and self-organization are discussed. It will become obvious that both domains partly overlap and sometimes no clear distinction is possible. A detailed discussion of bio-inspired techniques that are directly related to self-organization – especially in the domain of SANETs – is provided in Part V.

4.3.1 Exploring bio-inspired

Since the first papers were published by Ashby (1962) and by Eigen and Schuster (1979), many methods were called 'bio-inspired'. Not all these developments are based on observation of nature or biology. The buzzword 'bio-inspired' seems to have become an eye catcher. In general, the term 'bio-inspired' stands for all methods and techniques that originate in behavioral pattern, algorithms and structures that can be found in biological systems. It greatly differs from bioinformatics, which focuses on the application of computational efforts to solve biological questions.

Obviously, several methods and techniques are really bio-inspired, as they follow principles that have been studied in nature and that promise positive effects if applied to technical systems. Some examples are provided in the following subsection. There are three steps (we will discuss these in more detail in Chapter 6) that are always necessary for developing bio-inspired methods that have a remarkable impact in the domain of investigation:

1. Identification of analogies – which structures and methods seem to be similar.

2. Understanding – modeling of realistic biological behavior.

3. Engineering – model simplification and tuning for technical applications.

This list implies two very basic conclusions. First, the biological methods must be understood in order to work with them, to adapt to different domains and to develop helpful algorithms or tools. Secondly, it must be verified whether the new bio-inspired techniques may ever outperform previous solutions. This includes a careful investigation of the advantages and disadvantages of the bio-inspired mechanisms.

4.3.2   Bio-inspired techniques

There is a great number of different techniques and methods. An excerpt of these is provided, including a short description of the key ideas. This list is not intended to be complete. All-embracing domains such as Artificial Intelligence (AI) are explicitly excluded, since they already include many of the listed methods. Further information focusing on bio-inspired domains related to self-organization will be presented in Chapter 18.

- Cellular automata – Cellular automata represent a discrete model studied in computability theory, mathematics and theoretical biology. It consists of an infinite, regular grid of cells, each in one of a finite number of states. The grid can be in any finite number of dimensions. First, simple lattice networks were used as a model; then, the problem of self-replicating systems was studied (Neumann 1966). The best known example of a cellular automaton is the Game of Life. Cellular automata found broad application including self-learning systems and cryptography (Wolfram 2002).

- Attractor schemes – An attractor is a set to which the system evolves after a long enough time (Milnor 1985). In biology, foraging and reproduction represent effective attractor schemes. Usually, such attractors are used in fuzzy feedback loops. Unsupervised control can be introduced to a system by means of powerful attractors. Similarly, cells switch between various stable genetic programs (attractors) to accommodate environmental conditions (Kashiwagi *et al.* 2006).

- Artificial Neural Networks – Traditionally, a neural network is used to refer to a network of biological neurons. In modern usage, the term is often used to refer to an Artificial Neural Network (ANN). An ANN is an interconnected group of artificial neurons that uses a mathematical model or computational model for information processing based on a connectionist approach to computation. In most cases, an ANN is an adaptive system that changes its structure based on external or internal information that flows through the network (Bishop 1996). It can be used to model complex relationships between inputs and outputs or to find patterns in data.

- Swarm Intelligence – Swarm Intelligence (SI) is an AI technique based on the observations of the collective behavior in decentralized and self-organized systems (Bonabeau *et al.* 1999). Such systems are typically made up of a population of simple agents interacting locally with one another and with their environment. Although there is normally no centralized control structure, dictating how individual agents should behave, local interactions between such agents often lead to the emergence of global behavior. Examples can be found in nature, including ant colonies, bird flocking, animal herding, bacteria molding and fish schooling. The best known application of swarm principles is in the field of multi-robot systems (swarm robotics) (Dorigo *et al.* 2004).

- Artificial Immune System – An Artificial Immune System (AIS) is a type of optimization algorithm inspired by the principles and processes of the mammalian immune system (Hofmeyr and Forrest 2000). The algorithms typically exploit the immune system's characteristics of self-learning and memorization to solve a problem. The immune system is, in its simplest form, a cascade of detection and adaptation, culminating in a system that is remarkably effective. Data analysis and anomaly detection represent typical application domains (de Castro and Timmis 2002).

- Cellular signaling pathways – The intra- and inter-cellular communication between cells in tissues is an example for very efficient and specific communication. Basically, two communication paradigms can be differentiated. The regulation of the concentrations, e.g. of anions, allows diffuse messaging. On the other hand, specific reactions can be triggered by exchanging particles, e.g. proteins, that can be detected by the receiving cell and that may activate powerful signaling cascades. This behavior can be directly applied to different aspects in computer communications (Dressler and Krüger 2004; Krüger and Dressler 2005).

- Molecular computing – Molecular computing or Deoxyribonucleic Acid (DNA) computing is a form of computing which uses DNA and molecular biology, instead of the traditional silicon-based computer technologies. The first experimental use of the DNA as a computational system was demonstrated by Adleman (1994). He solved a seven-node instance of the Hamiltonian Graph problem – an $\mathcal{NP}$-complete problem similar to the traveling salesman problem. While the solution to a seven-node instance is trivial, it was the first known instance of the successful use of DNA to compute an algorithm. DNA computing has been shown to have potential as a means to solve several other large-scale combinatorial search problems.

- Evolutionary Algorithms – Evolutionary Algorithms (EAs) represent a search technique used in computing to find real or approximate solutions to optimization and search problems. A particular class of EAs are Genetic Algorithms (GAs) that use techniques inspired by evolutionary biology such as inheritance, mutation, selection and recombination (Goldberg 1989). GAs are implemented as a computer simulation in which a population of abstract representations (genotype) of candidate solutions (individual, phenotype) to an optimization problem evolves toward better solutions. Usually, the evolution starts from a population of randomly generated individuals and happens in generations. In each generation, the fitness of every individual in the population is evaluated, and multiple individuals are stochastically selected from the current population (based on their fitness) and modified (mutated or recombined) to form a new population. This generational process is repeated until a termination condition has been reached.

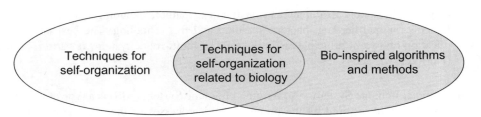

Figure 4.3 Distinction techniques for self-organization vs bio-inspired mechanisms

Table 4.1   General categorization of bio-inspired research domains

Objective	Bio-inspired research domains
Computation	DNA computing, cellular automata
Optimization	Genetic Algorithms, Artificial Neural Networks
Self-learning	Artificial Neural Networks, Artificial Immune System
Self-organization	Swarm Intelligence, Artificial Immune System, attractor schemes, cellular signaling pathways

4.3.3   Self-organization vs bio-inspired

Based on the list of the different bio-inspired research domains depicted in the previous subsection, we need to discuss the relationship between self-organization and bio-inspired. Biological self-organization is indeed a domain that brought us many great unforeseen concepts. Admittedly, there are also mechanisms used to coordinate massively distributed systems, i.e. self-organizing systems according to our definition, which do not relate to biology (at least to a given extent). Figure 4.3 depicts this classification. Considering all methods that we currently know and are employing for developing technical systems, only a part of them can be called bio-inspired. On the other hand, not all bio-inspired methods are supporting the self-organization of autonomous systems. Therefore, only the intersection of both domains represents bio-inspired methods for self-organization.

As can be seen, only a few of the bio-inspired research domains directly relate to issues of self-organization. Therefore, we need to distinguish them by means of primary objectives as well as methodologies. While a general categorization of bio-inspired research domains is not yet available, a rough overview can be found in Table 4.1. Besides self-organization, three other main domains are depicted: optimization, computation and self-learning. All these fields are only slightly related to self-organization.

5

Self-Organization in Technical Systems

In the last section, we briefly analyzed self-organization effects in nature. In order to summarize the introduction to self-organization, selected examples should be discussed for self-organization in technical systems. We start with the general applicability of self-organization methods and focus later on operations in WSNs and SANETs, respectively.

5.1 General applicability

If we speak about self-organization in technical systems, innumerable examples can be named and discussed as a matter of principle. Nevertheless, the intention of this section is just to give an impression what can be done with self-organization, which are typical application examples. The following summary can also be regarded as a list of self-organization requirements of technical systems, i.e. a description which systems really demand such a paradigm or which systems even rely on it, respectively.

5.1.1 Autonomous systems

The coordination and control of autonomous systems are facing challenging problems. Due to the – system inherent – lack of global control, the reliability and efficiency of the overall system are dominating questions. The research in this domain is basically driven by multi-agent systems – an AI control technique. Mataric (1995b) developed a survey on key terms in this research domain. The following terms were distinguished:

- Behavior and goals – The notion of behavior as a fundamental building block has been popularized in the AI, control and learning communities. As a control structure, we define 'behavior' as a control law for reaching and/or maintaining a particular goal. This definition specifies that a behavior is a type of an operator that guarantees a particular goal.

Self-Organization in Sensor and Actor Networks   Falko Dressler
© 2007 John Wiley & Sons, Ltd

- Communication and cooperation – Communication is the most common means of interaction ion among intelligent agents. Basically, we distinguish direct communication between neighboring entities. Messages can be destined to reach a single destination or multiple recipients. On the other hand, indirect communication is based on the observed behavior of other agents. This type of communication is referred to as stigmergic in biological literature. Cooperation is a form of interaction, usually based on some form of communication. Analogously to communication, explicit cooperation and implicit cooperation can be distinguished.

- Interference and conflict – Multi-agent control must deal with interference, i.e. any influence that opposes or blocks an agent's goal-driven behavior. In societies consisting of agents with identical goals, interference manifests itself as competition for shared resources. Resource competition manifests itself in homogeneous and heterogeneous groups of coexisting agents. In contrast, goal competition arises between agents with different goals.

In conclusion, it can be said that self-organization as a control paradigm is inherently demanded by cooperating autonomous systems. Task and resource allocation are typical application examples similar to communication and conflict resolution mechanisms.

5.1.2   Multi-robot systems

Multi-robot systems represent a specific domain of interest that highly correlates with the generic autonomous systems research. Mataric *et al.* (2003) formulated the problem as follows: 'Multiple cooperating robots hold the promise of improved performance and increased fault tolerance for large-scale problems such as planetary survey and habitat construction. Multi-robot coordination, however, is a complex problem. We cast this problem in the framework of multi-robot dynamic task allocation under uncertainty.'

Task allocation in heterogeneous multi-robot systems depends on various conditions. From a global perspective, in multi-robot coordination, action selection is based on the mapping from the combined robot state space to the combined robot action space. Several systems have been developed to address these issues (Gerkey 2003; Martin *et al.* 1999). However, the general problem of dynamically allocating tasks in a group of multiple robots satisfying multiple goals needs further investigation, as it represents one of the strongest multi-objective optimization problems.

The literature describes five distinct strategies for solving the multi-robot task allocation problem: auctions, motivation-based allocation, mutual inhibition, team consensus, and no allocation (Gage *et al.* 2004). In the first strategy – auctions – robots explicitly negotiate the task assignment through a bidding process. Auction-based approaches allow agents in the team to maximize utility or minimize cost that results from the task assignment. The second strategy – motivation-based task allocation – uses an internal motivation mechanism to cause behavior changes. Motivation-based cooperation distributes the task-allocation process equally among members of the team, and emergent team behavior results from simple control mechanisms within each agent. In task allocation through mutual inhibition, robots directly inhibit those around them from being chosen for a task. Task allocation by team consensus enables entire teams of robots to agree on a team strategy or formation. Such mechanisms are well studied in distributed systems research. Finally, some approaches use a fifth type of strategy – no allocation – to coordinate robot teams, and it is assumed that all robots cooperate on the same task.

Application examples of multi-robot systems include the analysis and synthesis of multi-robot coordination strategies for complex and uncertain domains, such as space exploration. Such coordination strategies represent a primary branch of research of self-organization studies.

5.1.3 Autonomic networking

The Internet did transform from a scientific network for specialists into an ubiquitously used network that provides the basis for a growing number of applications. Among the issues of highest importance for the future evolution of communication networks are functional adaptability or extensibility and robustness. The term 'autonomic networking' is used to describe communication networks consisting of self-managing elements capable to support self-configuration, self-healing and self-optimization. These desired properties require components for observation, for assessing the observed data, for representing and applying knowledge about constraints and goals. Due to the complexity of the network, and due to the multitude of administrative borders, decentralized self-organizing algorithms are required, with autonomous capabilities of individual nodes. The ultimate aim is to derive design paradigms for communication networks and distributed computing environments that are capable of providing rapidly adapting services and applications in scenarios in which networked devices and users interact in a highly dynamic manner.

Generally, autonomic networking refers to any self-managing communication networks including future versions of the Internet. Thus, the same context is addressed that is also in the focus of pervasive computing and ubiquitous computing. Weiser (1991) initiated the research domain of pervasive computing, as he stated: 'The most profound technologies are those that disappear. They weave themselves into the fabric of everyday life until they are indistinguishable from it.' Thus, a new research domain was created – the operation and control of massively distributed systems. Several years later, the term 'ubiquitous computing' was added in order to characterize the omnipresent computation facilities (Weiser 1993b).

New frontiers for pervasive computing were discussed by Zambonelli et al. (2004). They envisioned a future in which clouds of microcomputers can be sprayed in an environment to provide, by spontaneously networking with each other, an endless range of futuristic applications. The number of potential applications of the scenario is endless, ranging from smart and invisible clothes to intelligent interactive environments, self-assembly materials and self-repairing artifacts.

Estrin et al. (2002) summarized the most dominant challenges and the critical dimensions that must be considered during development and deployment. Immense scale refers to the vast number of interacting systems that will build future networks. Limited access means that most devices will be embedded in the environment in places that are inaccessible or expensive to connect with wires, making the individual system elements largely untethered, unattended and resource-constrained. Finally, extreme dynamics characterize the operational parameters and the environment. By design, the devices can sense their environment to provide inputs to higher-level tasks, and environmental changes directly affect their performance. The following dimensions have been identified, which concern the development of self-organizing pervasive environments:

- Scale – Scalability may be considered in different aspects. 'Sampling' is concerned with the frequency of separate actions. Basically, actions can stand for measurements or actuations. Depending on the sampling rate, different demands are made on the capabilities of the single system as well as of the communication infrastructure. 'Extend' depicts the spatial and temporal extend of systems. Either systems or behaviors last for a different length of time or the coverage can be different. 'Density' reflects the footprint of a system's installation. Density also affects the sampling, as higher densities tend to lead to increased sampling rates.

- Variability – Variability is concerned with the form and behavior of the specific system's aspects. The system 'structure' can be ad hoc or well engineered. It refers to the variability in system composition. 'Task' reflects the single mode of operation of a system that can be optimized in several directions. Variability in 'space' stands for system mobility. It applies to both system nodes and phenomena.

- Autonomy – The degree of autonomy has some of the most significant and varied long-term consequences for system design: the higher the overall system's autonomy, the lower the human involvement and the greater the need for extensive and sophisticated processing inside the system. Different 'modalities' provide noise resilience to one another. 'Complexity' is a measure for the real demand for autonomy. The greater the autonomy, the greater the system complexity that can be handled.

Self-organization is also a key objective for specific communication paradigms such as overlay networks. Self-organizing overlay networks are the keystone of modern peer to peer networks (Steinmetz and Wehrle 2005). The design of the overlay dramatically varies the efficiency of the network along multiple axes, including bandwidth, latency and robustness. Self-organization and self-adaptivity are the main characteristics of overlay networks (Leibnitz et al. 2006). Another specific property of overlay networks is resilience. For example, the Resilient Overlay Network (RON) architecture allows distributed Internet applications to detect and recover from path outages and periods of degraded performance within several seconds, improving today's wide-area routing protocols that take at least several minutes to recover (Andersen et al. 2001).

5.1.4   Mobile Ad Hoc Networks

A Mobile Ad Hoc Network (MANET) is a subtype of wireless ad hoc networks (Perkins 2000). It is a self-configuring network of mobile routers and associated hosts connected by wireless links, forming an arbitrary topology. Therefore, the main criteria relevant to MANETs are infrastructure-less operation, wireless communication and (possibly) mobility of the participating end systems. MANETs are becoming accepted as a valid commercial concept, which is confirmed by the creation of the MANET working group in the Internet Engineering Task Force (IETF) (MANET WG 2006). The research on MANETs is concerned with Medium Access Control (MAC) layer issues, routing and mobility. A special case of MANETs are underwater acoustic networks (Akyildiz et al. 2005a). The main difference is the communication channel that – while also wireless – has very different characteristics compared with other ad hoc networks.

A well known example of MANET research are the terminodes developed by Hubaux *et al.* (2000). Terminodes are personal devices that provide functionality of both the terminals and the nodes of the network. A network of terminodes is an autonomous, fully self-organized, wireless network, independent of any infrastructure. It must be able to scale up to millions of units, without any fixed backbone or server.

One special case of mobile ad hoc networks are Vehicular Ad Hoc Networks (VANETs). The communication between cars and trucks on the road is investigated for several reasons. The main motivation is the very specific type of mobility. It has been discovered that usual routing protocols fail in VANETs due to the high degree of node mobility and the tendency to build dense networks in traffic-jam situations (Sommer and Dressler 2007). In order to cope with these issues, solutions such as the Self-Organizing Traffic Information System (SOTIS) have been developed that follow the broadcast principle in communication networks and allow distributed data storage and processing (Wischhof *et al.* 2005, 2003).

All kinds of MANETs rely on a self-organizing infrastructure that is formed on demand. The addition or removal of nodes, their mobility and limited resources in the wireless medium must be supported by the developed communication mechanisms.

5.1.5   Sensor and Actor Networks

Other special cases of MANETs are WSNs and SANETs. Sensor nodes are considered to be small, embedded systems with limited processing capabilities, storage and energy. Sensors are attached to measure physical phenomena. A WSN may consist of hundreds or thousands of sensor nodes (Culler *et al.* 2004). The main differences between sensor networks and ad hoc networks are outlined below (Akyildiz *et al.* 2002b):

- The number of sensor nodes in a sensor network can be several orders of magnitude higher than the nodes in an ad hoc network.

- Sensor nodes are densely deployed.

- Sensor nodes are prone to failures.

- The topology of a sensor network changes very frequently.

- Sensor nodes mainly use broadcast communication paradigm, whereas most ad hoc networks are based on point-to-point communications.

- Sensor nodes are limited in power, computational capacities and memory.

- Sensor nodes may not have global identification (ID) because of the large overheads and large number of sensors.

SANETs refer to a group of sensors and actors linked over a wireless medium to perform distributed sensing and acting tasks (Akyildiz and Kasimoglu 2004). The realization of SANETs needs to satisfy the requirements introduced by the coexistence of sensors and actuators. Sensors gather information about the physical world, while actuators make decisions and then perform appropriate actions upon the environment, which allows a user to effectively sense and act from a distance. In order to provide effective sensing and acting, coordination mechanisms are required among sensors and actors.

Basically, WSNs and SANETs represent a class of massively distributed systems that emphasize communication as well as coordination aspects. For example, Low *et al.* (2005) investigated autonomic mobile sensor networks with self-coordinated task allocation and execution. Similar work was done by Buttyán and Hubaux (2003). We will depict selected aspects of WSNs and SANETs that relate to self-organization in Part II.

5.2   Operating Sensor and Actor Networks

Self-organization is referred to as the multitude of algorithms and methods that organize the global behavior of a system based on inter-system communication. Most networking algorithms work like that. Therefore, self-organization in this context is not a new research direction in this area. Nevertheless, most of these algorithms are based on global state information, e.g. routing tables. In the networking community, it is commonly agreed that such a global state is the primary source of scalability problems of the particular algorithms. Especially in the area of ad hoc networks, new solutions were discovered that show the properties of the new definition of self-organization. Most ad hoc routing algorithms as well as data-centric data-dissemination approaches are well known examples (Akkaya and Younis 2005). In addition to such communication issues, task and resource allocation is of particular interest in WSNs and SANETs (Low *et al.* 2005). Such coordination issues

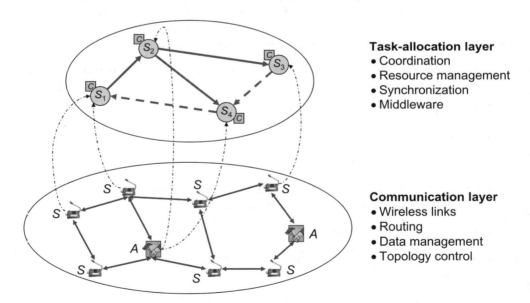

Figure 5.1   Problem spaces in SANETs. The communication layer comprises wireless transmission principles, routing and data management, and topology control. The task-allocation layer focuses on coordination, resource allocation and synchronization issues

require differentiated abstraction layers in order to be efficiently addressed in heterogeneous networks (Melodia *et al.* 2005). Cross-layer design approaches make it feasible to optimize task allocation and communication aspects simultaneously.

Basically, the problem space in SANETs can be divided into two layers. This principle is depicted in Figure 5.1. Communication aspects are addressed at a lower level. This makes it possible to exploit hardware-specific properties as well as topology information. Different routing functions in ad hoc sensor networks may be employed, depending on the scenario. The higher-level functions cover task-allocation issues. This includes mechanisms for coordination, resource allocation and synchronization. Typically, middleware architectures are used in this context.

In this book, we generally discuss self-organization aspects focusing on operations in SANETs. The basic methods and techniques that enable self-organization in massively distributed systems are outlined in the following section. Furthermore, we will use the discussed layering scheme in SANETs to discuss self-organization methods in more detail:

- Part II Networking Aspects: Ad Hoc and Sensor Networks – In this part, communication aspects will be presented that depict self-organization requirements and solutions in ad hoc and sensor networks. This chapter covers protocol issues (MAC, routing) as well as topology control and quality of service.

- Part III Coordination and Control: Sensor and Actor Networks – In this part, the second domain – coordination and control – will be covered. Self-organization techniques used for coordination and task and resource allocation are analyzed, including cross-layer concerns. Additionally, requirements on synchronization vs coordination are discussed.

6

Methods and Techniques

Self-organization as a process relies on a number of methods and techniques that enable the emergent behavior or the non-linear system properties. The purpose of this section is to outline these helper techniques and to discuss the development of self-organizing systems. This overview is generic for any self-organizing system. Nevertheless, we will discuss the design paradigms and available methods, focusing on communication networks in general and ad hoc networks in particular.

Often, questions on constructing and controlling self-organization arise in the research community. In our discussion of self-organization as a system inherent feature, we saw that this process depends on several factors: global state information is not available, individual systems derive their decisions locally and deterministic behavior is neither a requirement nor can be achieved without leaving the domain of self-organization. In this section, we will outline the basis methods that are underlying in all self-organizing systems, design paradigms, development and modeling techniques, which enable not only the understanding but also the usage of self-organization as a key paradigm for supporting massively distributed systems.

6.1 Basic methods

As already discussed in our definition of self-organization, self-organizing systems completely rely on localized decision processes. Three basic methods need to be distinguished. These are the building blocks for all approaches that should self-organize, as depicted by our definition of self-organization:

- positive and negative feedback;

- interactions among individuals and with the environment;

- probabilistic techniques.

All these mechanisms can be employed individually or in combination. Usually, quite complex control structures can be found in self-organizing systems. Nevertheless, deeper

Self-Organization in Sensor and Actor Networks   Falko Dressler
© 2007 John Wiley & Sons, Ltd

analysis will always discover multiple control units following one or more of the listed basic methods.

6.1.1   Positive and negative feedback

The use of feedback has several advantages. The local state of an individual system can be adapted to changing environments using adequate feedback from either the environment, a controlling system or neighboring systems. It is well known that feedback loops have been a well understood mechanism in engineering for decades. Typically, a single system analyzes its environmental conditions and updates its local state accordingly. The most simple form of feedback control is shown in Figure 6.1. Such simple feedback loops will not lead to an emergent system behavior. On the other hand, more complex systems that allow feedback between multiple systems and even feedback loops will always show a nonlinear system behavior.

Feedback can be divided into positive and negative feedback. Usually, system control will depend on positive feedback only. Positive feedback provides amplification capabilities, to act efficiently and in time on particular changes. Positive feedback generally promotes changes in a system. Self-enhancement, amplification, facilitation and autocatalysis are all terms used to describe positive feedback. Actually, positive feedback is one of the major mechanisms that make self-organization possible. The amplifying nature directly leads to the development of pattern. Such pattern can be the observable system state or even the environment changed by frequent modifications in the same direction. The amplifying nature of positive feedback means also that it has the potential to produce destructive explosions or implosions in any process in which it plays a role. Such effects are depicted in Figure 6.2. Two special cases are shown in this figure. The first one is the snowballing effect. This means that a single system might affect huge numbers of other systems, whether they need to be involved or not. The second case is the implosion effect. Many individual systems take notice of a small change in the environment. Since all these systems send feedback information, the resulting amplification can overload a single system and it can lead to over-reactions.

Positive feedback, i.e. the amplification of fluctuations, can be used to find a stable system state starting from any random initial state. If additional feedback is received, the system can leave this stable state in order to converge to another one. Thus, during the

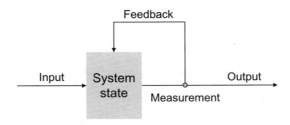

Figure 6.1 Simple feedback control consisting of a system that observes its own output in order to update its system parameters accordingly

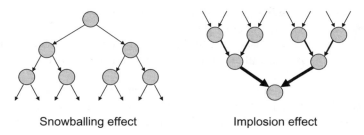

Snowballing effect   Implosion effect

Figure 6.2 Positive feedback can lead to snowballing effects (left) and overloading amplifications, i.e. implosions (right)

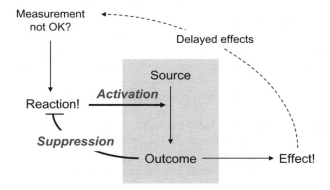

Figure 6.3 Complex system behavior driven by positive and negative feedback; the amplification effect is controlled by a simultaneously executed suppressor

system lifetime, the system will traverse a complex order of stable states. This behavior can be reflected by means of pattern generation.

Another kind of feedback is used for keeping positive feedback under control. Negative feedback is used for this purpose. Negative feedback can be created in the form of rules that have been developed to keep the system state within a given parameter range. In other cases, such an inhibition arises automatically, often simply from physical constraints.

Self-enhancing positive feedback coupled with antagonistic negative feedback provides a powerful mechanism for creating structure and patterns in many systems involving large numbers of components. Figure 6.3 depicts the complex behavior of a system caused by positive and negative feedback. The observation of an effect activates an internal process of the system, e.g. to produce some outcome from a particular source. This outcome changes either the system or its environment; thus, it results in an observable effect. If we consider a self-organizing massively distributed system, the time from inducing the effect until it can be measured by the system might be very long. Thus, an over-reaction of the system becomes highly predictable. This behavior is changed by using negative feedback. The same outcome that changes the environment is also used to suppress further reactions, i.e. measurements that activate the internal process.

Positive and negative feedback must be strictly coordinated. If these mechanisms are used uncontrolled, the system will either tend to do nothing, over-react or oscillate between both states.

Multiple terms are used in the literature for both positive feedback and negative feedback. In general, positive feedback is equal to reaction, amplification and promotion. Negative feedback is often described as diffusion, suppression and inhibition.

6.1.2 Interactions among individuals and with the environment

The use of feedback loops is directly linked to acquiring and actioning upon information. In our discussion of self-organization and the self-organizing system, we outlined as a main characteristic that the operation and control, i.e. the organization, arises entirely from multiple interactions among their components. Camazine *et al.* (2003) differentiate such interactions within self-organizing systems as interactions based on signals and those based on cues, respectively. While information transfer via signals tends to be conspicuous, communication via cues is often more subtle and based on incidental stimuli in the environment.

We have to distinguish between two kinds of information transfer:

- information transfer between individuals, i.e. direct communication between neighboring (in time and space) individuals via signals;

- interactions with the environment, i.e. indirect information flows via cues arising from work in progress (stigmergy).

Figure 6.4 depicts both ways of communication. Between individual systems, information can be exchanged directly via signals. This can be either a directed communication, as known from unicast connections in computer networks, or a diffuse transmission, such as a broadcast communication. Also depicted in Figure 6.4 is the indirect communication via the environment. Each system that participates in the complex global system may influence the environment, e.g. by changing surrounding conditions or leaving footprints. Other systems that observe the environment detect these changes and can act accordingly. Therefore, the indirect communication via the environment enables the information exchange over temporal barriers, i.e. using stigmergy as a control paradigm.

Concentrating on the interactions with the environment, two principles must be distinguished: stigmergy, i.e. indirect communication among individuals using the environment, and observation of the environment to detect changes for subsequent system adaptation.

Stigmergy, being an interesting research object, has been studied for many years (Abraham *et al.* 2006; Di Caro and Dorgio 1998; Theraulaz and Bonbeau 1999). Based on these findings, we can provide the following definition for stigmergy.

Definition 6.1.1 Stigmergy – Stigmergy refers to indirect communications among individuals using modifications of the local environment as a communication medium. Typically, previous work-in-progress is discussed as stigmergic communication.

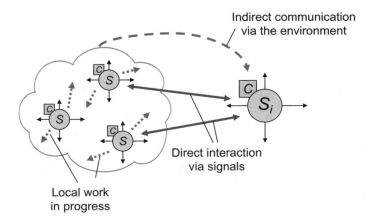

Figure 6.4   Information exchange between individual systems: directly via signals and indirectly via changes in the environment

6.1.3   Probabilistic techniques

A third component for successfully building self-organizing systems is probability. Basically, all self-organizing systems include probabilistic techniques. Such mechanisms can be used either to entirely organize the local behavior of a single system or at least for parameter settings of other deterministic algorithms. The use of probabilistic techniques has several advantages. First, the necessary state information can be drastically reduced and, secondly, it enables the break out of local optima. A well known example for pure probabilistic approaches is the nest-building process of termites (Resnick 1997).

The basic idea is shown in Algorithm 6.1. Starting in an environment with randomly deployed wood chips, the termites iteratively relocate a single chip in order to build a heap. After a number of iterations, the termites will finally be able to build a single heap.

Algorithm 6.1 Nest-building algorithm used by termites to collect wood chips

Ensure: Collection of wood chips on a single heap

1: initialize
2: **loop**
3:    **repeat**
4:       random walk
5:    **until** found wood chip
6:    **if** already carrying a wood chip **then**
7:       drop wood chip
8:    **else**
9:       take wood chip
10:   **end if**
11: **end loop**

Depending on the number of termites, this process will take some time. As only randomized operations are used, neither the completion time nor the final location of the heap is known in advance.

6.2   Design paradigms for self-organization

Besides the knowledge about the key properties and the basic methods for self-organization, further information is necessary to develop self-organizing systems. Design paradigms for self-organization in computer networks have been investigated by Prehofer and Bettstetter (2005). In this section, we follow the classification presented in this paper. Even though this discussion closely follows the requirements in communication networks, the listed design paradigms are transferable to any kind of self-organizing system.

6.2.1   Design process

The complete design process is depicted in Figure 6.5. We extended the scheme suggested by Prehofer and Bettstetter (2005) to reflect the operation and control mechanisms for self-organizing systems as described in the previous sections. This diagram gives a rough idea of how to proceed when designing a self-organizing network function. In general, it

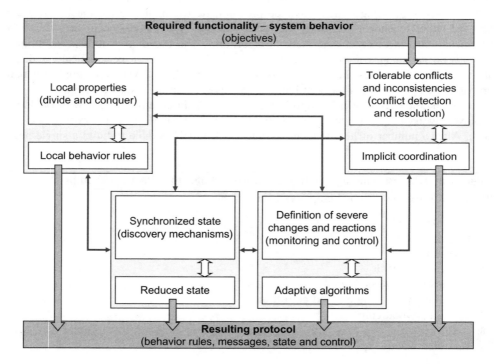

Figure 6.5   Design process for building self-organized systems

is still unknown how to define such a design process in detail – possibly, it is generally impossible.

The design process starts with a description of the overall system behavior that needs to be achieved. This global behavior description is reduced to local properties, e.g. using the divide-and-conquer strategy. Simultaneously, tolerance restrictions are determined from the global objectives. Internally, local behavior rules and implicit coordination mechanisms are designed.

Based on the acquired information, mechanisms to work on reduced state information will be selected. Additionally, an adaptive behavior of the resulting protocol is enabled. After a definition of expected severe changes and corresponding reactions, monitoring and control algorithms help to detect changes in the environment and to update the system behavior accordingly. Again, there is the need to exchange information about the selected algorithms between all four building blocks in order to optimize the final system behavior.

Prehofer and Bettstetter (2005) identified the following four paradigms that enable the design of a self-organizing system. A detailed discussion of these paradigms is provided in the following section.

1. Design local behavior rules that achieve global properties.

2. Do not aim for perfect coordination: exploit implicit coordination.

3. Minimize long-lived state information.

4. Design protocols that adapt to changes.

6.2.2 Discussion of the design paradigms

Paradigm #1: Design local behavior rules that achieve global properties – As previously discussed, centralized functions that establish a global property of the network will not be scalable to huge numbers of network nodes. Nevertheless, they are able to observe and control the process that leads to this property. An example is a centralized addressing scheme that is providing network-wide unique addresses to each participating node. Following the paradigm of self-organization, the central responsibilities have to be distributed among individual entities. No single entity is in charge of the overall organization. Instead, each node has to contribute to a collective behavior.

Following this paradigm, localized behavior rules must be designed that automatically lead to the intended global property. In this context, localized means that entities have only a local view to the overall system and interact only with their direct neighbors. Secondly, we have to relax the global goal, as an approximate solution will be more likely to be achieved.

The local view and interactions with neighboring nodes can be applied to many problems in communications, such as routing and congestion control. Topology control, enabling each node in an ad hoc network to maintain links to its k-th nearest neighbors is an example for localized behavior control. Another example is clustering, in which networks are partitioned into n clusters. Each cluster is controlled by a clusterhead. Several approaches can be found in the literature on how to achieve such clusters with purely local communication and without central control.

The key challenge in the design of such networking functions is to find local behavior rules that directly lead to the desired global behavior. One promising approach is to use the divide-and-conquer principle. Retrieved information is locally processed and aggregated in a way that enables other entities that receive this aggregated information to extract all necessary elements to compute globally valid solutions. Many routing protocols work like that. The most well known example is distance vector routing. Each node calculates the optimal paths towards all known destinations based on locally available information. Only the result is communicated to neighboring nodes. Thus, global optimizations will not be possible. Nevertheless, in many cases, the routing protocol performs as well as if the global state were available.

This design paradigm – reduction to a local view and local interactions – is the basis for typical self-organizing systems. This includes the property that locality also means that changes in the global system have only local consequences. Regarding the efficiency and performance, two notes have to be appended. First, such behavior helps to enable stability and robustness in larger systems, because failures will typically have only a local impact. Secondly, localized decision processes may lead to inconsistencies. We will discuss limitations of self-organizing systems later, in Section 3.7.

Paradigm #2: Do not aim for perfect coordination: exploit implicit coordination – Considering coordination among a number of nodes, two approaches can be distinguished: explicit and implicit coordination. Coordination issues in networks include, for example, the fair allocation of available resources, such as bandwidth and memory. Explicit coordination aims to use a centralized approach to avoid any conflicts and inconsistencies between the nodes. Such a scheme will always rely on particular signaling messages that are used to update global state information. The two main disadvantages of this solution are the low scalability (what about synchronizing thousands of nodes?) and the signaling overhead (what about dynamically changing conditions?).

The key idea used in self-organizing systems is to avoid explicit coordination and to tolerate a certain number of conflicts. For example, conflicts might be acceptable if they are localized, restricted in time or easily detectable and resolvable. Following this idea, the use of implicit coordination can be exploited. Implicit coordination refers to the use of locally available information, e.g. inferred from the local environment, for the decision process. A node observes other nodes in its neighborhood and uses this information to draw conclusions about the network state and to control its local behavior.

A typical example is the implicit allocation of the wireless channel in carrier sensing protocols with collision avoidance. In this case, nodes detect and analyze communications in their neighborhood. Based on these observations, each node can access the wireless channel with only a small collision probability.

An extreme form of implicit coordination is zero coordination that avoids coordination among the nodes at all. Obviously, resource conflicts can easily originate from such a behavior. The primary solution is randomization of the resource allocation process. Nodes may wait for a random time or randomly select other resources, e.g. use another wireless channel. In fact, many self-organizing systems in nature and technology use randomization to initialize the system or to recover from errors or deadlocks.

Exploiting implicit coordination represents a design paradigm that enables – in conjunction with conflict detection and resolution methods – resource- and time-efficient coordination among nodes in a self-organizing system.

Paradigm #3: Minimize long-lived state information – We are considering self-organizing systems that consist of a huge number of participating entities, e.g. nodes in a WSN. Many protocols and techniques used in modern networks require at least limited localized state maintenance. While paradigm #1 considered spatial restrictions of such state information, we now target the temporal properties of state information. Nodes will always keep configured state information, e.g. their address or location, which is either pre-configured or dynamically determined by a network management protocol. Additionally, a synchronized state might be needed that must be consistent among several nodes, e.g. routing information. The maintenance of a synchronized state is difficult in dynamic network environments.

To achieve a higher level of self-organization, the size of such a synchronized state must be minimized, e.g. limited in time. One approach to provide minimized state information is the use of discovery mechanisms. Nodes can on-demand query necessary information. Depending on the dynamics of the overall system, an optimum can be found for the query rate. In general, the more dynamic a system becomes, the less a synchronized state should be maintained.

The amount of maintained state information is related to the first two design paradigms. This means that higher localization and reduced coordination demand will also lead to less state that must be maintained. For example, if nodes use implicit observation-based coordination in a wireless network, they do not need to keep a long-lived state about this network. Similarly, if nodes use on-demand routing protocols, they will not need to store routing tables for a long time period – especially if the network is highly dynamic.

Paradigm #4: Design protocols that adapt to changes – The fourth paradigm is related to the configuration of individual systems as well as of groups of systems. Dynamics in terms of mobility, application demands and available resources generate the need for dynamic adaptations according to these changes. Considering the first three design paradigms, these adaptations must be performed locally, without global state information. Three levels of adaptation can be distinguished:

- Level 1: The simplest adaptivity is to cope with changes such as failures and mobility.

- Level 2: In order to react to changes and to optimize the system performance, protocols should be able to adapt their own parameters, e.g. timers and cluster size.

- Level 3: In case of dramatic changes, the complete change of the employed algorithms can be necessary, e.g. the exchange of a no longer converging clustering algorithm.

In most scenarios, it makes sense to develop protocols that allow all these three levels of adaptivity. Self-organization covers adaptivity as a main objective. In some cases, more complex adaptation might be required in order to provide system-inherent self-learning capabilities. The distributed change of algorithms as proposed by Level 3 is subject to further investigation. Differences in the interpretation of the current environment between neighboring nodes can easily lead to oscillating changes as well as to network partitioning.

6.3   Developing nature-inspired self-organizing systems

Many approaches have been proposed that claim to be either self-organizing or bio-inspired. Some of them fail to convince the academic community in this aspect. Basically, this is due to insufficiently conducted modeling. We will later discuss the necessary steps needed to transfer ideas and mechanisms of natural self-organization to technical applications. In short, at least the following three steps are needed:

(1) Identification of analogies – In order to transfer ideas from natural self-organization to a particular technical system, a detailed analysis of the systems under study as well as their internal system behavior is needed to identify similar structures and methods.

(2) Understanding – In a second step, the detailed modeling of realistic system behavior is recommended. This allows understanding of the underlying functions and mechanisms as well as their characteristics and limitations.

(3) Engineering – Finally, the model has to be simplified and tuned for application in the technical system. This step also includes the comparison with previous solutions that might still outperform the developed 'nature-inspired' self-organizing approach.

As can be seen, the modeling part has a great influence to the quality of the developed systems. In the next section, we give a brief overview of modeling techniques.

6.4   Modeling self-organizing systems

Most of the approaches for self-organization as described in the literature are the result of empirical studies. These studies provide a first glimpse into the benefits that might be expected from the developed methods and techniques. However, testing the accuracy and completeness of these algorithms requires a further state – the formulation of a rigorous model. In general, modeling requires the translation of a verbal understanding of the inter-actions among the individuals in the self-organizing system into a mathematical form such as a simulation or a set of equations. This process is a very useful exercise. It requires more precise definition of the system's behaviors.

6.4.1   Overview of modeling techniques

The development of such models follows the general approach of modeling. First, an in-depth study of the system is necessary. Then, the behavior can be reproduced in a model. Secondly, the model needs to be verified. In order to test the model, the behavior found in the system under study is compared to the output of the model. This process needs careful attention, such as verification under different environmental conditions. Ideally, the model predicts the system behavior correctly. If not, a third step is the further analysis of the system as well as the optimization of the model. This may include a complete change in the modeling technique.

In the following, selected modeling techniques are discussed that are commonly used in the context of studying complex and self-organizing systems. General recommendations to modeling have been described by Law and Kelton (2000). For example, they give a de-tailed classification and introduction to modeling and simulation. Roughly, system analysis using modeling techniques can be distinguished into physical models and mathematical models. Focusing on the latter ones, two solutions are recommended: analytical solution and simulation.

Besides the choice of the modeling technique, there are several common mistakes that can be made in analyzing the system and the use of simulation models. Jain (1991) provides a great overview to typical mistakes. The most common are listed below:

- No goals – When analyzing a system, regardless of which technique is being used, it must be clear which objectives are intended. Even though it might be an expensive act, once the problem is clearly formulated and the goals are written down, finding the solution is often easier. There is typically no general-purpose model available.

- Improper modeling technique – Using inadequate modeling techniques or improper languages can result in either too complex or even unmanageable models. Depending on the defined goals and the system that has to be analyzed, different modeling techniques are advisable.

- Biased goals or unsystematic approach – The selection of adequate metrics to evaluate the system's performance has a great influence on the results. The systematic approach is to identify a complete set of goals and metrics.

- Inappropriate level of detail – The level of detail used in modeling has a significant impact on problem formulation. It is clear that the goals of a study have a significant impact on what is modeled and how it is analyzed.

6.4.2 Differential equation models

Differential equations can be used to describe the behavior of the overall system over time. As an example, we consider a simple sensor model, as depicted in Figure 6.6. Each sensor can be in four different states. Considering a system consisting of a given number of sensors N, we can describe the overall system behavior by means of the number of sensors in idle state I, in measurement state M and in transmission state T, respectively, by means of differential equations, as depicted in Equation 6.1 (r_{IM} depicts the rate at which the sensor changes from I to M and so forth). Three states describe the local behavior: IDLE (resting state), MEASUREMENT (measurement and computation) and TRANSMISSION (communication with neighboring nodes). The transitions between the states are labeled with rates, e.g. the rate r_{IM} depicts the state change from IDLE to MEASUREMENT. I, M and T represent counters for sensor nodes in states IDLE, MEASUREMENT and TRANSMISSION, respectively:

$$\frac{\partial I}{\partial t} = r_{\text{MI}}M + r_{\text{TI}}T - r_{\text{IM}}I - r_{\text{IT}}I$$

$$\frac{\partial M}{\partial t} = r_{\text{IM}}I - r_{\text{MI}}M - r_{\text{MT}}M \qquad (6.1)$$

$$\frac{\partial T}{\partial t} = r_{\text{IT}}I + r_{\text{MT}}M - r_{\text{TI}}T.$$

These differential equations allow continuous modeling of the system behavior over time. Please note that the solution always provides an approximation of the system's state if discrete system states have to be considered. In this case, only fractional solutions are provided by the model. To find a stationary solution for this model, one can simply set all the derivatives $\dfrac{\partial I}{\partial t}$, $\dfrac{\partial M}{\partial t}$ and $\dfrac{\partial T}{\partial t}$ to zero and algebraically solve the system.

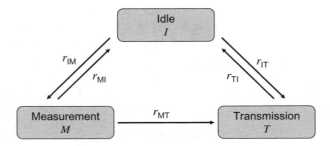

Figure 6.6 Simple state chart describing the behavior of a sensor node

6.4.3   Monte Carlo simulations

As differential equation models do not treat each subsystem, i.e. each sensor in our example, as an individual entity that is behaving in a stochastic manner, other modeling techniques are necessary if not only the global view is needed, but also the behavior of each participating system is of interest. Simulation models often follow the discrete approach. One particular approach is the Monte Carlo simulation that incorporates the randomness that one expects from complex and self-organizing systems. In such a simulation, each individual node is tracked separately, including its interactions with other nodes. Thus, it is possible to evaluate the statistical behavior of each subsystem as well as of the overall self-organizing system. Additionally, uncertainties can be included into the model.

The stationary solution from the differential equation model can be achieved by executing many runs of the simulation and measuring the average of the results.

6.4.4   Choosing the right modeling technique

We briefly presented two very different modeling techniques. There are many other modeling techniques available that can be used for performance evaluation (Jain 1991; Law and Kelton 2000). Regarding the depicted techniques, differential equation models can be used to characterize the quantitative behavior of the system over time, as well as equilibrium conditions. Furthermore, these models are more appropriate for describing continuous rather than discrete processes.

On the other hand, simulation models are often easier to develop than to deviate complex differential equation systems. Additionally, the simulation models allow the behavior of individual entities that participate in the global system to be observed. Nevertheless, detailed simulation models can demand huge amounts of computational power for processing and storage of state information of each individual entity.

Appendix I

Self-Organization – Further Reading

In the following, a selection of some interesting textbooks, papers and articles is provided that is recommended for further studies of the concepts of self-organization. Additionally, a list of major journals and conferences is provided that are either directly focused on self-organization and related aspects or broadly interested in publishing special issues or organizing dedicated workshops in this domain. Obviously, this list cannot be complete or comprehensive. It is intended to provide a starting point for further research.

Textbooks

- E. Bonabeau, M. Dorigo and G. Theraulaz, *Swarm Intelligence: From Natural to Artificial Systems*, Oxford University Press, 1999.

- S. Camazine, J.-L. Deneubourg, N. R. Franks, J. Sneyd, G. Theraula and E. Bonabeau, *Self-Organization in Biological Systems*, Princeton University Press, 2003.

- L. N. de Castro, *Fundamentals of Natural Computing: Basic Concepts, Algorithms, and Applications*, Chapman & Hall/CRC, 2006.

- M. Eigen and P. Schuster, *The Hypercycle: A Principle of Natural Self Organization*, Springer, 1979.

- J. H. Holland, *Adaptation in Natural and Artificial Systems*, 2nd edn, MIT Press, 1975.

- S. A. Kauffman, *The Origins of Order: Self-Organization and Selection in Evolution*, Oxford University Press, 1993.

- J. v. Neumann, *Theory of Self-Reproducing Automata*, University of Illionois Press, 1966.

Self-Organization in Sensor and Actor Networks   Falko Dressler
© 2007 John Wiley & Sons, Ltd

- M. Resnick, *Turtles, Termites, and Traffic Jams: Explorations in Massively Parallel Microworlds* MIT Press, 1997.

- H. von Foerster and G. W. Zopf (eds), *Principles of Self-Organization*, Pergamon Press, 1962.

- F. E. Yates, A. Garfinkel, D. O. Walter and G. B. Yates, *Self-Organizing Systems: The Emergence of Order*, Plenum Press, 1987.

Papers and articles

- G. Di Caro and M. Dorgio, 'AntNet: Distributed Stigmergetic Control for Communication Networks', *Journal of Artificial Intelligence Research*, vol. 9, pp. 317–365, December 1998.

- F. Dressler, 'Self-Organization in Ad Hoc Networks: Overview and Classification', University of Erlangen, Dept of Computer Science 7, Technical Report 02/06, March 2006.

- C. Gerhenson and F. Heylighen, 'When Can We Call a System Self-organizing?', Proceedings of 7th European Conference on Advances in Artificial Life (ECAL 2003), Dortmund, Germany, September 2003, pp. 606–614.

- F. Heylighen, 'The Science of Self-Organization and Adaptivity', *The Encyclopedia of Life Support Systems (EOLSS)*, 1999.

- J. O. Kephart and D. M. Chess, 'The Vision of Autonomic Computing', *IEEE Computer*, vol. 36(1), pp. 41–50, January 2003.

- C. Prehofer and C. Bettstetter, 'Self-Organization in Communication Networks: Principles and Design Paradigms', *IEEE Communications Magazine*, vol. 43(7), pp. 78–85, July 2005.

- K. Sohrabi, J. Gao, V. Ailawadhi and G. J. Pottie, 'Protocols for Self-Organization of a Wireless Sensor Network', *IEEE Personal Communications*, 2000.

- G. Theraulaz and E. Bonbeau, 'A Brief History of Stigmergy', *Artificial Life Journal*, vol. 5(2), pp. 97–116, 1999.

- M. Weiser, 'The Computer for the 21st Century', *Scientific American*, vol. 265(3), pp. 94–104, September 1991.

- M. Weiser, 'Ubiquitous Computing', *IEEE Computer*, vol. 26(10), pp. 71–72, October 1993.

- F. Zambonelli, M.-P. Gleizes, M. Mamei and R. Tolksdorf, 'Spray Computers: Frontiers of Self-Organization for Pervasive Computing', Proceedings of 13th IEEE International Workshops on Enabling Technologies: Infrastructure for Collaborative Enterprises (WETICE'04), 2004, pp. 403–408.

- C. Zimmer, 'Complex Systems: Life after Chaos', *Science*, vol. 284(5411), pp. 83–86, April 1999.

Journals

- *Complex Systems* (Complex Systems Publications)

- *Communications Magazine* (IEEE)

- *Computer* (IEEE)

- *Information Sciences* (Elsevier)

- *Intelligent Systems* (IEEE)

- *International Journal of Intelligent Systems* (Wiley Interscience)

- *International Journal of Systems Science* (Taylor & Francis)

- *International Journal of Unconventional Computing* (Old City Publishing)

- *Journal of Intelligent Information Systems* (Kluwer Academic Publishers)

- *Journal of Selected Areas in Communications* (IEEE)

- *Knowledge and Information Systems* (Springer)

- *Nature* (Nature Publishing Group)

- *Science* (AAAS)

- *Transactions on Computational Biology and Bioinformatics* (IEEE/ACM)

- *Transactions on Computational Systems Biology* (Springer)

- *Transactions on Systems, Man, and Cybernetics* (IEEE)

Conferences

- BIONETICS – IEEE/ACM International Conference on Bio-Inspired Models of Network, Information and Computing Systems

- CEC – IEEE Congress on Evolutionary Computation

- GECCO – Genetic and Evolutionary Computation Conference

- CIBCB - IEEE Symposium on Computational Intelligence in Bioinformatics and Computational Biology

- ECAL – European Conference on Advances in Artificial Life

- SMC – IEEE International Conference on Systems, Man, and Cybernetics

- SSCI – IEEE Symposium Series on Computational Intelligence

- WCCI – IEEE World Congress on Computational Intelligence

Part II

Networking Aspects: Ad Hoc and Sensor Networks

Self-organization techniques in communication networks can be seen as the key towards better scalability of networking techniques, especially in the context of wireless networks and low-resource sensor networks. The efficiency of such networks is usually not only measured in terms of bandwidth and throughput. Instead, energy considerations and scalability issues are investigated as primary objectives. After a brief introduction of mobile ad hoc and sensor networks, the relevant protocols and methods are discussed. The main focus is on the autonomous behavior of single network nodes that are self-organizing in a greater compound to a large-scale system. All studies are accompanied by case studies of typical protocols and techniques developed in the context of ad hoc and sensor network communication.

Outline

- Mobile ad hoc and sensor networks

  The main concepts and characteristics of ad hoc and sensor networks are discussed. This includes hardware and software constraints, application scenarios, and programming paradigms. Additionally, the most prominent research objectives are outlined.

- Self-Organization in sensor networks

  We revisit the basic methods of self-organization for the specific domain of sensor networks. In this context, we discuss the mapping of self-organization techniques to sensor network algorithms.

- Medium access control

  Wireless communication demands for careful resource management in order to reduce the probability of collisions and to satisfy the energy constraints of battery-powered sensor nodes. Primarily, we study busy–sleep cycles based on synchronization techniques and transmission–power control mechanisms.

Self-Organization in Sensor and Actor Networks   Falko Dressler
© 2007 John Wiley & Sons, Ltd

- Ad hoc routing

  First, the main differences between address-based ad hoc routing and data-centric forwarding are outlined. Then, based on several case studies, the ideas of proactive and reactive routing as well as the concept of link affinity are explained. Finally, dynamic address assignment techniques are introduced.

- Data-centric networking

  The basic methods of data-centric networking are introduced. This includes flooding-based techniques such as gossiping and optimized variants in particular. The presentation furthermore outlines agent-based techniques and diffusion routing as well as aggregation techniques and associated problems.

- Clustering

  Localized communication techniques are essential solutions to improve the scalability of sensor network algorithms. Therefore, clustering algorithms are investigated, including general clustering solutions, such as k-means and hierarchical clustering, and specialized solutions for sensor networks.

- Networking aspects – further reading

7

Mobile Ad Hoc and Sensor Networks

Wireless communication is an enabling technology for various applications. Starting with the ALOHAnet in 1970 (Abramson 1985), radio communication has already been used for decades. Nevertheless, only recently advances in wireless communications and electronics enabled the development of low-cost high-performance wireless networking technologies. In general, one has to distinguish between infrastructure-dependent and ad hoc, i.e. infrastructure-less, networks. The most prominent examples of wireless communication using a fixed infrastructure are cellular wireless networks (Wisniewski 2004), used particularly for mobile phones and wireless local area networks according to the IEEE 802.11 standard (IEEE 1999), better known as Wireless LAN (WLAN) working in infrastructure mode. In this book, we focus only on ad hoc wireless networks, which are operating without any pre-established infrastructure. The following sections will introduce ad hoc and sensor networks and elaborate on the commonalities and differences, accompanied by a summary of current research issues in these domains. The main goal is to introduce mobile ad hoc and sensor networks before studying selected communication aspects in detail.

7.1   Ad hoc networks

The name 'ad hoc' gives a rough idea about the principles and technology behind ad hoc networks. Instead of deploying a fixed network infrastructure, networking nodes are expected to establish network connections on demand in an ad hoc manner. Many application examples can be envisioned that rely on a communication network that must be established without any available infrastructure. Two primary reasons led to the development and support of ad hoc networks: the unavailability of a infrastructure in some cases and its costs. Examples of scenarios demanding an ad hoc network are described in the following.

- In several extreme cases, there is no infrastructure available by definition. This includes disaster areas and emergency operations. Consider, for example, hurricane

Self-Organization in Sensor and Actor Networks   Falko Dressler
© 2007 John Wiley & Sons, Ltd

Cathrina in 2005.[1] After destroying huge parts of New Orleans, emergency operations strongly required the quick setup of a communication network – which was possible thanks to the advances in ad hoc networking.

- Another case is the avoidance of unnecessary costs for a network infrastructure. For example, in rural sites or in construction areas, it is too expensive or inconvenient to set up an infrastructure-based communication network. A special form of ad hoc networks – Wireless Mesh Networks (WMNs) – are used in this case.

- Another scenario requiring ad hoc networking is collaborative and distributed computing. For quick data transfers between collaborating people, e.g. during a conference or in a classroom, a temporary network infrastructure must be set up among a group of people.

7.1.1 Basic properties of ad hoc networks

The basic requirements of ad hoc networks can be summarized as follows. The primary meaning of ad hoc networks is to construct a network without any infrastructure using only the abilities of the participants. Thus, ad hoc networks are constructed on demand and for a special purpose. In the simplest case, an ad hoc network consists of a number of computers building a single-hop network, i.e. all networking nodes are in the direct communication range of each other node. As the radio transmission range is limited, such a single-hop network will work in very few scenarios only. In all other cases, data must be delivered over a path involving multiple nodes. This multi-hop routing requires mechanisms for dynamic path identification and management. Besides the problem of the limited communication range of wireless communication, directed communication is often infeasible because of the distance, obstacles and specific physical phenomena disturbing the radio communication. Multi-hop networks can bypass these problems.

The multi-hop routing in ad hoc networks leads to a second major difference of ad hoc networks compared with infrastructure-based networks. Besides application-dependent data communication of end systems, all nodes must also provide network functions such as data forwarding and routing activities. This duality–each node participates as an end system and as a network element, i.e. a router–is unique to ad hoc networks.[2]

Studying the properties of ad hoc networks, we start with the simplest case–all participating nodes have a pre-configured unique identifier, i.e. address, and remain fixed in time and space. Then, the required network functionality includes mechanisms to detect the current network topology and to establish routes, i.e. communication paths, between all participating nodes. The best known example of such a network are Wireless Mesh Networks (WMNs). The application examples of WMNs are manifold. They range from community and neighborhood networking to spontaneous (emergency or disaster) networking (Akyildiz and Wang 2005; Akyildiz et al. 2005b). The key aspect for establishing such networks is the capability of the participating nodes to self-organize their network topology, i.e. to identify and to maintain wireless connections similar to wired ones in infrastructure-based networks.

---

[1]For details see, for example, http://www.telealtobut.it/katrina/.

[2]This duality can be observed also in peer-to-peer networks that have similar properties compared to ad hoc networks except for the wireless communication (Steinmetz and Wehrle 2005).

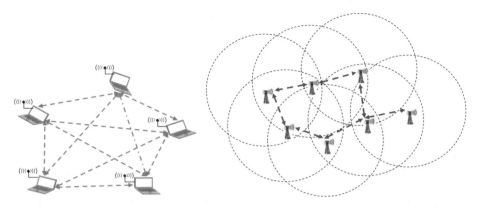

Figure 7.1 Typical examples of ad hoc networks; single-hop ad hoc communication be-tween portable computers (left) and multi-hop ad hoc networking without pre-deployed infrastructure (right)

Figure 7.1 depicts two typical examples of ad hoc networks. On the left-hand side, the collaboration between multiple users takes place. This example shows a special case of ad hoc networks—a fully meshed network. In such fully meshed networks, only single-hop communication is required, which simplifies the discovery and maintenance of the network topology. On the right-hand side, the normal case of multi-hop ad hoc networks is shown. The circles around the antennas depict the radio communication range, which is usually approximated in the form of circles. Only destinations within this range can be targeted in a single hop. The communication between arbitrary nodes relies on multi-hop forwarding of the data messages.

Without a pre-deployed network infrastructure, several aspects become more compli-cated compared with an infrastructure-based network. Essentially, there is no central entity available for the organization of network parameters. Additionally, as wireless communica-tion is used, the nodes have to deal with the limited range of the wireless radio transmissions as well as with much higher error rates, as collisions or physical effects like signal fading are quite common in any kind of wireless communication.

The lack of a central entity (like a base station) means that the participating nodes must organize themselves into a network. Thus, self-organization is the key to successful ad hoc networking or, in other words, it is an inherent capability of ad hoc networks. Besides other aspects, the following aspects must be considered in particular. The medium access control must be decided in a distributed fashion, as there is no base station that can assign transmission resources. Among others, the following issues must be addressed: distributed operation, synchronization, hidden terminal and exposed terminal problem, throughput, access delay, and fairness. On the network layer, the route and topology discovery is of highest interest. This includes route acquisition delay, loop-free routing, control overhead, scalability and others issues such as addressing and service discovery. For a more complete list of issues in ad hoc networks, please refer to Murthy and Manoj (2004).

Table 7.1   Comparison between infrastructure-based and ad hoc, i.e. infrastructure-less, networks

	Infrastructure-based network	Ad hoc network
Prerequisites	Pre-deployed infrastructure, e.g. routers, switches, base stations, servers	None
Node properties	End system only	Duality of end system and network functions
Connections	Wired or wireless	Usually wireless
Topology	Outlined by the pre-deployed infrastructure	Self-organized topology maintained by the nodes
Network functions	Provided by the infrastructure	Distributed to all participating nodes

Table 7.1 summarizes the properties of ad hoc networks compared to infrastructure-based networks. This list focuses on the most important capabilities but does not outline all possible commonalities and differences.

Based on the previous considerations, we can now define the term 'ad hoc network' more appropriately, focusing on the unique characteristics of wireless ad hoc networks.

Definition 7.1.1 Ad hoc network – An ad hoc network is created on demand and for a specific purpose without the need for any pre-deployed infrastructure. All nodes in an ad hoc network participate as end systems and as routers. The participating network nodes must self-organize to establish an operable network, i.e. they need to provide topology control, MAC, and routing functions.

7.1.2   Mobile Ad Hoc Networks

In more complex scenarios, the networked nodes are allowed to move in time and space. In this case, the quality of the network primarily depends on the speed to adapt to new topologies, i.e. to re-establish neighborship relations and communication paths through the network. In the last decade, a new research branch has emerged, concentrating on the study of mobility aspects in ad hoc networks. Within Internet research, the advantages of Mobile Ad Hoc Networks (MANETs) are well accepted. Therefore, the Internet Engineering Task Force (IETF) has also initiated a dedicated working group called MANET WG, focusing on different routing mechanisms in MANETs (Chakeres and Macker 2006).

In order to describe the behavior of MANETs, the key characteristic is the mobility model (Bai and Helmy 2004). Besides the need to self-organize the network topology in a timely manner, the kind and way of node mobility strongly influences all performance and quality aspects in such a network. While, usually, only spatial movement is considered in mobility models, in general, one has to distinguish between temporal and spatial mobility. Temporal changes refer to the appearance of new nodes or their outages, as well as to the temporary availability of nodes due to sleep cycles. According to Bai and Helmy (2004), spatial mobility can be categorized into random models, models with temporal dependency, models with spatial dependency and models with geographic restrictions.

MANETs can be used in several application examples. The most common scenarios are their use in emergency situations (as already cited for static ad hoc networks) and their application for inter-vehicle communication. While the establishment of an ad hoc network infrastructure in disaster areas does not usually rely on mobile nodes (while mobile scenarios can be envisioned as well), human actors are the main focus in emergency operations. These people, carrying networked nodes, will introduce a high amount of mobility.

Another example of MANETs' exhibiting a similar degree of mobility are Vehicular Ad Hoc Networks (VANETs) (Luo and Hubaux 2004). Such networks, also known as Inter-Vehicle Communication (IVC), enable the drivers (or their vehicles) to communicate with other drivers (or their vehicles) that are located out of direct radio communication range if a multi-hop network is built among several vehicles. As a result, information gathered through IVC can help to improve road traffic safety and efficiency (Luo and Hubaux 2004). However, due to mobility constraints, driver behavior and high mobility, VANETs exhibit characteristics that are dramatically different from many generic MANETs (Blum *et al.* 2004). The main applications of VANETs, as summarized by Reichardt *et al.* (2002), can be roughly categorized into three classes:

- Information and warning functions: Dissemination of road information (including incidents, congestion, surface condition, etc.) to vehicles distant from the subjected site.

- Communication-based longitudinal control: Exploiting the 'look-through' capability of IVC to help in avoiding accidents and platooning vehicles for improving road capacity.

- Cooperative assistance systems: Coordinating vehicles at critical points, such as blind crossings (a crossing without light control) and highway entries.

Besides the named issues in ad hoc networks, the mobility of nodes increases the need for efficient network operations even further. The route setup time and the associated overhead for topology control become dominating factors in MANETs. Depending on the amount and speed of temporal and spatial mobility, the topology of the ad hoc network can change very rapidly. As nodes are meant to self-organize the network state, the scalability of networking methods and techniques must be questioned in order to evaluate the performance of MANETs.

Let us consider the following example of an ad hoc network with one node oscillating between two positions. Figure 7.2 depicts this scenario. Obviously, the network topology changes with each move. The participating nodes will need some time to realize the move and to react to it by adapting to this new topology. If, for example, the reconfiguration time is in the same range as the oscillating frequency, the mobile node will never become able to fully participate in this ad hoc network. Thus, the mobility model must be reflected by the ad hoc operations by detecting it first and by adapting to adequate system parameters.

The following definition summarizes our considerations on MANETs. Besides all the properties and restrictions discussed for non-mobile ad hoc networks, MANETs allow networking nodes to move in time and space.

Definition 7.1.2 Mobile ad hoc network – Ad hoc networks that support node mobility in time and space, i.e. temporal and spatial changes of the network topology, are named

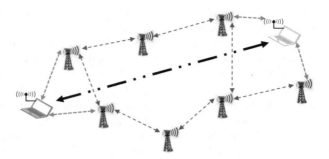

Figure 7.2 A mobile node oscillating between two positions in an ad hoc network; every time the node moves, a new network topology is created and must be adequately detected and distributed by members of the ad hoc network

mobile ad hoc networks. Such networks dynamically must adapt to changing topologies in a self-organizing manner.

Obviously, ad hoc and mobile ad hoc networks have to deal with a high degree of systems' dynamics. Nodes have to establish communication paths on demand, i.e. an ad hoc network must be created. This includes mechanisms for Medium Access Control, routing and general topology control. Additionally, MANETs require further dynamic, as changes of the network topology, i.e. node mobility in time and space, must be supported. Such dynamic systems are the target audience for self-organization methods. The kind and degree of self-organization strongly depends on the required adaptivity of the ad hoc network. Nevertheless, we can stick to self-organization as the main characteristic and requirement for efficient ad hoc networking (Dressler 2006b).

One last property of ad hoc networks needs to be mentioned to conclude our considerations on on-demand networking. The use case of an ad hoc network differs also in the degree of interactivity with human operators. Basically, the goal or the objective which the network is intended to solve is either pre-programmed or pre-defined, or determined by user interactions.

- No interactivity – The overall goal is provided either in the form of a precise description or in terms of a desired behavior; thus, the network must completely manage its topology and system parameters in a self-organizing way.

- Little interactivity – The system can be (partially) controlled through user interactions. This means that some basic network functions will be provided by the network while others are user-controlled.

- High interactivity – In this case, there is no pre-defined goal; the system behavior depends on regular user interactions. Perhaps all system parameters are user-defined.[3]

The higher the degree of interactivity, usually, the less self-organization of the network is required. On the other hand, if no manual control is needed (or if it is not possible at

---

[3] As a note, such networks may not really be called ad hoc networks, as, actually, most parameters are controlled manually and not in an automated 'ad hoc' manner.

all), then the system, i.e. the ad hoc network, must completely autonomously self-organize its parameters to adapt to changing environmental conditions. Such networks are the main focus of this book. A particular example are Wireless Sensor Networks (WSNs), which are described in the following section.

7.2 Wireless Sensor Networks

With the beginning of the 21st century, a new research domain increased the efforts in ad hoc network research: Wireless Sensor Networks (WSNs). The stringent resource restrictions of networked embedded systems that are used to build a WSN motivated several new research approaches (Akyildiz *et al*. 2002a,b; Culler *et al*. 2004). The basic idea of WSN is to combine the measurement of physical values, e.g. the temperature or the humidity, and to transport the measurement results to a base station for further processing. Thus, sensing and communication tasks have to be performed by so-called sensor nodes that collaborate to build an ad hoc network. Additionally, various kinds of pre-processing activities can be applied on the measured data. Nowadays, the scenarios for WSNs have become more challenging by introducing a high degree of heterogeneity, mobility and autonomy to the operation and management of sensor networks. This section reviews the main ideas behind sensor network technology and outlines properties, restrictions and typical application scenarios.

7.2.1 Basic properties of sensor networks

Sensor networks can be described as the composition of multiple sensor nodes that build an ad hoc network using wireless radio communication. While multiple properties and characteristics can be distinguished in such WSNs, the main differentiation criteria are the kind of sensor processing, the networking methods and paradigms, and the application scenarios (Margi 2003). A typical sensor network is depicted in Figure 7.3. In this example, several sensor nodes are depicted, equipped with radio communication facilities. As described for ad hoc networks, these wireless network interfaces are used to establish an ad hoc network, which is able to forward data in such a multi-hop network. On the right hand side, a base station is depicted. Usually, all sensor nodes deliver their data to such a base station, which, in turn, controls the sensor nodes (if necessary).

Obviously, multiple roles can be distinguished in such a sensor network. While all nodes equally participate in performing network-related operations, i.e. Medium Access Control (MAC) and routing functions, higher-layer functions differ depending on the role of a node. The following roles (or tasks) can be distinguished (Karl and Willig 2005):

- Sensors – Sensors or sources of data perform the measurement process, i.e. they use attached sensors to retrieve corresponding readings and report the results to 'somewhere'. Depending on the employed communication methods, a specific sink will be addressed or the data record will be transmitted through the network until an appropriate destination is found. Typically, the sensor nodes are equipped with different kinds of actual sensors.

- Base stations – Base stations, or more general sinks of data, represent nodes that are interested in receiving data from the WSN. Typically, such sinks are part of the

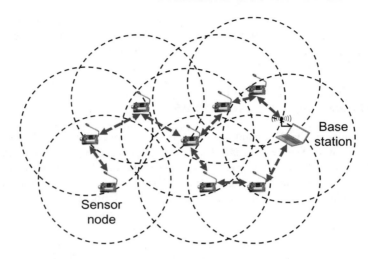

Figure 7.3 A typical WSN consisting of several sensor nodes and a base station; the communication network is created on demand using ad hoc networking mechanisms

sensor network. Nevertheless, also external sinks can be supported by introducing gateways to other networks.

- Actuators – Actuators control their environment by acting on received data. This can take place in the form of physical control by some kind of motor or electric lever, or in the form of modifications of logical system parameters. Usually, actuators also represent a sink. A more detailed study of actuators and their relationship to the sensor network can be found in Part III.

- Processing elements – Furthermore, processing elements can be found in sensor networks that pre-process received sensor data, such as by aggregating them. Currently, more sophisticated in-network processing methods are being developed for pure network-centric operation without external control. Basically, such processing elements represent sinks as well as sources of data, as received information is processed and the results are induced into the network again.

Thus, differently from other kinds of networks, there is usually no communication between arbitrary nodes in WSN. Instead, specific nodes generate data to be delivered to designated sinks. This assumption also holds for in-network data processing. In this case, these intermediate nodes represent sinks from an application point of view.

Besides discussing the potential roles which a sensor node may perform, it is necessary to analyze the system components of a typical sensor node. Basically, each node consists of a number of sensors, a processing unit with attached memory, a radio communication unit and a battery. In this textbook, we consider only embedded systems as sensor nodes, while, in principle, any system that can perform sensing and communication tasks may be called a sensor node. Thus, the processing unit will usually be a small micro controller.

From these observations, we can see that such sensor nodes, and therefore the entire sensor network, are highly resource-restricted. Extensive computations will not be possible

considering the available processing and storage capacities. Similarly, the radio communication unit will only be able to transmit messages with limited throughput to neighboring
nodes. Finally, the limited energy resources as provided by the attached battery will enable
the sensor node to operate only for a very limited time. We will highlight the capabilities
of a single sensor node in more detail in the next paragraph.

At the beginning of sensor network research, WSNs were considered homogeneous
and static. All sensor nodes periodically measure several sensor readings and transmit
them to a single base station (as depicted in Figure 7.3). Today, all these assumptions have
been dropped. First, mobility in time and space was introduced. Duty cycles have been
introduced in order to prolong the node lifetimes and, therefore, frequent topology changes
must be taken care of (Li *et al*. 2004; Wang and Xiao 2006). Similar requirements appear if
mobile nodes (in terms of spatial mobility) are considered. Secondly, in-network processing
allowed much better utilization of the available resources in the network (Dressler 2006a).
Nevertheless, such operation principles replace the single base station concept. Finally,
homogeneity in terms of hardware and software setups is usually replaced by variants
of heterogeneous sensor nodes, i.e. nodes equipped with different sensors and, possibly,
even different micro controllers or radio interfaces. Similarly, the installed applications
and operating systems become heterogeneous as system variants are developed that better
utilize the available resources, depending on the node's associated tasks. This, nowadays,
heterogeneity as a key characteristic of WSNs (Duarte-Melo and Liu 2002).

In summary, it can be said that WSNs are highly resource-restricted ad hoc networks.
Mobility and heterogeneity are becoming major concerns in this domain as application
scenarios approach that either directly demand these properties or emphasize on exploiting
mobility or heterogeneity aspects in order to perform sensor network applications more
efficiently (Lee *et al*. 2004; Wang *et al*. 2005).

Definition 7.2.1 Wireless Sensor Network – A sensor network is a specific type of
MANETs. It consists of sensors collecting particular measures, i.e. sensor readings, and
processing elements, which collect these measures for further processing. Usually, sensor networks are assumed to be strongly resource-restricted in terms of communication,
processing and storage capabilities, and in terms of available energy.

The simplest case of a WSN is a network consisting of a number of sensors and a
single base station. The other extreme is a network performing in-network processing, in
which all nodes may perform sensing and processing tasks (see also Section 14.4).

A particular idea needs to be highlighted here, as it was one of the main source of
inspiration for the sensor network community. In the late 1990s, the 'smart dust' idea
(Kahn *et al*. 1999, 2000) identified challenges and application scenarios for WSNs. The
dream was to distribute huge amounts of sensor nodes from an aircraft. These sensors
self-organize on the ground to form an ad hoc network and start monitoring the target
area according to the application requirements. First hardware solutions are also available
(Warneke *et al*. 2001, 2002). So far, the idea is still a dream, while many of the challenges
have been appropriately addressed or even solved.

Regarding the evaluation of WSNs, one has to keep in mind the application-dependent
objectives of a sensor network. Usually, it is not meant to simply move bits like another network. Rather, that it is about providing answers. Issues like geographic scoping

are natural requirements, absent from other networks (Karl and Willig 2005). Thus, WSN-specific performance characteristics are required compared with other networks. The highest impact on the application-level quality is the overall lifetime of the sensor network. Additionally, the application may have stringent requirements on coverage, i.e. the part of the network covered either by a specific sensor or by communication radii. We will discuss these properties and how to map classic performance metrics to the network lifetime in Section 17.3.

7.2.2   Composition of single-sensor nodes

Even though we are interested in the inter-networking between sensor nodes in a WSN, a brief description of the composition of a single-sensor node is helpful in order to understand the stringent resource restrictions of sensor nodes. Sensor nodes – as envisioned in this book – represent a specific class of embedded systems. A typical sensor node, as depicted in Figure 7.4, is composed of a processing unit, i.e. a micro controller, with some associated memory and, usually, permanent storage, such as a flash memory. For communication purposes, a radio transceiver is embedded into the device as well as several sensors. Finally, a battery module must be present to provide energy for all components (Akyildiz *et al.* 2002a; Estrin *et al.* 1999b).

Besides the specific hardware properties of such embedded systems, the software development process must incorporate the application characteristics of sensor networks. As previously described, sensor nodes are required to measure sensor readings, pre-process them and transmit the results towards a sink node. Thus, the typical process-oriented operation of computer systems is inappropriate for sensor nodes. Instead, event-driven architectures are needed. An event is triggered if a sensor reading is available, if a timeout is reached or if a data packet is received. All of these events are usually triggered by an interrupt. Usually, there is no other software-driven processing requirement on the node. Therefore, a node must only be activated by interrupts and the programming paradigms need to be adapted to this working behavior.

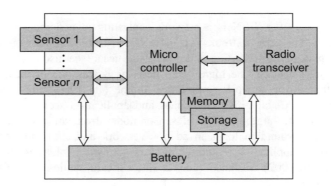

Figure 7.4 Schematic view of a typical sensor node; depicted are system components and their interconnections

Hardware components

The individual hardware components of sensor nodes are described in the following. This enumeration is intended to give examples of hardware modules as employed in typical sensor nodes – not as a complete and detailed evaluation of possible hardware installations.

- Processor (and memory) – A commonly employed micro controller for typical sensor nodes is the Atmel ATmega128. This 8-bit low-power system operating at 16 MHz provides 128 kByte flash memory and 4 kByte SRAM. Other systems manufactured by Altera, Cypress and others have similar capabilities. These data demonstrate the strongly limited processing capabilities of sensor nodes. The processing performance of such systems has been analyzed in a wide variety of tests. For example, the performance of computational intensive cryptographic operations has been evaluated by Passing and Dressler (2006).

- Radio transceiver – For wireless radio communication, multiple radio technologies and transceivers are used on sensor nodes. Commonly employed radio transceivers include the ultra-low-power single-chip RF transceiver Chipcon CC1000, operating in the 315/433/868/915 MHz SRD bands, and the single-chip 2.4 GHz IEEE 802.15.4 compliant (IEEE 2006) and ZigBee[4] ready RF transceiver Chipcon CC2400. Experimental throughput evaluations have, for example, been conducted by Anastasi *et al.* (2004).

- Battery – The most challenging constraints on the battery source are its limitation in size and capacity. The battery size is directly proportional to its capacity. Besides these constraints, additional aspects must be considered for their use in sensor nodes, including energy density, environmental impact, safety, cost, available supply voltage and charge/discharge characteristics. Battery management techniques help to prolong the lifetime of single-sensor nodes. This includes, for example, the exploitation of the inherent property of batteries to recover their charge when kept idle, as well as energy-scavenging techniques to recharge batteries from ambient energy (light, vibration and others). To estimate the energy consumption and, in conclusion, the node lifetime, experimental measurements have been conducted, for example, by Landsiedel *et al.* (2005).

- Sensors – As all micro controllers support a number of analog–digital converters, all kinds of sensors can easily be attached to a sensor node. Typical sensors can measure the temperature, light, motion, humidity, geographical positions and many other physical values.

In general, an innumerable variety of sensor nodes (composed of the mentioned components) exists, as sensor nodes are typically developed for a particular application. Nevertheless, three particular examples of commercially available sensor nodes should be mentioned, as these types are widely used in the scientific community to evaluate developed algorithms and methods in lab environments. Crossbow, Inc. continued the development and production of Mica and Mica2 motes[5] that were originally developed at the Berkeley University. A similar sensor node developed at the ETH Zurich is the BTnode.[6] Both nodes, as shown

---

[4]ZigBee Alliance – http://www.zigbee.org/.
[5]http://www.xbow.com/.
[6]http://btnode.ethz.ch/.

Figure 7.5 Typical sensor node bases that are commercially available: BTnode from ETH Zurich (left) and Mica2 from Crossbow, Inc. (right)

in Figure 7.5, employ the ATmega128 micro controllers and the Chipcon CC1000 radio transceiver, and can be equipped with various sensors.

Operating system and programming paradigms

The operating systems and programming paradigms are entirely different compared with other computer systems. The main reason lies in the event-driven operating principle of sensor nodes, which must be supported by the operating systems and, in the optimal case, also by the programming languages. As depicted in Figure 7.6, each sensor node completely relies on reacting to events, such as sensor readings, timers and data messages, being received by the radio transceiver. These events must immediately be processed when they happen. The rest of the time, i.e. when no events have to be processed, the node is idle. This behavior is called an event-based programming model, which is basically an interrupt handler. The problem is that the system must not remain in interrupt handler too long in order not to lose events.

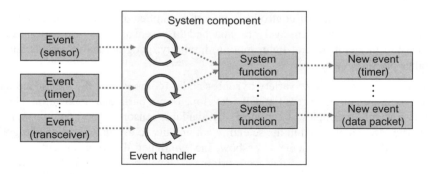

Figure 7.6 Event-driven operating principle of sensor nodes; the input consists purely of events or interrupts, which are processed by the system in the form of event handlers and system functions; the output, if any, is again a series of created events

The best known example of such an operating system is TinyOS[7] (Girod *et al.* 2004). TinyOS was developed at the Berkeley University as the runtime environment for their Mica motes (Handziski *et al.* 2005). A specific programming language–nesC (a C-like dialect)–was designed to directly support the event driven programming model (Levis *et al.* 2004). The immediate goal behind TinyOS was to show a small memory footprint, i.e. to be able to operate on micro controllers with predominant memory restrictions. The primary design criteria were to develop a component-based system. These components may interact by exchanging asynchronous events. In order to build a program, the components are wired together in the compilation step. While TinyOS supports the concept of a task, there is no scheduler for preemptive task switching available. A task is meant to continue until completion, interrupted only by asynchronous events. A specific simulation environment–TOSSIM–was developed to support TinyOS applications (Levis *et al.* 2003). The original nesC programs can be evaluated using this environment, which provides a great basis for first stage debugging of sensor network applications.

Another operating system is Nut/OS,[8] which is frequently used, for example, on the BTnode hardware. Nut/OS is an intentionally simple Real-Time Operating System (RTOS) for the ATmega128 micro controller. It provides similar mechanisms to those found in TinyOS, such as asynchronous events, periodic and one-shot timers, and interrupt-driven streaming I/O. Additionally, the basic concepts of non-preemptive multi-threading is supported. Thus, concurrent threads can be used to perform multiple logical threads of execution in parallel. Nut/OS is programmed in C, which simplifies the porting of available software libraries. Nevertheless, only limited support of the event-driven operating paradigms is available.

7.2.3 Communication in sensor networks

WSNs are supposed to build and maintain their network infrastructure on demand using ad hoc networking techniques. As current sensor networks must support node mobility in time and space, one possible solution is to employ MANET routing algorithms. Nevertheless, the primary objectives of WSNs must be regarded. Sensor data are measured according to an application-specific scheme and transported to a given base station (or pre-processing node). Thus, different communication paradigms might be appropriate in the context of WSN. In general, the following two paradigms can be distinguished: address-based and data-centric communication (Akkaya and Younis 2005; Akyildiz *et al.* 2002a).

- Address-based routing – All classic communication networks rely on a global network-wide addressing scheme, providing unique node identifiers. If a data message needs to be routed through the network, routing tables provide mechanisms for finding the destination node using its address information and forwarding the message along the identified path. This kind of routing is called id-centric or address-based (Akkaya and Younis 2005). The main advantage is the inherent possibility of transmitting data messages to a given well identified destination node. Obviously, mechanisms are required to set up such unique addresses in advance. Usually, this is done by manual configurations but there are also dynamic address-allocation mechanisms

---

[7]http://www.tinyos.net/.
[8]http://www.ethernut.de/en/software.html.

available (Sun and Belding-Royer 2004; Weniger and Zitterbart 2004). This kind of address-based routing is discussed in more detail in Chapter 10.

• Data-centric communication – In contrast to routing methods using well known identifiers, especially in WSNs, there is rarely the need to uniquely identify a specific
  node but to get selected information from a particular area under observation. Thus,
  typically, data messages are encoded in a way that other systems that receive the
  message can either recognize the message and process the contained data or forward
  the message according to the data description towards appropriate sink nodes. Such
  data-centric communication schemes are also known as application-centric routing
  methods. While data messages can obviously not be delivered to well identified
  destinations, sink nodes can still be identified according to their interest in specific
  message types (Intanagonwiwat et al. 2000). The main advantage is the simplified
  application of aggregation (Krishnamachari et al. 2002; Rajagopalan and Varshney
  2006) and in-network data-processing mechanisms (Dressler 2006a). Data-centric
  communication is discussed in Chapter 11.

Culler et al. (2004) discussed several issues for self-organization in WSNs. Primarily,
they focused on connectivity issues and dissemination and data collection. By exchanging
information, nodes can discover their neighbors and perform a distributed algorithm to
determine how to route data according to the application's needs. Various radio communication-related issues, e.g. interference and antenna orientation, as well as mobility make
determining connectivity a priori difficult. Instead, the network discovers and adapts to
whatever connectivity is present. The networking capabilities of sensor networks require
the inter-operation of multiple protocol layers.

This demand is also discussed by Akyildiz et al. (2002b). The resulting protocol stack
is shown in Figure 7.7. In addition to the standard layered protocol processing, multiple
management planes are required to coordinate the layers according to the available energy,
mobility patterns and application requirements.

A main aspect of WSN communication is the localized operation. Being also a primary self-organization method, localized operation prevents unnecessary network-wide
message exchanges, thus limiting possible network congestion and increasing the scalability
of the communication methods. Two alternatives exist to enable such location-dependent
techniques. First, the relative position to neighboring nodes or within a specific part of
the network can be exploited, needed for directed communication (Estrin et al. 1999a,b).
Secondly, geographical positions of destination nodes can be used for another routing category – geographical routing, which became an important research issue in ad hoc and sensor
network routing (Mauve and Widmer 2001; Navas and Imielinski 1997).

Data-aggregation techniques can clearly reduce the amount of data to be transported
through the sensor network. Rajagopalan and Varshney (2006) summarized techniques
and issues for data aggregation in WSNs. The authors distinguish between the following
different approaches: chain-based, tree-based and grid-based aggregation. We discuss these
aggregation concepts in detail in Section 11.5. The main issues targeted by aggregation
mechanisms are energy constraints, Quality of Service (QoS) awareness, reliability and
security. Obviously, there is always a trade-off between energy, accuracy and latency. If
a huge number of sensor readings is aggregated into a single measure, e.g. calculating
the mean or median, the aggregation procedure has to wait until all single messages are

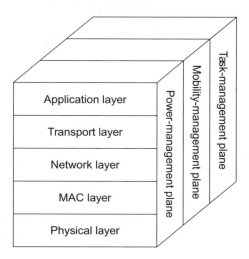

Figure 7.7   The protocol stack of typical WSNs; besides the classical layered protocols, multiple control planes exist that are used to configure system parameters at all layers

available, thus increasing the end-to-end latency. While the mean or median is a less accurate measure than the single measurements, the network load is essentially reduced and, therefore, also the energy that is required for transmitting the messages through the network.

Finally, the summary of communication-related issues in WSNs needs to be concluded by discussing the primary message creation paradigms push vs pull:

- Push relies on periodic measurements and directed communication of the retrieved data towards a base station. Thus, there is no need for controlling the sensor nodes in terms of requesting information or even detecting available nodes. On the other hand, sensor data are transmitted, even if no base station is available or the measurement results are not needed. Typically, either pre-programmed base station addresses in conjunction with address-based routing schemes are employed or various flooding techniques (see Section 11.2) for data-centric communication (Kwon and Gerla 2002; Kyasanur *et al*. 2006; Pleisch *et al*. 2006).

- Pull is a reactive strategy that requires an active request before starting any transmission. Compared with push, the main cost is in operation and control, i.e. in locating needed sensor nodes and sensing appropriate requests. On the other hand, the pull strategy ensures that only measurements that are really required are transmitted. Additional advantages are the possibility to update the application parameters on the sensor node with the request, e.g. the sampling rate, and the possible assistance of the routing protocols, as the request message can be used to update topology information in the involved nodes. Supported are address-based routing approaches, e.g. when using database-like requests (Madden *et al*. 2005; Sadagopan *et al*. 2003) as well as data-centric approaches such as directed diffusion (see Section 11.4) (Intanagonwiwat *et al*. 2000).

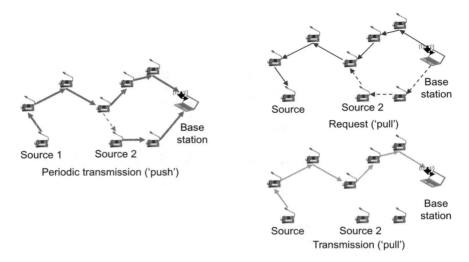

Figure 7.8 Push (left) requires only unidirectional communication – while usually more bandwidth is needed. In contrast, pull (right) needs bidirectional links but only requested data are transmitted. The dashed lines represent alternative paths

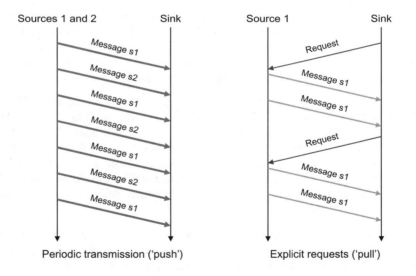

Figure 7.9 The behavior of the push (left) and pull (right) strategies is depicted in the form of the time scales indicating the individual messages

Figures 7.8 and 7.9 depict the main differences between both techniques. The push strategy (left) only relies on the periodic transmission of sensor data. Both sources (even if only one is needed) transmit their readings to the base station. The pull strategy requires a first request to finally initiate the transmission of sensor data. This (possibly expensive) request ensures that only the needed data are transmitted over the network.

7.2.4 Energy aspects

Sensor nodes usually draw energy from attached batteries. Thus, even if energy-harvesting methods like solar panels are available, energy-saving techniques are strongly required in order to maintain the node operable for as long as possible. The desired long runtime must be distinguished into the runtime of individual devices and the runtime of the network as a whole. Usually, application demands do not bother with individual nodes, as long as the global application-dependent objective can still be fulfilled. As the network lifetime is possibly the most important characteristic of sensor networks – actually, we will show that it is, as all other performance aspects can be represented as lifetime aspects – a more specific discussion of network lifetime will be presented in Section 17.3.

Three topics represent research domains focusing on energy aspects in WSNs. Even though they are pointing in different directions, all of them must be considered when constructing energy-aware and energy-efficient sensor network algorithms and applications:

- Energy-efficient networking protocols either rely on reducing the required energy for transmitting sensor data to the base station, i.e. reducing the energy consumption (e.g. energy/bit), or they optimize the system behavior of a single node by taking the battery characteristics and status into account (Adamou *et al.* 2001; Akkaya and Younis 2004). Solutions are, for example, the optimized selection of different paths through the network, as long single hops require more energy for a transmission compared with multiple short hops (Chiasserini and Rao 2001; Toh 2001), or the exploitation of the recovery effects of typical batteries (Barsukov 2004; Lahiri *et al.* 2002).

- If energy savings should be achieved, different optimizations are often possible. Thus, conflicts can occur and must be resolved. A simple example is the effort to exploit the battery-recovery effects that conflicts with algorithms for node synchronization or low-latency data transmissions (Karp *et al.* 2003; Wang and Xiao 2006).

- Last, but not least, some form of recharging or energy scavenging from the environment is needed in order to increase the available energy for sensor applications. Various techniques have been developed over the last few years and technology progresses quickly in terms of miniaturization and improved efficiency (Roundy *et al.* 2003).

7.2.5 Coverage and deployment

One of the most important characteristics in terms of applicability and energy consumption in WSNs is coverage. As this parameter is mainly influenced by the sensor network deployment, both characteristics are explained in conjunction.

Coverage

The sensor coverage problem centers on the fundamental question about the ability of a sensor network to observe a physical space. Such coverage aspects are the main research focus of the sensor network community (Cardei and Wu 2004). Two forms of coverage should be distinguished here – radio coverage, i.e. connectivity, and sensor coverage:

- Radio coverage – In most papers on sensor networks, the authors refer to coverage as the radio coverage, i.e. the wireless connectivity. Usually, the coverage area is approximated by circles, as physical barriers are hard to model and failure-prone. Thus, if two nodes are within their coverage range, a successful wireless connection is assumed.

- Sensor coverage – In its original meaning, coverage represented the sensing area. This definition is still used (and, unfortunately, sometimes mixed up with radio coverage) in the literature. As sensors have very different characteristics, a circular approximation may lead to severe estimation errors. Considering the nature and purpose of WSNs, sensor coverage is the more important aspect, while connectivity is required at the same time to transmit measured sensor data to the base station.

In the following, we will use the term 'coverage' in the sense of sensor coverage as it is usually stated in the literature. Discussing coverage problems, we also need to distinguish between area coverage and target coverage. Area coverage depicts the ability of a sensor network to cover (i.e. to monitor) a geographical area. Thus, each point in the physical space of interest must lie within the sensing range of at least one sensor. Target coverage is about monitoring a set of targets that are located within a domain of interest by at least one sensor. As these targets can be mobile, target coverage closely relates to target tracking.

Both area coverage and target coverage can be extended to k-coverage. In this case, at least k sensors need to be able to monitor a point in space (area coverage) or a moving object (target coverage).

Deployment

Compared with traditional computer networks, there are a number of possible deployment techniques for WSNs. First of all, regular deployment is usually done by human operators (Shen et al. 2006). It follows a well planned structure with fixed node positions. Node failures are either handled by network mechanisms or by replacing failed nodes on demand (again, by human operators). Such regular deployment does not necessarily follow a geometric structure. Nevertheless, this is often a convenient assumption. Secondly, random deployment is often considered in the context of sensor networks, e.g. by dropping sensor nodes from an aircraft (Kahn et al. 2000). Usually, uniform random distribution of nodes over a finite area is assumed. Unfortunately, this is often not achieved (Chellappan et al. 2005; Wu and Yan 2005). Practical experiments have shown that dropped sensor nodes tend to concentrate in heaps instead of uniformly covering the ground. Figure 7.10 depicts the two deployment strategies. As can be seen, random deployment tends to produce areas with different degrees of coverage (dense, sparse and even uncovered).

Resulting problems of both deployment strategies (regular and random) are targeted to be solved by mobile sensor nodes (Wang et al. 2004; Wu and Yan 2005). Such mobile sensor nodes are intended to compensate for deployment shortcomings. They can either be passively moved around by some external force (wind, water), be actively moved by human beings or robot systems, or they can move on their own, e.g. by using a built-in carriage. Exploiting any of these mobility solutions, sensor nodes can actively seek out

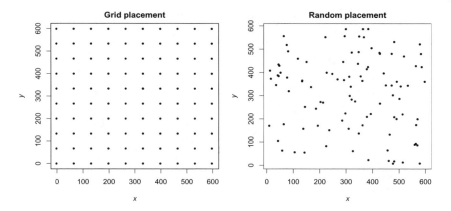

Figure 7.10 Sensor network deployment following a regular pattern (left) and a random distribution (right), respectively

'interesting' positions according to a pre-programmed global objective, e.g. to increase the coverage of the sensor network (Liu *et al.* 2005).

As regular deployment is considered expensive in terms of required resources and the necessary time for the deployment, and random deployment often leads to coverage holes, several improvements have been discussed, focusing on mobility-supported deployment or specific deployment strategies (Bai *et al.* 2006; Meguerdichian *et al.* 2001a). Additionally, heterogeneity properties have been exploited to increase the sensor coverage in WSNs (Lee *et al.* 2004).

7.2.6   Comparison between MANETs and WSNs

Comparing MANETs and WSNs, we can find many commonalities. Both networks strongly rely on self-organization mechanisms to maintain neighborship relations and network topologies. The available energy is limited in both cases and, therefore, the energy efficiency of employed algorithms and methods is of the highest importance. Additionally, both networks often use wireless multi-hop communications. Available resources and properties of both networks are summarized in Table 7.2.

On the other hand, many differences can be detected by closely analyzing the characteristics of MANETs and WSNs. Considering the applications and the available equipment, usually, MANETs employ more powerful equipment. This includes the availability of 'human in the loop'-type applications, higher data rates and more resources, such as memory and storage. From an application point of view, WSNs depend much more on application specifics, i.e. they are built for specific applications. In MANETs, a number of similar devices (in terms of functionality) are collaborating, using changing applications. Different mobility models can be distinguished for MANETs and WSNs, as mobility is the core concern of MANETs but only a kind of additional property in WSNs (and not all nodes will be mobile).

Table 7.2   Resources and properties of MANETs compared
with WSNs

Resources and properties	MANET	WSN
Available energy	High	Low
Processing power	High	Low
Memory and storage	High	Low
Density and scale	Low	High
Mobility	High	Limited*
Heterogeneity	Medium*	Low*
Varying user demands	High	Low

* Depending on the application scenario

The main differences between both technologies lie in the following four domains. First, the interaction with the environment is the core of sensor networks (sensor readings have to be collected) but completely absent in MANETs. Secondly, both networks strongly differ in their typical size. WSNs are considered much larger; therefore, employed mechanisms must be highly scalable. Similarly, the energy constraints are much higher in sensor networks, as small embedded systems are used compared with relatively rich mobile PCs in MANETs. Finally, both technologies differ in the primarily used communication paradigms. While MANETs always rely on address-based routing, WSNs often employ data-centric communication mechanisms.

The predominant research objectives can be summarized as follows. Ad hoc networks rely on efficient routing while considering mobility and – sometimes – QoS. In contrast, WSNs need to achieve optimal network lifetimes (depending on energy efficiency and application layer requirements) and scalability. In selected scenarios, WSNs are have to care about mobility issues and response times. Both ad hoc and sensor networks demand a high degree of autonomy; thus, self-organization methods must be employed to prevent centralized and human oriented control.

7.2.7   Application examples

In order to conclude the section on WSNs, typical application examples should be reviewed to give a comprehensive understanding of sensor networks and their characteristics. Mark Weiser initiated the ubiquitous and pervasive computing concept in 1991 (Weiser 1991). He already envisioned scenarios of networked embedded systems that are ubiquitously surrounding us to support our daily life. Sensor networks can be regarded as such pervasive environments, as they show the same characteristics, such as scale (sampling, temporal and spatial extent, density), variability (structure, space) and autonomy (Estrin *et al.* 2002). More practical application examples have been surveyed, for example, by Akyildiz *et al.* (2002a,b) and Karl and Willig (2005). In the following, we will outline selected application examples with practical backgrounds. These examples have been chosen so as to demonstrate the different requirements on WSNs, depending on the application scenario.

Emergency operations

In emergency situations, sensor nodes can be dropped from an aircraft over a critical region. These sensors periodically sense their environment, e.g. for human beings in the local neighborhood or to derive a temperature map. In such applications, nodes must be able to work in hazardous environments for a limited period of time. Sensor networks are envisioned to play an important role in a variety of disaster and emergency operations (Lorincz et al. 2004). The most compelling application is wireless vital sign monitoring of multiple victims at a disaster scene. First responders would place wireless, low-power vital sign sensors on each patient. Usually, a given number of base stations will collect the gathered data. In some cases, stationary sensor nodes are supposed to be supported by mobile robot systems (Kumar et al. 2004). In this example, sensor nodes are used to assist first responders in emergency situations. The sensors are employed to detect living human beings and general environmental conditions in hazardous situations in order to protect firefighters during their tasks. The demands on the reliability of the network are very high, while the network must not be operable for a long period of time.

Habitat monitoring

An obvious application scenario for WSNs is habitat monitoring. Depending on the installed sensor hardware, sensor nodes can be employed to detect animals wandering around in their natural habitat or to analyze environmental conditions in the jungle. Several aspects of the hardware and software for such sensors must be considered (Tanenbaum et al. 2006). For example, these sensor nodes will be stationary in most cases but assumed to work for a very long time (usually, the biologists will not be able to replace batteries or even complete sensor nodes on a daily basis). Two projects gained some interest in the sensor network community—Great Duck Island[9] and ZebraNet.[10]

In the Great Duck Island project, the behavior of breeding ducks was analyzed. Sensor nodes were distributed around the nests to observe arriving and leaving ducks (Mainwaring et al. 2002; Szewczyk et al. 2004). All sensor nodes were stationary, thus providing a rather simple environment for establishing an ad hoc network. Unfortunately, the impact of the people maintaining the sensor nodes was still too high (failing nodes must be replaced similarly to frequent battery changes).

In the ZebraNet project, much larger sensor nodes were used according to the capabilities of zebras to carry items and according to the technical requirements (GPS receivers, WLAN communication) (Juang et al. 2002). In this project, the inter-species interactions were analyzed in particular. From the sensor network perspective, all nodes were mobile and highly energy-restricted. On the other hand, computational performance and storage were available to a high degree. In retrospect, the most challenging properties were the especially sparse network (how to achieve successful transmissions to a base station) and the maintenance of the sensor nodes (mobile software updates, how to 'reboot' a zebra).

---

[9]http://www.greatduckisland.net/.
[10]http://www.princeton.edu/~mrm/zebranet.html.

Precision agriculture

A couple of application examples have been reported in the domain of precision agriculture (Baggio 2005). The observed and controlled processes include the bringing out of fertilizer, pesticides and irrigations only where needed. In traditional agricultural solutions, single weather stations are employed to obtain large-scale averages from a given region. Remote sensing of the environmental conditions across different parts of the region under observation, e.g. a farmland, can greatly improve the quality of the measurements that build the basis for the activities of the farmers and, therefore, improve the final product and maximize benefits. Such sensor applications mostly focus in the optimized control and maintenance of vineyards (Beckwith *et al.* 2004). In such highly sensitive plantations, for instance, accurate monitoring of temperature can reveal significant variations in heat accumulation (hot and cold spots), and thus in grape maturity, even in small sections of a vineyard (Galmés 2006). Burrell *et al.* (2004) demonstrated that this kind of vineyard computing can achieve the best quality of wine by adapting the right harvest dates to measured variations (a process known as precision harvesting). Galmés (2006) reported on the lifetime demands in such application scenarios. As all deployed sensor nodes are stationary and only infrequent measurements are needed, the research primarily focuses on increasing the network lifetime to a maximum degree.

Home automation and health care

Smart home environments are considered to support our daily life in many ways. Application scenarios range from home automation (automatically controlled windows, heating or air-conditioning systems) to assistance in various health care applications (monitoring of vital signs, support of elderly people). Smart sensor nodes and actuators in appliances will learn how to provide needed services (Schulzrinne *et al.* 2003). Tele-monitoring of physiologic data for post-operative or intensive care and long-term surveillance of chronically ill patients or elderly people have already become a reality in selected projects (Noury *et al.* 2001). This includes tracking and monitoring mechanisms as provided by WSNs in particular.

The challenge regarding smart homes, especially for supporting elderly and handicapped people, is to compensate for handicaps and support the individual in order to give them a more independent life for as long as possible. For example, Dengler *et al.* (2007) developed a common architecture for a smart home environment, which was verified in an experimental lab environment. The focus of this work was on helping elderly or handicapped people to live a more independent life for as long as possible. Elderly or handicapped people are an important target group concerning smart home environments, so special requirements have to be taken into account. For example, a health-monitoring and emergency help system has to be established. The requirements made to control infrastructure and interfaces have to be easy and self-explanatory. The user should be integrated and feeling well in the new environment. When developing smart home environments for this target group, the main focus is on compensating handicaps and limitations. The first objective was the development of a sensing and monitoring system. In detail, the architecture must take care of domestic systems, air conditioning, lights and heating, as well as control the basic functions of home entertainment and security systems. To interact with the physician and to verify collected

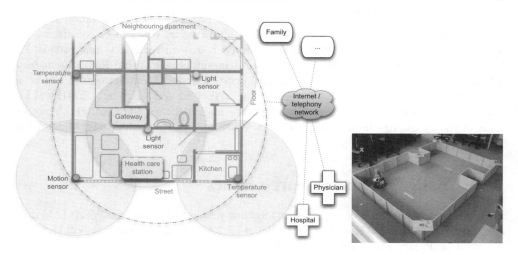

Figure 7.11 Smart home scenario based on sensor network technology. The system architecture (left) was developed to support heterogeneous sensor nodes and networking technologies; the applicability was demonstrated in a lab environment (right)

vital signs, as an additional goal, the architecture must be able to communicate with external services, e.g. by placing emergency calls. These objectives require a remarkable amount of technology to be used. But the fact that this technology must be transparent and easy to use is one of the most important points. All improvements and technology are useless if they are not accepted by the user. Thus, the user–system interface must be as simple and powerful as possible. And the system must basically operate in a self-organized way.

Figure 7.11 depicts the smart home architecture developed by Dengler *et al.* (2007). Several sensor nodes are considered to collect environmental information, such as light and temperature information, in addition to motion sensors and temperature sensors at the stove. Both can detect incorrect usage of potentially dangerous systems. A gateway is used to interconnect different networking technologies and to transmit status information (for patient monitoring and emergency operations). The architecture was tested in an artificial lab environment.

Logistics

Today, the production of complex products such as cars or electronic devices is a strongly distributed process. Multiple manufacturers participate on an in-time production. Therefore, high demands emerged on the logistics that are coordinating the complete process. It is necessary to obtain information on individual parts, their status and their position in real-time. Sensor networks are considered to help achieving these demands by equipping products or parts with sensor nodes. WSNs can be used to localize parts and to store information about the production process. Additionally, the sensor nodes can obtain and store environmental parameters such as the humidity and temperature (for example, in some cases, specific parts must not be exposed to high temperatures). The ability to locate sensor nodes within a sensor network can also be exploited for anti-theft protection. The application scenario for WSNs in logistics therefore includes mobile sensor nodes that can be localized. Additionally, part-specific information must be stored and – on request – transmitted to a base station. In some cases, passive readouts might suffice, e.g. using Radio Frequency ID (RFID) technology. In recent studies, the importance of WSNs for logistics has been highlighted. The term 'product quality monitoring' (Akyildiz and Kasimoglu 2004) was introduced and characterized as the forthcoming mass market (Hatler and Chi 2005). Another study explicitly emphasizes the applicability of WSNs technology for logistics (Mindbranch 2004).

7.3   Challenges and research issues

Several survey papers have been published targeting research challenges in WSNs (Akyildiz et al. 2002a,b; Chong and Kumar 2003; Culler et al. 2004). In this section, we will highlight the most important constraints and challenges in WSNs, as well as important research aspects currently studied in the sensor network community.

7.3.1   Required functionality and constraints

Networking aspects in WSNs include the following functions that must be provided by any sensor network, regardless of the scenario-dependent constraints and restrictions. The MAC is considered to maintain wireless connections between neighboring nodes. This includes all aspects of synchronization, collision avoidance and channel management. If a single-hop connection is not possible, network protocols have to be involved for efficient multi-hop communication. Independently of the communication paradigm (address-based vs data-centric), the network must be able to find a way to transmit data messages from source to sink. Topology control mechanisms support the management and control in the MAC and network layer protocols by identifying and maintaining topological structures of the network. Finally, localization techniques are often required to locate single nodes within the network. This information can be used for network maintenance issues, but also by the applications for searching or tracking systems. Other issues such as (time) synchronization, collaboration and task allocation will be discussed in Part III. As no manual control is possible in ad hoc and sensor networks, all these functions strongly rely on self-organization methods, which provide techniques and solutions for autonomous operation in such distributed environments. The following list summarizes the most important constraints in WSNs regarding the mentioned objectives:

- Available energy – Sensor nodes are operated by batteries that provide limited energy for all electrical parts of the sensor node (Lahiri et al. 2002). Energy-harvesting methods can prolong the node lifetime. Unfortunately, such methods cannot be used in all application scenarios (Rahimi et al. 2003).

- Processing power – Micro controllers employed for sensor nodes usually provide very limited processing performance (see size and energy restrictions). It cannot be expected to get a similar computational performance compared with other computer systems.

- Memory and storage – Typical micro controllers bring their own memory (RAM and flash). Thus, the characteristics of the available memory usually correlate with the size of the micro controller, i.e. only a few kBytes will be available to the application programs.

- Bandwidth and throughput – Wireless radio transceivers as used by most sensor nodes are optimized for low-energy operation. The main result is a relatively small bandwidth available to the application. Due to unreliable radio communication, the achievable throughput is even smaller than the nominal bandwidth.

- Reliability – Depending on the application scenario, the demands of the reliability of the WSN can strongly differ. This includes reliable communication (the use of fully reliable transport protocols is usually impractical (Anantharaman *et al.* 2004; Holland and Vaidya 1999)) as well as the error-proneness of the hardware (which can be high in hazardous environments).

- Deployment – Constraints on the deployment must be reviewed in two dimensions. First, depending on the application scenario (and the costs), networks can be either sparse or dense. Secondly, manual deployment is not always possible or feasible. Thus, automated, random or manual deployment strategies must be distinguished (Howard *et al.* 2002; Zou and Chakrabarty 2003).

- Mobility – The degree and kind of mobility in WSNs depends on the application requirements. While most sensor networks are considered to be stationary in space, duty cycles induce temporal variations in the topology. In specific scenarios, spatial mobility is also considered, while different mobility models must be applied (Bettstetter 2001; Camp *et al.* 2002).

- Addressing – Typically, off-the-shelf sensor nodes do not have a globally unique address pre-programmed into the system like WLAN interfaces of mobile PCs. Thus, the availability of globally unique Ids or addresses is limited. Thus, networking mechanisms must either dynamically allocate unique addresses or even abandon address-based techniques (Sun and Belding-Royer 2004; Weniger and Zitterbart 2004).

- Scalability – A primary constraint in sensor networks is the scalability of employed methods and algorithms. While most solutions work fine in small-scale networks, e.g. consisting of 10–20 nodes, the same algorithms might not be able to terminate (due to transmission latencies, collisions, memory restrictions) in large-scale networks, e.g. consisting of multiple hundreds or even thousands of nodes (Akyildiz *et al.* 2002a).

- Costs – Economical constraints usually restrict any project. In sensor networks, this constraint especially holds, as the costs of the overall network explode with the number of nodes.

- Environmental conditions – The required degree of self-organization in WSNs primarily depends on the accessibility of the environment in which the sensor networks have to be deployed. If it is inaccessible, either because it is hazardous or just remote, operation and management must be organized by the sensor nodes themselves. Possibly, the replacement of failing nodes is not possible (or is too expensive).

- Miniaturization – Finally, the size and shape of sensor nodes are a challenging constraint that affects the costs and deployment strategies. Thanks to advances in microelectronics and engineering, sensor nodes are becoming smaller and smaller. The best examples are the smart dust devices as described by Kahn *et al.* (2000).

7.3.2 Research objectives

According to the operating principles and the identified constraints of sensor networks, a number of current research objectives can be derived. The presented list focuses on topics that have not yet been studied in depth in the sensor network community. The following topics are explicitly excluded, as they are discussed in detail in this book: self-organization, data-centric networking, synchronization and coordination.

Network lifetime

The network lifetime is perhaps the most important metric for the evaluation of sensor networks (Dietrich and Dressler 2006). Of course, in a resource-constrained environment, the consumption of every limited resource must be considered. However, network lifetime as a measure of energy consumption occupies the exceptional position in that it forms an upper bound for the utility of the sensor network (Giridhar and Kumar 2005). The network can only fulfill its purpose as long as it is considered 'alive', but not after that. It is therefore an indicator of the maximum utility that a sensor network can provide. If the metric is used in an analysis preceding a real-life deployment, the estimated network lifetime can also contribute to justifying the cost of the deployment. Lifetime is also considered a fundamental parameter in the context of availability and security in networks (Khan and Misic 2007). In conclusion, it can be said that a sensor network should fulfill its task for as long as possible. Thus, the definition of 'lifetime' depends on the application. In general, the lifetime of individual nodes is relatively unimportant, as long as the network as a whole is still functional (Olariu and Stojmenovic 2006). Unfortunately, network and sensor lifetime are often treated equivalently in the literature. Dietrich and Dressler (2006) initiated a discussion on a more concise definition of network lifetime (see also Section 17.3).

Robustness and fault tolerance

WSNs must be able to adapt to changes in their environment, such as severe environmental modification (e.g. after an earthquake or hurricane) and node failures. Self-monitoring algorithms and adaptive operations are employed to self-maintain the complete sensor network. Such auto-configuration is necessary, as manual configuration is often not an option. The solution space ranges from parameter changes to incorporating additional resources, e.g. newly deployed nodes. The robustness of sensor network techniques against node failures is one of the main requirements. Nodes may run out of energy, get physically destructed or become disconnected due to irregularities in the radio communication (Ding *et al.* 2005; Zhou *et al.* 2005). As a wide range of densities must be supported by the sensor network algorithms, depending on the application, a large or small number of nodes per unit area will be available. Thus, spare nodes are often available for exploiting this redundancy (Shah *et al.* 2005a).

In-network processing

The first sensor networks followed the simple design principle of distributed sensor nodes, collecting sensor readings and transmitting the results over an ad hoc network to a base station. The base station is responsible for evaluating the received data and, possibly, for reconfiguring the sensor nodes if necessary. Basically, all nodes in the network collaborate towards a joint goal. A new trend is to pre-process data in the network (as opposed to at the edge). This can greatly improve the performance and efficiency (response time, lifetime). Data aggregation attempts to conserve energy and bandwidth (Rajagopalan and Varshney 2006). It represents one major domain of in-network processing algorithms. Sensor data is collected by intermediate nodes and condensed according to different aggregation algorithms. While additional processing power is required, network bandwidth and transmission energy is saved (which is much more expensive) (Krishnamachari *et al*. 2002; Mhatre and Rosenberg 2004). Another domain of in-network processing is network-centric operation and control. Dressler (2006a) proposed a concept for rule-based data processing in WSNs. It is based on three principles: data-centric operation, specific reaction on received data and simple local behavior control using a policy-based state machine. The flexible rule descriptions allow the expression of aggregation patterns, and also routing and actuation commands for pure data-centric sensor networks.

Quality of Service

Traditional QoS metrics do not directly apply to typical sensor networks. Examples include the transmission delay or the delay variation, as multimedia communication is typically not considered to take place in WSNs. Still, the service of WSNs must be 'good', i.e. the application expects the right answers at the right time. In this sense, each sensor network represents a distributed real-time system. In such a system, the quality of the received data messages correlates with their age, i.e. the significance of the measured data degrades over the time. Real-time communication architectures are needed that support such application demands (Lu *et al*. 2002). Additionally, QoS metrics like the packet loss ratio can be used to dynamically evaluate the network quality and to adapt the application behavior accordingly. For comparing WSN algorithms, e.g. routing protocols, QoS measures are used to show the differences between the single techniques, e.g. the end-to-end delay. Nevertheless, even in such evaluations, the focus is more on energy consumption than on pure delay measurements (Anastasi *et al*. 2004; Chiasserini and Garetto 2004).

Software management

Recently, the list of research objectives in sensor networks was extended to cover software-management aspects in WSNs as well (Handziski *et al*. 2005; Hurler *et al*. 2004). Due to the heterogeneity of employed hardware platforms and the low resources in terms of processing power, available memory and networking capacities (sensor nodes are usually able to run a single task only) (Margi 2003), new approaches for efficient software engineering and software management are needed. Han *et al*. (2005) depict a model consisting of three fundamental components: the execution environment at the sensor node, the software distribution protocol in the network and the optimization of transmitted updates. Reprogramming of nodes in the field might be necessary, for various reasons. The best known examples are improved flexibility (switching between applications) as well as possible failure treatment (relocating tasks to other nodes). Several approaches for multi-hop network-based node reprogramming have been discussed in the literature. The best known system is Deluge (Hui and Culler 2004), which was developed for reprogramming Mica2 sensor nodes. Deluge propagates software updates over a multi-hop ad hoc network. In this example, all sensor nodes are assumed to run the same piece of software. To optimize the reprogramming procedure, incremental network re-programming was introduced (Jeong and Culler 2004). Nevertheless, in many application scenarios, heterogeneous sensor nodes in terms of hardware and software heterogeneity are required. Thus, the flexible exchange of software components or even dynamic software composition (Schröder-Preikschat et al. 2007) according to the current application requirements or to changes in the environment are thought to be possible solutions. Profile-matching techniques have been proposed (Truchat et al. 2006) in combination with dynamic code composition and node reprogramming (Dressler et al. 2007b; Fuchs et al. 2006).

8

Self-Organization in Sensor Networks

Sensor networks are inherently relying on self-organization and adaptive algorithms. Previously, we outlined the primary concepts and properties of WSNs. In this section, we will investigate objectives and methods for self-organization in sensor networks. It will become clear that new protocols and paradigms are needed compared with conventional approaches. In several case studies, we show that first solutions are already available. These examples also help in understanding the main principles of the different self-organization mechanisms and their interactions.

8.1 Properties and objectives

Communication in wireless networks was first inspired by well known approaches from wired standards such as the Internet. Later, issues such as mobility (Bai and Helmy 2004; Bhatt *et al.* 2003), high dynamics in terms of new devices joining a network and other devices leaving, and rapid changes in the environment have been identified. Such challenges made a distributed configuration necessary. This configuration should be completely decentralized and organized based on local information only. Thus, it evolves autonomously (Bulushu *et al.* 2001; Low *et al.* 2005) – we call such mechanisms 'self-organizing' (Dressler 2006b; Gerhenson and Heylighen 2003).

In Part I, we discussed the evolution of system control from classical centralized control over distributed systems to pure self-organizing systems. In this context, we refer to centralized control in terms of external or manual coordination. Obviously, the scalability increases towards self-organized systems. More and more single devices can be put together to collaborate on a given (global) goal. One striking property of self-organization (corresponding to the provided definition) is the missing determinism of the algorithms.

The overall objective of any sensor network is to fulfill a specific application (or any number of concurrently installed applications). At the moment, we do not want to focus on the particular objective or goal. We can reduce the problem of achieving an overall goal

Self-Organization in Sensor and Actor Networks   Falko Dressler
© 2007 John Wiley & Sons, Ltd

(regardless of the degree of interactivity with human beings) to the question of how to control all the individual nodes in the sensor network.

Focusing on networking aspects (issues on coordination and control will be discussed in Part III), we need several functions for wireless ad hoc communication. Regarding the number of nodes, their distribution and communication characteristics, self-organization is needed to achieve several objectives. Primarily, these methodologies try to mitigate the need for centralized control or even to abandon it. The listed objectives must be addressed, keeping the already mentioned constraints in mind, such as energy, processing power, storage and mobility. The most prominent issues to be solved using self-organization methods in sensor networks are briefly introduced below. We discuss these issues in more detail in Chapter 17.

- Scalability – The scalability parameter describes how many interacting nodes can be supported with, at worst, a linear increase in the resource requirements. Scalability must be regarded as an overall measure for how well the different protocols and techniques cooperate with each other. This also affects the reliability of the communication, as connections might be influenced by the methodology itself. Depending on the application scenario, scalability is often the limiting factor in sensor networks, as, typically, networks with huge amounts of nodes are considered.

- Lifetime – The lifetime of a sensor network is perhaps the most important parameter, as it ultimately describes the capability of the network to fulfill the application requirements. Network lifetime focuses on the network as a complex system, ignoring the fact that single nodes may fail as long as the overall goal is still fulfilled. Besides pure energy concerns, most characteristics describing the quality of an application, e.g. sensor coverage or transmission delay, can be reduced to lifetime discussions. Additionally, it also describes the availability of the network, i.e. the question of whether the sensor network is always available or if are there special protocol-inherent maintenance periods.

The quality of all networking algorithms developed for sensor networks can be measured in these three categories. Besides these dominant targets, a number of further goals must be addressed by particular solutions. Multi-objective optimizations are needed in most algorithms, e.g. latency vs energy. This also addresses the named objectives, e.g. to achieve a high degree of reliability might lead to a reduced scalability.

Figure 8.1 depicts the approaches that are used to maintain the main networking functions, i.e. global connectivity, in conventional and self-organizing networks. On the right-hand side, the organization of conventional networks based on global state is shown. This allows explicit coordination among network nodes according to deterministic optimization algorithms. On the left-hand side, the behavior in self-organizing networks is depicted. In this case, implicit coordination is achieved by using local state, interactions to neighboring nodes and probabilistic schemes. Global optimization is no longer possible while the scalability in the network is increased. In the literature, multiple (while similar) schemata for this categorization are available, e.g. provided by Prehofer and Bettstetter (2005) and by Dressler (2006b). Based on the available state information and the classification in the network protocol stack, we develop a two-dimensional categorization in the following section.

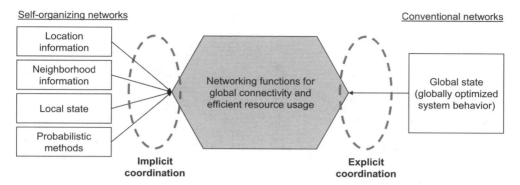

Figure 8.1 In contrast to conventional networks, adaptive on-demand networks do not rely on global state information. Instead, location information, neighborhood information, local state and probabilistic methods are employed. Thus, implicit coordination has to be used instead of explicit control

8.2 Categorization in two dimensions

Methodologies for self-organization in ad hoc and sensor networks can be categorized in multiple dimensions. In this section, we first provide a categorization of self-organization methods and, secondly, try to classify some well known mechanisms and on-going developments. More detailed case studies are provided in the following sections.

The categorization of self-organization methods in ad hoc and sensor networks opens a multidimensional space. In general, the methodologies can be grouped horizontally by their use of state information and vertically by their function in the protocol stack. For both horizontal and vertical differentiation, selected methodologies available for self-organization are discussed and summarized in Section 8.3.

8.2.1 Horizontal dimension

Figure 8.2 depicts the horizontal dimension. Reading the figure from the left, necessary state information to perform the particular algorithm is decreasing. As already mentioned, mechanisms in ad hoc networks try to avoid global state information in order to increase the scalability of the particular approach. The different categories are discussed below. Most of the given examples will be discussed in separate case studies at the end of this chapter.

While the required state is reduced towards the probabilistic methods, the determinism or predictability of the algorithms is reduced as well. Therefore, the best solution for a particular application scenario must be chosen carefully by comparing all application requirements at the same time.

Also depicted in Figure 8.2 is the class of feedback loops. We will discuss the relationship between the primary self-organization methods and the techniques shown in the categorization in the next section.

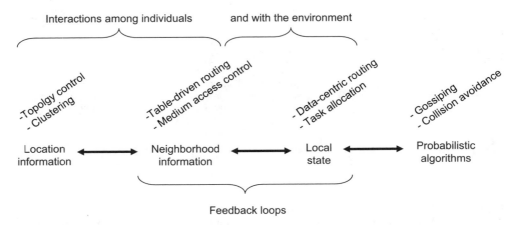

Figure 8.2 Horizontal categorization of self-organization mechanisms in ad hoc and sensor networks

Location information

Geographical positions, i.e. absolute or relative positions in a given area of interest, or affiliation to a group of surrounding nodes, e.g. clustering mechanisms, are used to reduce necessary state information for performing routing decisions or synchronizations. Usually, similar methods to those known for global state operations can be employed in this context. Depending on the size of active clusters or the complexity to perform localization methods, such location-based mechanisms vary in communication and processing overhead. Named examples are topology control and clustering. Both techniques rely on the finding of information about the absolute or relative position of nodes in the network. Regardless of the specific algorithms, location information must be collected and maintained. In-band mechanisms such as network-centric localization can be used as well as out-band systems like the Global Positioning System (GPS).

Neighborhood information

Further state reduction can be achieved by decreasing the size of previously mentioned clusters to a one-hop diameter. In this case, only neighborhood information is available to perform necessary decisions. Usually, so called 'hello' messages are exchanged in regular time periods. This keeps the neighborhood information up to date and allows the exchange of performance measures such as the current load of a system. Any collision-aware MAC protocol is an example of the use of neighborhood information. Typically, the nodes coordinate among all neighboring systems who is allowed to transmit in the next time frame. Similarly, table-driven routing protocols rely on neighborhood information in the first level. Additionally, this information is exchanged with other neighbors to finally obtain global routing paths.

Local state

Local system state is always available. Especially in sensor networks, system parameters can be combined with environmental conditions as observed by attached sensors. Local state can be manipulated by different events. For example, previous message exchanges can influence the node behavior as well as the current time. Examples for local state techniques are data-centric routing and – in some sense – specific task-allocation schemes. Data-centric routing algorithms do not work on global topology information, but apply simple local programs to received data messages. According to the rules, these messages are handled appropriately, e.g. by forwarding them or by initiating necessary local actions. In addition to message-driven actions, local tasks may be selected according to sensor readings or a fixed time schedule.

Probabilistic algorithms

In some cases, it is useful to store no state information at all. For example, if messages are very infrequently exchanged or in the case of high mobility, pure probabilistic methods can lead to comparably good results. Statistical measures can be used to describe the behavior of the overall system in terms of predicted load and performed operations. Obviously, no guarantee can be given that a desired goal will be reached. The best known example in ad hoc and sensor networks is gossip-based routing. Instead of flooding messages through the whole network (creating a huge amount of overhead due to duplicates), probabilistic schemes are used to decide whether to send or to drop a message. Depending on the probabilities and the network topology, gossiping can achieve comparably good performance. At the MAC layer, collision-avoidance techniques employ probabilities as well. If stations start to send in a randomized scheme, the probability of collisions can be reduced to a certain degree.

8.2.2 Vertical dimension

In contrast, Figure 8.3 shows the layered system architecture. Depending on the particular layer, there are different aspects of self-organization issues. A common control plane co-ordinates and controls mobility questions and some additional cross-layer or cross-service issues have to be considered. In the given figure, some layers are highlighted, as they pose the primary base for research in sensor networks. Therefore, we concentrate on the following mechanisms in ad hoc and sensor networks: MAC layer, network layer and application layer.

Several support functions are provided by additional control planes. According to Aky-ildiz *et al.* (2002a), at least mobility, energy and task control must be supported in the context of sensor networks. These communication-external properties have a strong influence on the efficiency and performance of the network functions. Additionally, cross-layer optimizations are of specific relevance in sensor networks. For example, neighborship management is usually provided by the MAC layer. The same information is required at the network layer. Similarly, security solutions must not be applied at multiple levels due to the strict resource restrictions in WSNs. Based on the given application scenario, particular mechanisms from different layers might interact to achieve a common goal, e.g. to reduce the necessary amount of energy, but they might also interfere with one another, e.g. by defining different sleep cycles at different layers to reduce the energy consumption.

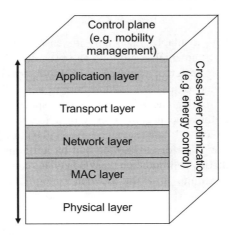

Figure 8.3 Vertical categorization of self-organization mechanisms in ad hoc and sensor networks

MAC layer

The primary objective of the Medium Access Control layer is to manage access to the wireless radio link and to exchange message exchange between directly connected nodes, i.e. between nodes within the wireless transmission range. Historically, contention-based mechanisms dominate this layer. Thus, all neighboring nodes equally contend for the channel. Carrier-sensing techniques help to reduce collisions and randomized sleep intervals ensure that two nodes cannot starve each other. Additionally, synchronization between neighboring nodes might be used to optimize the link sharing. For example, Time Division Multiplexing (TDM) schemes ensure exclusive access in selected time slots. While overhearing techniques enable the nodes to transmit messages at arbitrary times (obviously, the receiving node will be 'always on'), duty cycles can be used to reduce the energy consumption of the radio receiver as well as to optimize the overall performance. Such sleep cycles must be managed in the local neighborhood. Self-organization mechanisms help to perform concurrent access, to synchronize nodes and to maintain duty cycles in a distributed manner without the need for central management and pre-configuration or nodes and algorithms.

Network layer

End-to-end forwarding of data packets is provided by the network layer. Two different tasks must be provided by this layer: routing and data forwarding. Routing cares about identification of the current network topology. Typically, routing tables are maintained by exchanging routing information among neighboring nodes. We can distinguish between different techniques, i.e. table-driven and link-state routing, as well as between different procedures, i.e. proactive and on-demand routing. Obviously, all of these mechanisms ultimately rely on state information's being synchronized between many (or even all) nodes in the network. On the other hand, data forwarding describes the actual 'routing' process.

Received messages are examined and forwarded towards the destination by examining the destination address and matching this address with entries in the routing table. While routing might involve a huge amount of global state information, forwarding mainly addresses performance optimization, error control and (sometimes) congestion handling. Many approaches have been developed that perform both tasks in a self-organized manner. Today, data-centric routing couples both tasks. Messages are exchanged, processed and forwarded according to their 'meaning', i.e. to the semantics of the message.

Application layer

Besides the contents of the networked application itself, many coordination tasks must be organized: which node should act as a master, which nodes should perform which part of the global task, etc. Task-allocation schemes have been studied in various kinds of ad hoc networks that are based on self-organizing systems or pure probabilistic approaches. Besides such coordination issues, which we will discuss in more detail in Part III, the application layer defines the basic requirements of any communication in the network. In WSNs, coverage, connectivity and availability can be regarded as the main characteristics. All of them can be described in terms of the lifetime of the complete network. Besides requirements, the application layer provides necessary information about the communication patterns that can be expected from the particular application running in the sensor network.

8.3   Methods and application examples

In Part I, we discussed the primary methods of self-organization in massively distributed systems. The presented horizontal and vertical categorization of self-organization as used in ad hoc and sensor networks (focusing on networking aspects) follows a different scheme. Therefore, some words are needed to explain this relationship.

In this section, first, a mapping with the primary self-organization methods is provided. Secondly, methodologies developed for communication in ad hoc and sensor networks are classified in terms of employed self-organization mechanisms as defined in the depicted classification scheme. We follow the horizontal classification – including global state algorithms, as these build the starting point for optimizations in distributed systems – and discuss the specific techniques and algorithms in vertical order.

8.3.1   Mapping with primary self-organization methods

Self-organization can be achieved by employing positive and negative feedback loops, interactions among individuals and with the environment, and using probabilistic techniques. These primary methods can be used in any context of complex and massively distributed systems. The application for networking aspects in ad hoc and sensor networks is a very specific domain. For better characterization, we developed the two-dimensional classification as presented in the last section. The following list maps both concepts (it is also summarized in Table 8.1):

- Positive and negative feedback loops – Feedback loops are used to update operational parameters of individual systems and groups of multiple systems according to the

Table 8.1 Mapping of primary self-organization methods to the horizontal classification in ad hoc and sensor networks

	Location information	Neighborhood information	Local state	Probabilistic
Feedback loops		Feedback is provided by observing and evaluating system parameters; this can be done either by local means (sensor readings) or with external help of neighboring systems		
Interactions	Information exchange among remote nodes using routing techniques	Local interaction among direct neighbors within their wireless radio communication range	Interactions with the environment or indirect interactions with other nodes using environmental changes (stigmergy)	
Probabilistic techniques	Randomness is often exploited to prevent unwanted synchronization effects, e.g. for retrial attempts			Stochastic methods

current state. Thus, adaptive algorithms can be built using such approaches. Positive feedback is used for amplification and negative feedback for the smooth control of the adaptation process. In sensor networks, feedback is provided by observing and evaluating various system parameters. Sensor nodes can employ locally attached sensors to measure environmental conditions – focusing on changes in these conditions. Thus, sensor readings are analyzed periodically and appropriate parameter changes are initiated. Feedback can also be provided by external helpers, i.e. neighboring nodes that measured or detected environmental changes to which the system (or the group of systems) should adapt. In summary, neighborhood information and local state are combined for executing adaptive algorithms that control the behavior of the individual sensor node.

• Interactions among individuals and with the environment – Actually, such interactions can be divided into three categories. First, interactions with remote nodes might be necessary in order to synchronize on a common objective. In ad hoc and sensor networks, for example, routing algorithms fall into this category. However, other mechanisms also rely on remote node interactions such as coverage optimizations. All these techniques rely on location (or topology) information. The second category concentrates on interactions among direct neighbors. In wireless networks, such interactions are highly needed for efficient use of the radio resources. For example, collision-avoidance schemes employ RTS/CTS handshakes or even synchronization schemes. For higher layers, localized updates can be used to reduce unnecessary redundancies and to increase the fault tolerance. Stigmergic actions, i.e. indirect interactions by changing the environment, build category number three. Especially in sensor networks, nodes can exploit their capability to observe the environment to synchronize themselves in a particular area. Local state can be adapted by measuring physical phenomena and selecting operating parameters and even algorithms accordingly.

- Probabilistic techniques – Probabilistic techniques have been described in the context of basic self-organization methods as well as in the context of self-organization in sensor networks. Such techniques allow stateless operation in any scenario. While the operation costs are reduced to an absolute minimum, there is no state maintenance procedure that must be cared for; the determinism and predictability are reduced as well. In the context of sensor networks, probabilistic routing, e.g. gossiping, and probabilistic task allocation are typical examples that have proven their feasibility in multiple scenarios. Additionally, randomness is often exploited to prevent unwanted synchronization effects in various mechanisms and protocols. For example, retransmissions due to collisions are artificially and randomly delayed in order to prevent starvation of single nodes. Other solutions employ the random selection of paths through the network to improve network utilization and to prevent bottlenecks.

8.3.2   Global state

While the maintenance of global state is not part of the class of self-organizing mechanisms, we briefly describe typical mechanisms relying on global state information in ad hoc and sensor networks to further demonstrate the advantages and limitations of self-organization. Using global state information, optimal solutions can either be directly calculated or at least approximated in the case of multi-objective optimizations. The primary purpose of global state algorithms in communication networks is twofold. First, this global state must be collected and maintained, and, secondly, the specific optimization algorithms must be performed. Depending on the size of the network in terms of participating nodes, both steps may require unsuitably high amounts of time and memory.

In MAC protocols, the energy efficiency as well as the overall performance can be optimized using synchronized sleep cycles. The optimal scheduling of sleep cycles can be calculated using distributed global state information. For example, mechanisms for delay efficient sleep scheduling (Lu *et al.* 2005) and energy-efficient real-time medium access control (Adamou *et al.* 2001) have been developed. Nevertheless, the maintenance of the perfect synchronization induces eminent scalability problems (Dewasurendra and Mishra 2004).

Similar to most Internet routing protocols, pro-active routing mechanisms have been developed for ad hoc networks. All these protocols are based on periodic state exchange, as they rely on globally synchronized topology information. Basically, such pro-active routing mechanisms can be grouped into distance vector-based or table-driven algorithms, e.g. the Destination Sequenced Distance Vector (DSDV) routing protocol (Perkins and Bhagwat 1994), link-state mechanisms, e.g. Optimized Link State Routing (OLSR) (Clausen and Jacquet 2003), and hierarchical approaches, e.g. Hierarchical State Routing (HSR) (Iwata *et al.* 1999). Considering the behavior of these algorithms, all of them always allow the find of an optimal path through the network according to specific routing metrics, e.g. the shortest path in terms of hops. The different protocols differ in the principles used for state maintenance and, thus, in the convergence speed and the necessary amount of maintenance overhead.

Mobility management and energy control can be incorporated into global state algorithms. Basically, such additional information can be integrated into the optimization algorithms. Nevertheless, the state maintenance requirements are drastically increasing. As every node in the network needs up-to-date global state information, each state change, e.g. movements or energy drains, must be propagated to all other nodes. In large networks or in ad hoc networks with high mobility, the state maintenance might be much too expensive. Similarly, networks with infrequent data transmissions will tend to perform state maintenance operations only and wasting all their energy for these application-independent actions.

8.3.3 Location information

Regionalizing global optimization algorithms is a first approach to improve the efficiency of operations in ad hoc networks (Hollos *et al.* 2004). This idea can be exploited in two directions, both relying on location information. First, geographical information can be used in various topology control issues, e.g. for stateless setup of routing paths. Sometimes, the retrieval of such geographic information can be expensive, especially in mobile networks (Shah *et al.* 2005a,b). Therefore, local position information within a limited area is often used to identify location information, e.g. for clustering techniques.

The explicit calculation of a route towards a destination can be prevented if the positions of the source, the neighboring nodes and the destination are known (Mauve and Widmer 2001). Thus, global state can be prevented and the routing problem in ad hoc networks can be reduced to the identification of accurate position information of (at least) the sender, the receiver and the neighboring nodes. First solutions for location-aided routing in mobile ad hoc networks were proposed by Ko and Vaidya (1998) and Navas and Imielinski (1997). Meanwhile, various alternatives have been developed using geographic positions for optimized routing decisions (Caruso *et al.* 2005). Optimized versions of geographical routing incorporate histories and meta-information (Xu *et al.* 2001). Problems may occur due to inconsistencies in the position information (Kim *et al.* 2004). Especially in phases with high mobility of the communicating nodes, the update of position information might become a bottleneck. While all geographical routing approaches follow similar approaches to select a next hop node during data transmission, i.e. a node that is 'closer' or 'closest' to the destination, different solutions have been proposed to counteract dead ends.

Clustering techniques allow the logical grouping of nodes according to a given distance metric. The state information is reduced, as complete state must be maintained only by the group or cluster members. The primary idea is to group nodes around a clusterhead that is responsible for state maintenance and inter-cluster connectivity. Clustering is a crosscutting technology that can be used in nearly all layers of the protocol stack. While clustering algorithms have already been used in various domains of networking, the use of standard algorithms like k-clustering and others have also been successfully demonstrated in ad hoc networks (Fernandess and Malkhi 2002). Besides the reduction in state information, the primary objective of using such techniques in ad hoc and sensor networks is to reduce the necessary amount of energy required for maintenance and control operations (Chen *et al.* 2004). Examples of efficient clustering algorithms are passive clustering (Kwon and Gerla 2002), which reduces the necessary overhead for maintaining the structure of the clusters, and on-demand clustering (Cramer *et al.* 2004), which mitigates the need for permanent maintenance of clusters by creating them on demand.

Application examples in ad hoc networks especially focus on efficient routing algorithms that make use of such clustering mechanisms (Ibriq and Mahgoub 2004). A typical example is the Zone Routing Protocol (ZRP) (Haas and Pearlman 2001). Other cluster-based routing protocols (Chiu *et al.* 2003) have been investigated, as well as optimized control loops (Banerjee and Khuller 2001). In the context of sensor networks, an additional issue is data management. Low-Energy Adaptive Clustering Hierarchy (LEACH) (Heinzelman *et al.* 2000) and its competitor Hybrid Energy-Efficient Distributed Clustering Approach (HEED) (Younis and Fahmy 2004) are examples for power-aware cluster-based communication approaches for ad hoc and sensor networks.

8.3.4 Neighborhood information

Compared with cluster maintenance or even global state, neighborhood information can be gathered quite easily. Actually, it is also used as a starting point for maintaining clusters or global state. The basic idea is to periodically exchange information between directly connected neighbors. In wireless networks – as studied in the context of ad hoc and sensor networks – a single data message will suffice to transmit local state to all neighbors. Usually, 'hello' or 'sync' messages are employed, which include all the necessary information for the particular algorithm to take decisions based on the local state and the state of its neighbors. The overhead induced by these additional messages can be reduced by piggybacking state information to other 'normal' data messages.

Obviously, the overhead for state maintenance is drastically reduced compared with location information-related approaches. Nevertheless, optimal solutions cannot be computed based on this very spatially restricted state information, e.g. for end-to-end communication paths or global allocation schemes. In this context, we speak of weak synchronization between coordinating nodes. Such weak synchronization is employed by many MAC protocols to reduce the probability of collisions or to implement energy-saving duty cycles. In the network layer, re-active routing protocols counteract the need to create and maintain global state for routing decisions.

The most popular mechanism for coordination at the MAC layer is the RTS/CTS handshake (see also Chapter 9). Rather short messages Request To Send (RTS) and Clear To Send (CTS) regulate access to the shared wireless medium. The collision probability for these short messages is quite low compared with typical data messages. On the other hand, 'sync' messages are employed to synchronize all neighboring nodes to a common sleep cycle (Lu *et al.* 2005). Nodes are allowed to sleep as long as there is no scheduled local action pending and if there is no message to receive, respectively. Obviously, such busy–sleep schedules must be synchronized in the local neighborhood. The same message exchange can also be used for further, more sophisticated performance improvements. The most prominent MAC protocol in the wireless ad hoc domain is IEEE 802.11 (IEEE 1999), which is also known as WLAN or WiFi. It features RTS/CTS based solutions for the hidden/exposed terminal problems, adaptive sleep cycles and energy control using overhearing techniques. More specialized for wireless sensor networks is the Sensor MAC (S-MAC) protocol (Ye *et al.* 2002, 2004). In addition to the mechanisms established in IEEE 802.11, a mechanism called adaptive listening has been integrated. It allows enhanced performance while enabling maximized sleep periods at the same time.

Reactive routing, also known as on-demand routing, is an example of network layer mechanisms employing neighborhood information. Actually, reactive routing intends to set up global state information. Nevertheless, it does so only if triggered by the local application or by direct neighbors. Reactive routing protocols do not keep global routing tables up to date. Instead, they only manage neighborhood relationships. On-demand, i.e. if messages have to be transmitted, routing information is gathered by flooding route requests through the network in order to find a suitable path towards the destination. Several optimizations in terms of adjustable caches for previously determined route information allow a fine-tuning of the algorithms, depending on the application scenario. Well known examples are the Ad Hoc on Demand Distance Vector (AODV) (Perkins and Royer 1999; Perkins *et al.* 2003) and its successor, Dynamic MANET on Demand (DYMO) (Chakeres and Perkins 2007). In contrast, diffusion techniques prevent the calculation of routing paths and employ interest distributions instead. The class of diffusion algorithms can also be classified into neighborhood information. While multiple diffusion algorithms have been developed during the last few years, directed diffusion (Estrin *et al.* 1999b; Intanagonwiwat *et al.* 2000) is still the best known approach. Other variants of directed diffusion try to optimize particular aspects, such as the minimization of energy consumption (Durvy and Thiran 2005) or the inclusion of geographical information in Geographical end Energy-Aware Routing (GEAR) (Yu *et al.* 2001). Constrained Anisotropic Diffusion Routing (CADR) (Chu *et al.* 2002) is a form of diffusion algorithms that optimizes both energy and transmission latency.

8.3.5   Local state

In most circumstances, the local system state can be maintained very easily. In this context, we consider the update of primary system parameters, the reception of messages and the update of sensor readings to examine environmental conditions. In summary, any algorithm relying on the value of one or many local system variables only falls into this category. Obviously, we need to clarify which events may change or update local system variables to describe the behavior of the local state algorithms:

- Internal events, e.g. timers – Various timers may be used to modify system behavior over the time. For example, retrials of algorithms, e.g. the message retransmission after a timeout, may use a binary back-off. This means that the waiting time is doubled with every try – resulting in a changing system behavior as observable from other nodes.

- Sensor readings – Attached sensors are intentionally used to observe the local environment. If changes are detected, the local system behavior may be adapted accordingly. For example, a system might want to update its duty cycle according to the value of the physical measures under observation. If a predefined threshold is exceeded, the message rate should be increased and, therefore, the system should contend more frequently for the wireless channel.

- Messages – The reception of data messages can be seen as an external event. The receiving node may want to update its local behavior, depending on the message content. As we will see, not only the content may be used to determine optimal system parameters, but also associated meta information such as the received signal strength.

Many MAC protocols employ local state information for various purposes. We already mentioned back-off techniques (which are used in the network, the transport and the application layer as well). Additionally, meta information, e.g. the received signal strength, from received messages is exploited. The Power-Control MAC (PCM) protocol (Jung and Vaidya 2002) is an energy-aware extension to typical contention-based MAC protocols. It adapts the transmission power to the current needs in the local network. For this purpose, the RTS/CTS handshake between adjunct nodes is used to estimate the necessary transmission energy (signal strength) to reach the destination node. As the required energy for a successful transmission over a distance of r increases – depending on the environment – with r^2 up to r^4, a huge amount of energy can be saved using this method.

On the network layer, a number of data centric routing algorithms have been proposed over the last few years. The main idea is to process received messages according to their internal 'meaning'. In most cases, no routing tables are needed for data routing. If a message is unknown to a particular node or if the node decides not to process the message for other reasons, it will forward the message to directly connected neighbors. The best known examples for data-centric message processing are various aggregation algorithms (Krishnamachari et al. 2002). Additionally, complete message processing, including message modification, data aggregation, actuation control and routing, is possible using rule-based approaches (Dressler 2006a).

8.3.6 Probabilistic techniques

Techniques in the category of probabilistic algorithms intend to keep no state information at all. Therefore, they show the best behavior if very few messages per time have to be transmitted because the overhead due to state maintenance is negligible (or even zero). Nevertheless, the overhead for actual transmitting messages can be much higher. Interestingly, probabilistic algorithms are employed in most ad hoc and sensor network protocols. Usually, we find randomized waiting times or similar approaches to prevent synchronization effects, e.g. if two neighboring nodes want to transmit a message simultaneously, a collision will occur. If both nodes start again after the same waiting time, the collision will be repeated with a high probability. Thus, in MAC protocols and congestion-aware communication mechanisms, stochastic distributions and random delays are employed to prevent such global synchronization effects. For routing purposes, obviously, probabilistic algorithms will not be able to find an optimal path from a source towards a sink. Instead of flooding messages through the entire network, probabilistic algorithms are often used for routing and data dissemination in ad hoc and sensor networks. A comparison of data dissemination protocols in ad hoc networks is, for example, provided in Bokareva et al. (2004) and Boukerche and Nikoletseas (2004).

Without routing tables, information exchange in communication networks can be organized by flooding the messages through the entire network. Optimized flooding strategies (Kwon and Gerla 2002) try to prevent the forwarding of duplicates of the packet by using a limited time-to-live field or sequence numbers. The probability that a message will arrive at a destination is very high, even in the case of mobility and error-prone wireless channels.

On the other hand, the overhead due to message transmissions into unessential parts of the network increases with the network size. Similarly, every duplicate that is transmitted to the destination must be regarded as an unnecessary overhead. Gossiping (Haas *et al.* 2002a) and rumor routing (Braginsky and Estrin 2002), as alternatives to flooding, have been developed to cope with this problem. Here, the main idea is to transmit messages to selected neighbors (based on a random scheme) instead of to all of them. The advantage is the drastically reduced overhead. Nevertheless, the probability for successful message transmissions decreases rapidly with the distance between sender and receiver. Specific network topologies, e.g. a number of nodes in a row, contribute to reducing the delivery ratio. Haas *et al.* (2002a) investigated this effect and suggested the first solutions. Probabilistic parametric routing (Barrett *et al.* 2003) and weighted probabilistic data-dissemination schemes (Dressler 2005) further improve the behavior of these algorithms. The optimization goal is the reduction of the overhead due to unnecessary messages compared with the probability of reaching the final destination. This group of algorithms has been extended to probabilistic lightweight group communication (Luo *et al.* 2004).

9

Medium Access Control

The primary objective of MAC protocols is to manage the communication between directly connected neighbors. This especially includes the management of the wireless communication channel. In wireless networks, medium access is rather difficult to realize due to several issues. First, it is not possible to send and to receive at the same time – at least if only a single transceiver can be used. Secondly, from the sender's point of view, it is hard to estimate the interference situation at the receiver, while only the channel status at the receiver side counts for successful transmission. Thirdly, high error rates make it rather hard to establish a well coordinated communication link between two nodes over the wireless medium. As all wireless networks rely on a distributed coordination function, appropriate scheduling techniques must be employed. Wireless networks inherently lack any centralized coordination due to their distributed nature. Coordination can be achieved by exchanging control information, while, obviously, such control packets must not consume too many resources.

The typical requirements of MAC protocols for wireless networks and, therefore, also for WSN include the following items. As for any network protocol, high throughput and low error rates are envisioned. Nevertheless, the primary measures are the protocol overhead and the energy efficiency.

Overhead is primarily generated by protocol actions that focus on the coordination among neighboring nodes. Obviously, the protocol overhead also contributes to the energy consumption. The primary approach to reduce energy requirements is to periodically switch on and off the participating devices (duty cycles). For coordinated duty cycles, the protocol overhead increases and vice versa. Thus, the MAC protocols for wireless communication demand optimization algorithms between energy wastage due to overhead and due to coordination. Additional effects that also contribute to energy wastage are the following problems that are unique to wireless communication protocols:

- Collisions – Whenever two simultaneous transmissions result in a collision, the transmitted packets will be lost. Thus, a retransmission must take place, which requires additional energy (and the detection of and reaction to collisions need further efforts).

Self-Organization in Sensor and Actor Networks Falko Dressler

- Overhearing – If a node is receiving a packet that is destined for another receiver in the same wireless transmission range, the required energy for receiving the packet (or at least parts of it) and the detection that it is destined to another node are waste efforts.

- Idle listening – As the receiving node has no knowledge (in non-time-schedule scenarios) when a sender may begin a transmission, it must be idle, waiting for a possible packet reception. Thus, energy is wasted for doing nothing.

Another issue appearing in the context of MANETs is the mobility of nodes. Node mobility can essentially affect the performance (throughput) of the protocol. For example, bandwidth reservations or exchanged control information may become useless if nodes are moving quickly, or, more precisely, if the topology changes rapidly. Therefore, protocol design must take this mobility factor into consideration, as system performance should not be significantly affected due to node mobility.

Basically, we can distinguish three classes of MAC protocols used in wireless radio networks. First, contention-based protocols must be named, as this class represents the biggest part. Protocols from this class do not share any time or resource information among the neighboring nodes. Thus, no a priori resource-reservation mechanisms are available. Secondly, contention-based protocols with reservation mechanisms include specific support for real-time traffic using QoS guarantees. In order to enable such guarantees, topology and resource information must be shared among all involved nodes in the network. Finally, contention-based protocols with scheduling mechanisms employ synchronization mechanisms to reduce the probability of collisions.

In the following, we will primarily focus on pure contention-based protocols. Nevertheless, some aspects of resource management and synchronization essentially influenced all state-of-the-art MAC protocols – even those used in sensor networks.

9.1   Contention-based protocols

Pure contention-based protocols do not rely on an a priori resource reservation. Whenever a packet should be transmitted, the node contends with its neighbors for access to the shared channel. Obviously, strict QoS guarantees cannot be provided. The contention period can be regarded as a conflict resolution. If multiple nodes want to access the channel at the very same time, collisions cannot be avoided. Thus, contention resolution has to be provided by the protocols. Two approaches can be distinguished: sender-initiated protocols, i.e. the packet transmissions are initiated by the sender node, and receiver-initiated protocols, i.e. the receiver node initiates the contention-resolution protocol.

The main problems in contention resolution are hidden and exposed terminals. These problems are unique in wireless networks. Figure 9.1 depicts the characteristics of these problems:

- Hidden terminal problem – This problem refers to possible collisions of packets due to the simultaneous transmission of those nodes that are not within the direct transmission range of each other but within the transmission range of a common receiver. This scenario is shown in Figure 9.1 (left). Two senders, S1 and S2, are

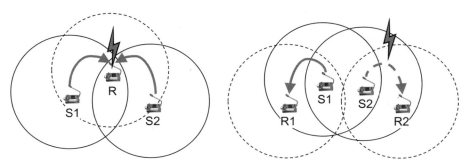

Figure 9.1   Basic principles of the hidden (left) and exposed (right) terminal problems

simultaneously transmitting packets to the common receiver R. This transmission results in a collision at R, while this collision cannot be recognized at S1 and S2.

- Exposed terminal problem – On the other hand, the exposed terminal problem depicts the possible starvation of single transmissions, i.e. the inability of a node, which is blocked due to transmission by a nearby transmitting node, to transmit to another node. In Figure 9.1 (right), the transmission from S1 to R1 blocks the other sender S2. As S2 can recognize the ongoing transmission, it will wait for a free channel – which, in some cases, might be not possible for a long time.

In order to control transmissions and to counteract the named two problems, the following two mechanisms can be employed. Receivers may inform potential interferers while a reception is ongoing. For example, a signal indicating the ongoing transmission can be sent out. Unfortunately, the same channel on which actual reception takes place cannot be used for this signal. Thus, separate signaling channels must be used. A popular example is the busy-tone protocol, e.g. the Dual Busy Tone Multiple Access (DBTMA) protocol (Deng and Haas 1998). On the other hand, receiver may inform potential interferers before a reception is ongoing. In this case, the same channel can be used for signaling control information and for data transmission. Nevertheless, the receiver itself needs to be informed, by the sender, about the impending transmission. Potential interferers need to be aware of such information. The best known approach in this category is the Multiple Access with Collision Avoidance (MACA) protocol (Karn 1990), which will be outlined below.

The primary idea of MACA is the use of additional signaling packets. The primary protocol behavior is depicted in Figure 9.2. The following protocol actions are specified:

1. The 'Sender' requests a permission to send a packet by broadcasting a Request To Send (RTS). This procedure should ensure that, first, the receiver is ready to receive the transmission and, secondly, no other nodes will interfere with the transmission.

2. The 'Receiver' sends out a Clear To Send (CTS) if it agrees upon the transmission of the forthcoming data packet. The CTS is broadcasted within the transmission range of the receiver.

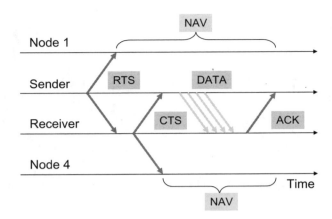

Figure 9.2 RTS/CTS handshake as used in the MACA protocol

3. If the RTS/CTS handshake was successful, i.e. if both the RTS and the CTS are successfully received by the receiver and the sender, respectively, the sender starts its data transmission.

4. In order to conclude the data transmission, the receiver sends an Acknowledgment (ACK) to the receiver, denoting the successful reception. Using the ACK message, error detection is provided by the MAC layer, as the original error detection in the transport layer is slow and introduces a lot of overhead in ad hoc and sensor networks.

All potential interferers overhear the RTS/CTS handshake. In our example, as shown in Figure 9.2, 'Node 1' can overhear the RTS and 'Node 4' the CTS. All RTS/CTS packets carry the expected duration of the data transmission. The MACA protocol requires that all participating nodes store this information in a local Network Allocation Vector (NAV). This NAV can be seen as a description of the current usage of the wireless medium. Transmissions may only be initiated if the local NAV declares the medium to be free.

Collisions may still occur, as there is an enforcement procedure that synchronizes the NAV of neighboring nodes. Thus, the RTS/CTS handshake ameliorates, but does not solve, the hidden/exposed terminal problems. Inconsistencies in the NAV of neighboring nodes may occur, e.g. due to colliding RTS/CTS packets, as shown in Figure 9.3, and, due to mobility of nodes, may enter the transmission range of ongoing data transfers at any time. If a packet is lost, i.e. if a collision took place, the node uses the Binary Exponential Back-off (BEB) algorithm to back-off for a random time interval before retrying. Each time a collision is detected, the node doubles its maximum back-off window. The principles of the MACA protocol are used in many modern MAC protocols, e.g. in IEEE 802.11 WLAN (IEEE 1999), and in S-MAC, which is described later.

Multiple issues have been identified in the original MACA protocol. Therefore, several extensions have been proposed. One example is the Multiple Access with Collision Avoidance for Wireless (MACAW) protocol (Bharghavan et al. 1994). This approach primarily addresses the back-off mechanism as specified in MACA. First, the binary back-off mechanism can lead to starvation of flows. As an example, please consider two nodes S1

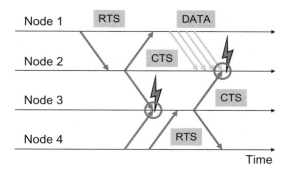

Figure 9.3 Situation in which the RTS/CTS handshake will fail due to colliding RTS/CTS packets

and S2, which are generating a high volume of traffic. If one node, let us assume node S1, starts sending, the packets transmitted by S2 collide. According to the back-off algorithm, S2 backs off and increases its back-off window. Thus, the probability of node S2's acquiring the channel is constantly decreasing. As a first solution, each packet can carry the current back-off window of the sender. The receiving nodes will use this value for their own back-off. Secondly, large variations in the back-off values are typical for the BEB algorithm, as the back-off window increases very rapidly and is reset after each successful transmission. MACAW proposes to replace the BEB by using the Multiplicative Increase and Linear Decrease (MILD) back-off mechanism, which increases the back-off window by a factor of 1.5. Considering the fairness in the utilization of the wireless medium, basically, the MACA protocol provides a per node fairness while MACAW introduces a per flow fairness by supporting one back-off value per flow. Further extensions to the original MACA have been developed such as the MACA By Invitation (MACA-BI) protocol by Talucci and Gerla (1997).

9.2 Sensor MAC

The raising interest in sensor networks and the increasing requirements on energy efficiency forced the network community to reconsider the developed MAC protocols for wireless radio networks. As described in Section 7.2, the primary objectives deal with the energy and lifetime of the sensor nodes. To cope with these goals and to retain the flexibility of contention-based protocols while improving the energy efficiency in multi-hop networks, a new protocol, Sensor MAC (S-MAC), has been proposed by Ye *et al.* (2002). According to the findings that the idle listening principle of MACA is particularly unsuitable if the average data rate is low (nothing happens most of the time), the primary idea is to switch nodes off, and ensure that neighboring nodes turn on simultaneously to allow packet exchange ('rendez-vous'). Especially in typical sensor networks, the nodes tend to send very infrequently, as the transmission of packets is highly energy-intensive.

The basic approach to save energy is to introduce a coarse-grained listen/sleep cycle. The basic principles of this behavior are depicted in Figure 9.4. All networked nodes follow

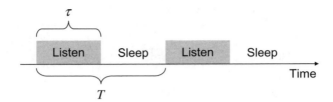

Figure 9.4 Coarse-grained listen/sleep cycle

a common cycle of consecutive listen and sleep periods. In the listen period, i.e. when the nodes are awake, essentially, the RTS/CTS handshake is performed. If this handshake is successful, the packet transmission can occur. During the sleep period, no packet exchange can take place. Obviously, the listen/sleep cycle must be synchronized between neighboring nodes, i.e. an exchange of scheduling information is needed.

Also shown in Figure 9.4 are the common denotations for the involved time periods. τ refers to the listen period and T describes the time for one complete cycle. Based on these measures, the duty-cycle D can be calculated as depicted in Equation 9.1. Thus, the duty-cycle represents a measure for the energy efficiency of one node, i.e. the percentage of active time (listen) to the total time:

$$D = \frac{\tau}{T}. \tag{9.1}$$

S-MAC has been developed to cope with the following two issues (Ye and Heidemann 2003; Ye *et al.* 2002). Today, S-MAC is considered a reference measure when comparing MAC protocols in the sensor network community.

- Synchronization of the sleep cycle of multiple sensor nodes. This synchronization procedure also supports to work in a multi-hop network.

- Optimization of the transmission delay and throughput in low duty-cycle networks.

9.2.1 Synchronized listen/sleep cycles

S-MAC enhances the presented coarse-grained listen/sleep cycle, as shown in Figure 9.5. Actually, the listen period is divided to support synchronization between neighboring nodes

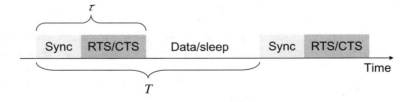

Figure 9.5 S-MAC duty cycle consisting of sync, RTS/CTS and data/sleep periods

as well as the contention for the wireless channel using the RTS/CTS handshake mechanism. The explicit synchronization using well defined SYNC messages was introduced to support low duty-cycle operations of 1–10% (Li *et al.* 2004). Basically, all nodes choose their own listen/sleep schedules. These schedules are then shared with their neighbors to make communication possible between all nodes in the surrounding neighborhood. Each node periodically broadcasts its schedule in a SYNC packet, which provides simple time synchronization. To reduce the overhead, S-MAC encourages neighboring nodes to adopt identical schedules.

The principles of the synchronization process of S-MAC are depicted in Figure 9.6. Shown are four nodes starting at arbitrary times. All nodes try to pick up schedule synchronization from neighboring nodes, i.e. if a node can successfully receive a SYNC message during the initial 'waiting time', it takes over this schedule (in the given example, nodes 2 and 3 receive a SYNC from their neighbors 1 and 4, respectively). If no neighbors can be found, the nodes pick some schedule to start with (nodes 1 and 4 in our example). If additional nodes join, some node might learn about two different schedules from different nodes (node 3 in Figure 9.6). These nodes provide bridging functionality between synchronized islands. According to Ye *et al.* (2002), the complete procedure for maintaining the schedule table works as follows.

1. Each node first listens for a fixed amount of time ('waiting time', which is at least the synchronization period). If it does not hear a schedule from another node, it immediately chooses its own schedule and starts to follow it. This includes the

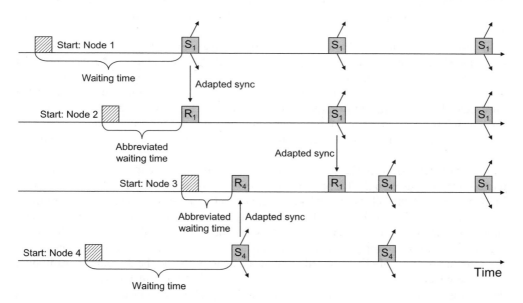

Figure 9.6 Synchronization process of S-MAC. Shown are four nodes starting at arbitrary times. If a node can successfully receive a SYNC message during the initial 'waiting time', it takes over this schedule (nodes 2 and 3). If no neighboring node is sending a SYNC, the node selects its own schedule and follows it (nodes 1 and 4)

periodic broadcast of a SYNC packet to the surrounding neighborhood. As SYNC messages may be sent by two nodes simultaneously, a randomized carrier sense time is used to reduce the chance of collisions on SYNC packets.

2. If a node receives a schedule from a neighbor before choosing or announcing its own schedule, it adapts it by setting its schedule to be the same. Then, the node continues as if it would have chosen its own schedule, i.e. it will periodically announce its schedule to all neighboring nodes.

3. If a node receives a new schedule after agreeing on a (different) common schedule with one or more neighbors, it adopts both schedules by waking up at the listen intervals of the two schedules. This way, already synchronized S-MAC islands can be interconnected. On the other hand, if the node follows its own schedule (without any synchronized neighbors) and receives a SYNC packet, it will discard its current schedule and follow the new one.

9.2.2   Performance aspects

The basic scheme of periodic listen and sleep essentially helps to increase the lifetime of a single sensor node according to the selected duty-cycle D. On the other hand, it strictly controls the communication of radio packets. There is exactly one chance for each node to either send or receive one data packet in a frame, i.e. in one time schedule. Thus, in a multi-hop network, the end-to-end latency gets accumulated at each hop. Thus, the optimization of the energy consumption may result in a large delay in multi-hop transmissions as a side effect.

Figure 9.7 explains this effect. Let us consider a linear topology of four sensor nodes (A, B, C, and D). In order to transmit a packet from A to D, the packet must be relayed by B and C, respectively. Thus, in the first time slot, A will forward the packet to B, which, in turn, will forward it to C in the second time slot. Finally, C will forward the packet to D in the third time slot. Assuming a low duty-cycle, only small parts of the schedules will be needed for transmission and, most of the time, the nodes will sleep. Thus, three listen/sleep cycles will be needed for this single packet, resulting in low bandwidth utilization, depending on the duty-cycle.

Ye *et al.* (2004) proposed an enhancement for this performance penalty, which was named adaptive listening. The primary intention was to improve the performance of S-MAC in terms of the achieved end-to-end delay and the possible throughput. The primary idea is very simple. At the end of one transmission, another listen period is appended, called the adaptive listening period. As this period takes place within the scheduled sleep period, all transmission-involved nodes obtain an extra transmission window. The active listening period can only be triggered at nodes that can overhead the ongoing transmission, i.e. those which are able to determine the scheduled end of the transmission as denoted in the RTS and CTS packets. These nodes will wake up after the transmission and contend again for the channel using a RTS/CTS handshake.

Figure 9.8 demonstrates the use of adaptive listening. Again, we consider a four-hop linear topology and an end-to-end communication between nodes A and D. In the first time slot, A transmits its data packet to B. C is able to overhead the CTS packet and sched-ules a new contention period after the announced end of transmission. After finishing the

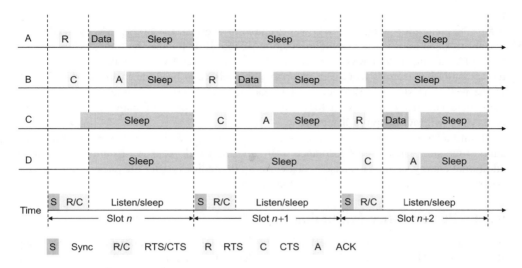

Figure 9.7 Transmission of a single data packet using S-MAC over a four-hop linear topology; three complete time slots, i.e. listen/sleep cycles, are needed to complete the end-to-end transmission

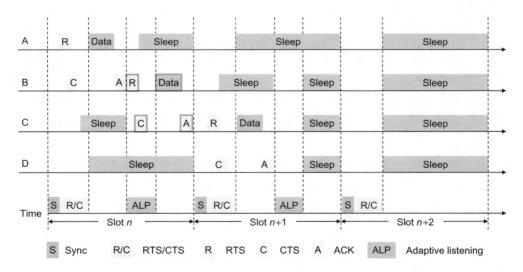

Figure 9.8 Transmission of a single data packet over a four-hop linear topology using the adaptive listening technique in S-MAC; only two complete time slots, i.e. listen/sleep cycles, are needed to complete the end-to-end transmission compared with the three time slots without adaptive listening

transmission from A to B, a new active listening period is started. A, B and C are participating in the contention window. As D was not able to overhear the first RTS/CTS handshake, it went to sleep until the next synchronization period. B announces its transmission request in a RTS packet, to which A responds by falling asleep and C by sending a CTS. Thus, the transmission from B to C can occur. Nothing else can be done in this first time slot. In the next time slot, C will send the data packet to D. Thus, in summary, only two time slots were necessary to transmit the data packet from A to D compared with the version of S-MAC without adaptive listening.

Using adaptive listening, it becomes possible to pass a data packet over two hops in a single time slot. Thus, the performance is greatly improved. Adaptive listening also allows additional energy savings because nodes wake up immediately after the packet transmission completes for immediate contention for the channel.

9.2.3   Performance evaluation

The performance of S-MAC was tested by Ye *et al.* (2002, 2004) in a lab measurement using Rene motes.[1] The same experiment was reproduced by Chen *et al.* (2006) in a simulation environment using the network simulator *ns-2* (Fall and Varadhan 2007). The following selected performance characteristics of S-MAC are reviewed to demonstrate its efficiency:

- Mean energy consumption per byte – The total energy consumed by all nodes divided by the total number of bytes received by the sink.

- Mean end-to-end delay – The sum of all end-to-end delays divided by the total number of packets.

- End-to-end goodput – The total number of bytes received by the sink divided by the time from the first packet generated at the source until the last packet was received by the sink.

In all the experiments, the same network topology is considered. Figure 9.9 depicts this topology. Ten nodes are placed in a linear network with one source (node 1) and one sink (node 10). In the simulation experiments, three S-MAC protocol modes have been compared:

- Mode 1: No periodic sleep – full active mode, sleep only to avoid overhearing;

- Mode 2: 10% duty-cycle without adaptive listening;

- Mode 3: 10% duty-cycle with adaptive listening.

Source   Sink

Figure 9.9 Scenario used for the performance evaluation of S-MAC: 10 nodes in a linear network with one source (node 1) and one sink (node 10)

---

[1]http://www.xbow.com/.

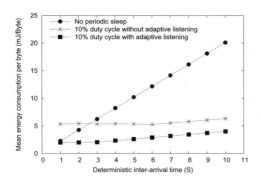 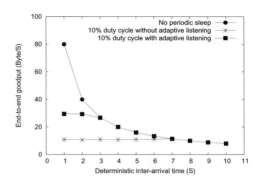

Figure 9.10 Mean energy consumption per payload byte (left) and mean end-to-end goodput (right), both using a CBR traffic source [Reproduced by permission of Falko Dressler (Chen *et al.* 2006); © 2006 ACM]

Figure 9.10 (left) depicts the mean energy consumed in transmitting a single byte from the source to the sink. The simulation was repeated for different traffic loads. As used by Ye *et al.* (2002) and Chen *et al.* (2006), the inter-arrival time is plotted on the x-axis, which means that the traffic load decreases with increasing values of the x-parameter. The energy consumption without periodic sleep is being reduced with higher traffic load. The primary reason is the highly increasing number of bytes, which can be transmitted in the same time. If lower duty-cycles are used, the energy consumption is almost equal for all data rates – while obviously improved with adaptive listening, as, in this mode, more data (about twice as much) can be transmitted in the same listen/sleep period compared with the scenario without adaptive listening.

The effects on the end-to-end goodput are shown in Figure 9.10 (right). With increasing data rates, the goodput increases almost linearly without periodic listen/sleep cycles. For S-MAC without adaptive listening, the goodput is limited by the duty cycle of 10%. Using adaptive listening doubles the possible goodput.

In Figure 9.11, the end-to-end delay as induced by S-MAC is analyzed for two different traffic sources – CBR and On/Off traffic. With increasing traffic rates, the end-to-end delay is increasing as well. The primary reason is the queuing in all intermediate nodes. As S-MAC with adaptive listening and the simple mode without any listen/sleep cycles supports higher data rates, the queuing effects become visible only at comparably high data rates.

From the above results, we can see that periodic sleep helps S-MAC to achieve a good energy performance, i.e. an adequate ratio of consumed energy to the communication performance, under light traffic loads. Under heavy traffic load, S-MAC with adaptive listening greatly improves the problems of large latency and low bandwidth utilization that periodic sleep introduced.

9.3 Power-Control MAC protocol

Another approach to reduce the energy for radio communication is followed by the Power-Control MAC (PCM) protocol (Jung and Vaidya 2002, 2005). It provides a MAC layer

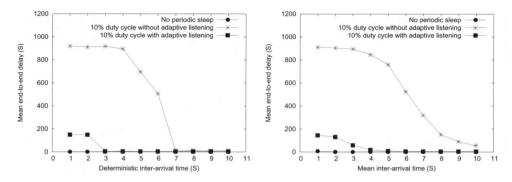

Figure 9.11 Mean end-to-end delay using a CBR traffic source (left) and an On/Off traffic source (right), respectively [Reproduced by permission of Falko Dressler (Chen *et al.* 2006); © 2006 ACM]

solution for localized power control by varying the transmission power to reduce the overall energy consumption. In addition to providing energy saving, the PCM protocol can improve the spatial reuse of the wireless channel.

PCM is an extension to typical contention-based MAC protocols. It suggests transmitting the RTS/CTS handshake messages with the maximum available power p_{max}. The RTS/CTS handshake is used to determine the required transmission power $p_{desired}$ that is used for the subsequent DATA/ACK transfer. The principles are shown in detail in Figure 9.12. In this example, node 4 is about to send a packet to node 5. It sends its RTS packet with the maximum available transmission power p_{max}. The same power is used by node 5 for the corresponding CTS. Finally, both nodes agree on the transmission power $p_{desired}$ for the DATA/ACK exchange.

PCM exploits the signal level of the received RTS for the calculation of $p_{desired}$ in combination with some well known minimum threshold for the received signal strength Rx_{thresh} that is necessary for correctly decoding the messages. The transmission power is calculated using Equation 9.2 (whereas p_r denotes the received power level). As the radio signal is assumed to be optimal, i.e. to be distributed with the same quality within a circular area, a constant parameter c is used to increase $p_{desired}$ according to the environmental conditions. Additionally, it is assumed that the signal attenuation between source and destination nodes is the same in both directions, and that the noise level at the receiver is below some predefined threshold (encoded in Rx_{thresh}). These assumptions may be wrong in certain scenarios. However, it is likely to be true with a fairly high probability (Jung and Vaidya 2002).

$$p_{desired} = \frac{p_{max}}{p_r} Rx_{thresh} \times c. \tag{9.2}$$

Unfortunately, collisions cannot be completely avoided using this scheme, as there may be nodes which can sense the signal of the RTS/CTS exchange but cannot decode it because the signal level is too weak (nodes 1 and 8 in Figure 9.12). During the DATA/ACK period, these nodes do not sense a signal any longer. Therefore, they may initiate their own

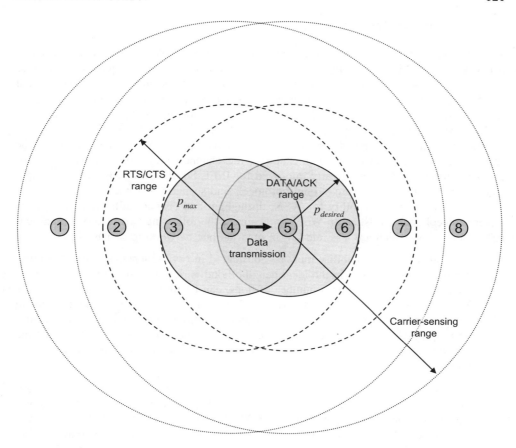

Figure 9.12 Transmission ranges used by PCM for RTS/CTS and DATA/ACK, respectively; also shown is the maximum carrier sense range

RTS/CTS exchange, which results in a collision with the still ongoing data transmission. PCM addresses this issue by varying the signal level of the data transfer. Periodically, the transmission power is increased to p_{max}, thus allowing distant nodes to sense the signal of the ongoing transmission. The period can be adapted to the carrier sensing algorithm in order to optimize the behavior of the protocol.

In summary, it can be said that the PCM protocol is a good example of achieving optimal throughput while reducing the necessary transmission energy to a minimum. The necessary calculations are based on locally available information only by observing the neighboring environment, i.e. PCM extracts the knowledge from monitoring and analyzing the received RTS/CTS messages, which represent the behavior and conditions of the surrounding environment. To some extent, node mobility is also supported as well as changes in the network topology.

9.4   Conclusion

There is a broad set of well established MAC protocols available in the ad hoc and sensor network domain. Most of the protocols rely on contention-based techniques. Due to space limitations, we only discussed a small set of protocols to outline some specific aspects that can be found in other approaches as well. Basically, all typical MAC protocols employ similar solutions for the hidden and exposed terminal problem, respectively, as introduced by MACA and MACAW and improved by IEEE 802.11 Wireless LAN.

The performance of the different MAC protocols strongly varies, depending on the specific protocol characteristics. For example, the channel utilization is rather poor for unsynchronized MACA and greatly improved for IEEE 802.11. Similarly, the throughput and the end-to-end goodput are strongly protocol- and scenario-dependent.

Considering the applicability for WSN, throughput and channel utilization are not the major protocol characteristics. In WSN applications, the following two factors are particularly dominating, which are less relevant in other ad hoc networking scenarios:

- Scalability – The scalability must be considered in two dimensions. First, the number of supported nodes in the local neighborhood varies, and, secondly, the network diameter, i.e. the maximum number of hops, can be limited by the behavior of the MAC protocol. For example, IEEE 802.11 networks need a global synchronization while S-MAC supports the interconnection of multiple 'synchronized islands'. Similarly, protocols with a fixed schedule do not work well in large networks, while the adaptive listening mechanism provides an adaptive solution even for low duty-cycle protocols.

- Energy efficiency – Throughput-optimized MAC protocols are usually less energy-saving. For example, the MACA variants (including IEEE 802.11) are rather energy-exhausting compared with S-MAC or PCM. The adaptive adjustment of the transmission power in PCM according to the current distance between two communicating nodes significantly reduces the energy consumption of a single data packet. Additional savings can be achieved by employing listen/sleep periods with a low duty-cycle. S-MAC proposes such a scheme in combination with adaptive adjustments of the protocol behavior according to the current network utilization.

In our discussions of MAC protocols for sensor networks, we outlined two specific developments. S-MAC supports listen/sleep periods with multiple syncs as well as low duty-cycles (i.e. long sleep periods) with adaptive listening. PCM includes an adaptive approach for adjusting the transmission power in a well controlled manner.

As can be seen from the analyzed MAC solutions, there is a strong need for dynamic networking approaches. These dynamics is usually handled by means of self-organization techniques. Adaptive algorithms for scheduling and synchronization can be developed based on locally available information only. This includes schedule initialization, synchronization maintenance and periodical neighbor discovery. None of the analyzed protocols is relying on a global schedule algorithm – while this is, of course, possible and typically used in small, fixed-sized, real-time systems.

More advanced research on MAC protocols focuses on cross-layer techniques exploiting common functionalities between the MAC, the network and the application layer, e.g. for neighborship management and duty-cycle adjustment.

10

Ad Hoc Routing

In general, routing in communication networks refers to the path-finding processes to interconnect two nodes over a multi-hop network. Thus, routing supports the end-to-end data communication. Often, the routing process is mixed up with the forwarding process. Forwarding means the actual packet delivery to a next hop node towards its destination. In this book, we discuss several means of routing, starting with its original purpose, the maintenance of so-called routing tables, but also completely different paradigms, i.e. data-centric data forwarding.

An ideal routing protocol should show the following main characteristics. This list is appropriate for any kind of routing, including the application in MANETs and WSNs:

- Fully distributed – Centralized routing processes have several limitations. The most important one is the missing scalability to support huge numbers of nodes. Thus, the routing decisions should be taken fully distributed.

- Adaptive – The routing protocol should be aware of topology changes. Depending on the kind of network, such topology changes can be the result of node failures or node mobility (Hong *et al.* 2002).

- Minimized coordination – The number of nodes involved for route computation should be minimal. In the ideal case, only the nodes on a particular path should be involved in detecting this path. At least, the overhead should be reduced by incorporating as few nodes as possible in the path-finding process.

- Localized state – In many routing protocols, global state is always available to some extent. Nevertheless, the design of protocols working with reduced global state information is highly demanded, as it is hard to maintain and synchronize the global state information.

- Loop-free, free from stale routes – Routing loops, i.e. paths though the network ending in a circular route, and stale routes, i.e. useless routes ending in a dead end, may result in unnecessary data transmissions without any possibility of reaching the final destination. Thus, energy and bandwidth are consumed without purpose.

Self-Organization in Sensor and Actor Networks   Falko Dressler
© 2007 John Wiley & Sons, Ltd

- Limited number of broadcasts – The protocol itself should limit the number of broad-cast packets (which are usually used for path finding). The main reason is to avoid collisions in the wireless medium (as a remark: even in wired networks, routing protocols try to prevent broadcasts, as these must be processed by every connected computer; multicast transmissions are used instead).

- Quick and stable convergence – Convergence refers to the state of the network in which all necessary paths have been determined by the routing protocol. This system state should be reached as soon as possible. Before reaching convergence, routing loops and stale routes may appear due to unsynchronized routing information.

- Optimal resource utilization – First routing protocols only used the hop count, i.e. the distance in the number of hops between source and destination, as a metric to evaluate the quality of a particular route. More enhanced protocols also evaluate the available bandwidth, processing capabilities, memory and battery statistics for route determination.

- Localized updates – In the case of failures or other topology changes, the necessary updates should be limited to the local environment, i.e. the neighborhood surrounding the source of the topology change. The protocol should prevent interactions with unrelated parts of the system to reduce the routing overhead.

- Provision of QoS – Application demands on the service quality should be supported by the routing protocol, e.g. by selecting only paths with a given minimum available bandwidth or a maximum end-to-end delay (De *et al.* 2003). Thus, additional traffic profiles to the typical best-effort communication should be supported.

Regarding the applicability of routing protocols in MANETs and WSNs, some additional requirements and characteristics must be considered to successfully develop efficient routing protocols (Akkaya and Younis 2005). As already discussed in Chapter 9, the use of wireless radio communication introduces several challenges to all network protocols, especially including MAC and network layer.

Gupta and Kumar (2000) have shown especially that the capacity of wireless networks, and thus of WSNs and MANETs, is limited and upper bounds can be formally given. For example, when n identical randomly located nodes, each capable of transmitting at W bits per second and using a fixed range, form a wireless network, the throughput $\lambda(n)$ obtainable by each node for a randomly chosen destination is $\Theta\left(\frac{W}{\sqrt{n \log n}}\right)$ bits per second under a noninterference protocol.

In MANETs, the most challenging issue is the demanded awareness for rapid topology changes caused by node mobility. Nodes may be mobile in time and space and the routing protocol must be able to detect such changes and to react appropriately by updating the local routing state. Obviously, the protocol overhead for such detections and resulting reactions should be minimized. Additionally, the unreliable radio communication must be considered, as routing information can easily get lost and must be retransmitted. Perhaps, specific routing paradigms can be successfully employed. Fortunately, additional information is available in some cases, which can be exploited for routing decisions. The most prominent example is geographic information that is used by several geographical routing approaches (Mauve and Widmer 2001).

Considering the properties of WSNs, the energy efficiency is perhaps the most important aspect for application in sensor networks (Chang and Tassiulas 2004). Fortunately, according to the working principles of WSNs, other communication paradigms can be thought of, which may be exploited for efficient routing and forwarding (Brooks *et al*. 2003b). So, the content of data packets can be analyzed for routing decisions instead of using the identifications of nodes in the network. Additionally, flooding strategies often help in the domain of WSNs – at least if well controlled and coordinated.

Focusing on the application in WSNs, the primary measures for evaluating the efficiency of routing schemes are the following:

- Route setup time – How long does it take to discover a path from a source to a destination?

- Convergence time – What time is required for the routing protocol to converge after a topology change?

- Protocol overhead and energy efficiency – How much energy and bandwidth are required for protocol inherent operations?

In the following sections, we first discuss the two main categories of ad hoc routing, and their requirements and limitations. Then, we focus on address-based routing approaches and outline selected case studies to explain the working principles and application scenarios.

10.1   Overview and categorization

So far, we have discussed the main requirements and characteristics for general-purpose ad hoc routing protocols. Additionally, we have already outlined the primary measures for performance evaluation, focusing on WSNs in particular. In this section, the two main categories of routing techniques in WSNs are distinguished and a more detailed classification of address-based ad hoc routing protocols is provided.

10.1.1   Address-based routing vs data-centric forwarding

The first routing approaches for ad hoc and sensor networks were based on well known techniques known from Internet protocols. In this domain, every routing protocol requires each node to have a network-wide unique address or identifier, which is used for routing purposes. Each packet that is forwarded though the network carries a specific destination address pointing towards its final destination. The category of routing protocols using these unique addresses or similar identifiers is called address-based routing. A typical example is depicted in Figure 10.1. All nodes in the network carry a network-wide unique address. This address is used in all transmitted data packets as an identifier of its final destination. A survey of ad hoc routing protocols for WSNs is, for example, provided by Akkaya and Younis (2005) and by Hong *et al*. (2002).

The destination address used for address-based routing can point towards a single system, i.e. unicast communication, to a group of systems, i.e. multicast, or to all systems in the network, i.e. broadcast to all nodes. The obvious advantage of such address-based routing algorithms is the possibility of identifying specific (unique) nodes and sending messages to them. As unique addresses are not available in all networks, an essential amount

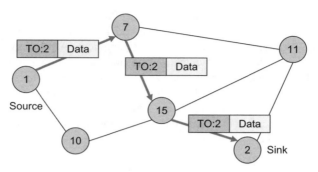

Figure 10.1 Principles of address-based routing; in this example, a data packet carrying the destination address 'TO:2' is forwarded along an established path towards its final destination: node 2

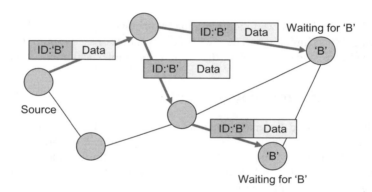

Figure 10.2 Principles of data-centric forwarding. Messages are forwarded according to their internal meaning; in this example, messages of type 'B' are transmitted to two sinks, which requested exactly messages of type 'B'

of work may be needed to assign such unique addresses to all nodes in a given network (Sun and Belding-Royer 2004; Weniger and Zitterbart 2004). We discuss this issue later, in Section 10.4.

In some cases – especially in WSNs – unique addresses are not demanded by the application requirements but only by the employed routing protocols. Therefore, a new message-forwarding approach has been introduced, which is called data-centric forwarding (Dressler 2006a; Krishnamachari et al. 2002). The main idea is to remove any addressing scheme and replace it with the specific semantics of the transmitted messages (Figure 10.2). As can be seen, the node addresses are replaced by a kind of interest of particular nodes in specific message types. The messages have to be encoded in a way that the destination address is replaced by an identifier explaining the message type. Thus, payload information (data) is used to forward messages towards an appropriate destination.

Table 10.1   Comparison between address-based routing and data-centric forwarding techniques

	Address-based routing	Data-centric forwarding
Routing approach	Identification of a path according to the destination address of the data message	Determination of the destination of a data message according to the content of the packet
Prerequisites	Network-wide unique node addresses	Pre-defined message types and semantics
Routing techniques	Proactive routing (continuous state maintenance) or reactive routing (on-demand path finding)	(Probabilistic) flooding schemes or interest-based reverse routing
Advantages	Usually low delays in connection setup and data dissemination	No address information required and simplified self-management and redundancy
Disadvantages	Network-wide unique address identifiers required	Increased overhead for single transmissions

Table 10.1 outlines the main differences between and objectives of address-based routing and data-centric forwarding. In the context of this chapter, we will discuss address-based schemes in more detail and refer to Chapter 11 for further information about data-centric forwarding.

10.1.2   Classification of ad hoc routing protocols

There exists a great number of approaches for address-based routing (Murthy and Manoj 2004; Rajaraman 2002). Figure 10.3 depicts a common classification scheme for ad hoc routing. The highlighted boxes in this scheme will be stressed in more detail in the following section. Basically, routing protocols can be distinguished in several dimensions and categories. In the following, we outline some of these dimensions and provide specific examples of routing protocols that use the specific information or employ the particular techniques.

Routing information update mechanism

Routing protocols can be differentiated according to their method to acquire and to distribute routing information. Basically, three techniques can be distinguished: proactive, reactive and hybrid protocols. The basic working principles will be outlined below.

Proactive routing   Proactive routing protocols rely on the periodic collection and exchange of topology information. Depending on the specific protocol, minor changes of the overall protocol behavior can be denoted. Nevertheless, all these protocols essentially always provide up-to-date routing information, which can be used to forward data packets

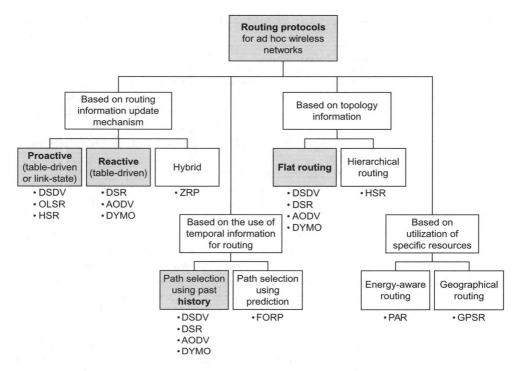

Figure 10.3 Classification of routing protocols according to Murthy and Manoj (2004)

though the network towards their destination. These periodic protocol updates will require a remarkable amount of network overhead for state (topology) maintenance, basically depending on the specific protocol and the amount of mobility. On the other hand, data packets can be forwarded at any time to any destination within the network.

Such proactive routing techniques can be either table-driven routing protocols or link-state mechanisms. The primary difference between these two groups is the topology maintenance. Table-driven protocols such as Destination Sequenced Distance Vector (DSDV) (Perkins and Bhagwat 1994) or Hierarchical State Routing (HSR) (Iwata *et al.* 1999) periodically exchange routing tables between neighboring nodes. This way, globally valid routing tables can be maintained. On the other hand, link-state techniques such as Optimized Link State Routing (OLSR) (Clausen and Jacquet 2003) distribute topology information so that each node can calculate optimal paths on their own, e.g. using the Dijkstra shortest-path algorithm (Dijkstra 1959).

Reactive or on-demand routing   Reactive routing protocols have been introduced in the MANET community to prevent periodic routing exchange, which may consume an essential amount of the available network resources. The principle of reactive protocols is to only search for paths though the network if data packets need to be transmitted. Thus, the routing protocol overhead is reduced to an absolute minimum while an additional delay is introduced for the setup of routing information.

The first protocol proposed in this context was Dynamic Source Routing (DSR) (Johnson 1994; Johnson and Maltz 1996), to which some additional enhancements have been developed over several years (Johnson *et al.* 2007, 2001). A more current approach is the Ad Hoc on Demand Distance Vector (AODV) (Perkins and Royer 1999; Perkins *et al.* 2003), which is probably the best known protocol in the ad hoc networking community. This protocol will be described in detail in Section 10.2.3 based on specific implementation guidelines (Chakeres and Royer 2004; Royer and Perkins 2000). The successor of AODV is Dynamic MANET on Demand (DYMO) (Chakeres and Perkins 2007), which has been designed to optimize specific aspects of the AODV routing protocol.

Hybrid routing   The advantages of proactive routing protocols, i.e. the fast delivery of packets, as routing information is always available, and of reactive routing protocols, i.e. the reduced costs in terms of network overhead for state maintenance, can be combined in some way. A typical example is the Zone Routing Protocol (ZRP) (Haas and Pearlman 2001; Haas *et al.* 2002b).

Use of temporal information for routing

Usually, routing protocols rely on past temporal information for maintaining routing tables. Received topology information, collected either in a proactive information exchange or during an ongoing reactive path determination, reflects the state of nodes and their topological neighborhood as observed when the routing update happened. All these protocols work in this way: DSDV, DSR, AODV and DYMO.

Optimizations can be expected if a protocol would be able to guess or to predict the topology of the network at some point in time, i.e. if the routing protocols can use future temporal information for the routing process. Flow Oriented Routing Protocol (FORP) (Garcia-Luna-Aceves 2002) is such an example, which relies on flow information to determine routing topologies.

Topology information

Most MANET and WSN routing protocols use a flat topology because no special addressing schemes exist that inherently structure the overall network. This is a major difference to the original Internet routing protocols, which benefit from a well structured address space. Flat routing protocols are, for example, DSDV, DSR, AODV and DYMO.

Hierarchical topology-based routing protocols have been tried in the ad hoc networking domain as well. The most prominent example is HSR. Such hierarchical topology protocols are comparable to cluster-based approaches that are frequently used in WSNs for various purposes. In contrast to hierarchical addressing schemes, clustering methods are highly dynamic and usually do not rely on fixed node addresses. Such clustering schemes will be discussed in Chapter 12.

Utilization of specific resources

Other resources can be used for more efficient routing based on meta-information or to maximize network lifetime by applying special power metrics to the routing process. Two specific examples are briefly introduced below. Nevertheless, there is a wide spectrum of further approaches available, which cannot be discussed here.

Geographical information-assisted routing Explicitly constructed routing tables contain information as to which hop a packet should be forwarded to next. As an alternative, this information can be derived implicitly from the physical placement of nodes. If the positions of the current node, its current neighbors and of the destination for the data packet are known, this geographical meta-information can be used to simplify the routing process by forwarding data packets to the next hop that is closest to the final destination, i.e. to search for a next routing hop which maximizes the progress in a single step (Mauve and Widmer 2001). Possible options include the transmission of a message to a given area, which is called geocasting, or the utilization of position information to aid other routing techniques, i.e. position-based routing. Possibly, location services are needed to map node IDs to node positions.

The best known protocol in this domain is Greedy Perimeter Stateless Routing (GPSR) (Karp and Kung 2000). It exploits geographical location information for routing in ad hoc networks. Basically, this protocol encourages the sender to transmit the message to the neighbor that realizes the most forward progress towards the final destination. This simple strategy might send a packet into a dead end. Thus, right-hand rules are needed to get out of a dead end: 'Put right hand to the wall, follow the wall.' In GPSR, two strategies are implemented. Primarily, the 'greedy' strategy forwards the packet as far as possible towards the destination. If no progress is possible, the strategy is switched to 'face routing', i.e. to send the packet around the face using the right-hand rule.

Power-aware routing Energy is one of the primary (and most restricted) resources in WSNs. Thus, energy-aware (or power-aware) protocols are strongly in demand in this domain (Singh *et al.* 1998). A number of power-aware routing metrics have been developed to assist other routing protocols (Salhieh and Schwiebert 2002). For example, the following metrics can be employed:

- Minimized energy consumption per packet – The energy needed for transmitting a packet through the network is the sum of the processing costs at each hop plus the power consumption for all radio transmissions. The path with the minimum transmission energy should be selected by the routing protocol.

- Minimized variance in node power levels – In order to increase the whole network lifetime, the energy distribution among all nodes can be used as a metric. Nodes with higher remaining power can be preferred.

- Minimized maximum node costs – This metric minimizes the maximum costs per nodes by preferring or blocking particular nodes.

Further considerations on energy efficiency and network lifetime can be found in Section 17.3.

10.2 Principles of ad hoc routing protocols

The main purpose of this section is to introduce some basic properties of ad hoc routing protocols based on selected examples. We explicitly do not go into the core details of the selected protocols. For this information, we refer to the literature. These case studies have been selected to introduce and to discuss specific mechanisms of self-organization in the domain of ad hoc routing.

10.2.1 Destination Sequenced Distance Vector

The basic idea of DSDV (Perkins and Bhagwat 1994) was to start with a standard routing protocol as used in the Internet. In this special case, one of the best known table-driven protocols, Routing Information Protocol (RIP) (Hedrick 1998), was used for inspiration. Some improvements have been added to the standard algorithm according to the special needs of ad hoc networks. Basically, DSDV periodically sends full route updates to all neighboring nodes. To prevent too costly updating on topology changes, incremental route updates are supported by DSDV.

The propagation of an (incremental) update is depicted in Figure 10.4. Node 1 detects a new neighbor (node 5) and distributes this knowledge to all its neighbors. These nodes forward the message again until all nodes update their routing tables. In each message, the distance towards node 1 is denoted. Thus, all nodes which received the update can conclude a path with the minimal number of hops towards node 5. As an example, the routing table of node 15 is outlined in Figure 10.4. In our case, node 5 can be reached via the neighboring node 7 in three hops.

In order to prevent routing loops, aging information is added to the route information propagated by distance vector exchanges. Using this time information, a received routing update can be analyzed, whether it is a new one or it was received over a longer path. This aging information is provided by sequence numbers, which are stored together with a particular route in the local routing table. Furthermore, DSDV suggests updating unstable routes with some delay to prevent fluctuations.

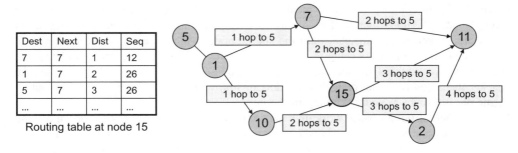

Dest	Next	Dist	Seq
7	7	1	12
1	7	2	26
5	7	3	26
...	...	...	...

Routing table at node 15

Figure 10.4 Principles of distance vector routing as used by DSDV; in this example, the route to node 5 is flooded through the network and all nodes update their routing tables accordingly

Advantages   The main advantage of DSDV is the availability of routes to all destinations at all times. This implies much less delay in route setup compared with on-demand routing approaches. The incremental updates with sequence number tags allow adaptating of existing wired network protocols. DSDV was one of the first algorithms available for routing in ad hoc networks. In general, it is quite suitable for creating ad hoc networks consisting of a small number of nodes.

Disadvantages   Being a proactive routing protocol, DSDV relies on periodic updates of the routing tables of all nodes in the network. Thus, the induced overhead greatly reduces the bandwidth efficiency. Updates due to broken links lead to an enormous control overhead during high mobility. Therefore, even small networks with high mobility or larger networks with low mobility can completely choke the available bandwidth – the scalability of DSDV is not suitable for large networks because the exhaustive control overhead is directly proportional to the number of nodes.

Whenever the topology of the network changes, DSDV makes the network unstable for a period of time until the route updates converged. Thus, the protocol is also unsuitable for highly dynamic networks. Similarly, to obtain information about a particular destination node, a node has to wait for a table update message initiated by the same destination node. This may lead to delayed updates and could result in stale routing information, e.g. as a particular node joins and leaves the network periodically.

10.2.2   Dynamic Source Routing

In reactive routing protocols, the main idea is to prevent periodic exchange of routing information to maintain a globally valid routing table in each node in the network. Thus, usually, no routing information exists at the time of packet forwarding, or, more precisely, initially, no information about next hop is available at all. A possible recourse is to send the data packet (or a special route-discovery packet) to all neighbors, i.e. flooding the network. Hopefully, at some point, the packet will reach the destination and an answer is sent back. This reply is used for backward learning the route from the destination to the source.

One of the first reactive ad hoc routing protocols was DSR (Johnson 1994; Johnson and Maltz 1996; Johnson et al. 2001), which was finally standardized by the IETF (Johnson et al. 2007). In contrast to other typical routing protocols, such as the outlined DSDV or AODV and DYMO, which will be discussed later, DSR does not maintain routing tables in all networked nodes. Instead, only the source node, which is creating the packet data, establishes and maintains a route towards the destination. This path, i.e. the complete route through the network, is included in each data packet and forwarded accordingly. Thus, the packet itself describes the path that it wants to follow through the network. This routing principle is known as source routing in the networking area.

This source-routing-based working behavior represents a visible optimization for routing in ad hoc networks. There is no need to maintain globally valid routing tables and the state information to be stored in each node, i.e. the routing information, is minimized. Thus, the resource requirements for intermediate (forwarding) nodes are quite low, as less memory, processing power and bandwidth is needed for state maintenance.

DSR uses separate Route Request (RREQ) and Route Reply (RREP) packets to discover a route from the source node to the destination. Data packets are only sent once a route

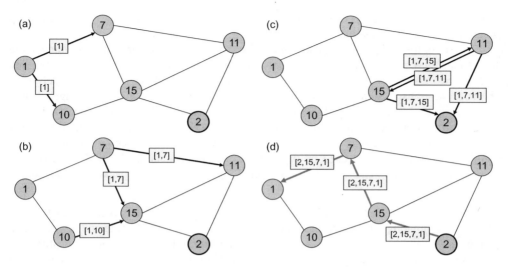

Figure 10.5   Route determination for source routing using DSR. First, a Route Request (RREQ) is flooded through the network (a, b, c) until the destination is found. All intermediate nodes include their address into this route request. Finally, the destination sends a Route Reply (RREP) back on the reverse path, i.e. using source routing

has been established. The key objective is to employ much smaller discovery packets compared with the data packets for route determination. The routing information is stored in the discovery packets. This route discovery procedure is depicted in Figure 10.5. In this example, node 1 searches a path towards node 2. Thus, a RREQ is flooded though the network until it reaches the destination node (steps a–c). Finally, the destination extracts the routing information from the discovery packet and unicasts the RREP to the source node using the reverse path as denoted in the RREQ message (step d). Routing loops can be easily prevented by analyzing the already listed path in the discovery packets and scanning for the local node address.

In order to reduce the necessary communication overhead due to frequent RREQs's being flooded through the entire network, DSR encouraged all nodes to employ route caches. These caches are used to store all possible information extracted from the source route contained in a data packet or in a RREP. The cached information is used to optimize the route construction phase, as intermediate nodes may answer the RREQ on behalf of the destination.

The optimization of DSR's using route caches may unfortunately lead to stale route caches, i.e. the storage of out-dated routing information. Fast topology changes, e.g. due to high mobility, may degrade the quality of the route caches essentially. Route maintenance procedures take care of such out-dated routes. If a link breaks, a Route Error (RERR) message is sent towards the source and the route construction is re-initiated.

Regarding the performance of DSR, the protocol behaves well for smaller, less saturated wireless networks. With an increasing network load, on the one hand, DSR is able to passively retrieve much routing information from other RREQs. On the other hand, the probability of lost RREQ packets increases and, therefore, the overhead and the route setup delay due to new RREQs increase as well.

Several optimizations and enhancements have been proposed for DSR, of which some found their way into the final protocol specification (Johnson *et al.* 2007). Primarily, they address two shortcomings of DSR: possible reply storms and the route setup delay for short-lasting transmissions. The first problem (reply storms) can happen is a huge amount of nodes have cached route information towards a searched destination node. All these nodes will send a RREP on behalf of the destination. The use of an exponential back-off has been proposed to avoid frequent route discovery packets. The second problem (route setup delay) has been addressed by piggy-backing data packets on the RREQ. This way, short-lasting connections, especially those consisting of a single data packet, are privileged.

Advantages   Compared with all proactive routing protocols, DSR's being a reactive approach is eliminating the need to periodically flood the network with table update messages. Routes are only searched if a data packet has to be transmitted to a particular destination. The use of source routing techniques, i.e. the pre-definition of the path at the source, also eliminates the need to store routing information in all intermediate nodes in the networks. Thus, DSR has very few storage and maintenance requirements. Caches can be used to reduce the route setup delay (to the cost of higher storage and maintenance requirements). Tests have shown that the connection performs well in static and low-mobility environments.

Disadvantages   The primary disadvantages of DSR include the route setup delay, the costly and failure-prone route maintenance, and the routing overhead. As in all reactive routing protocols, the connection setup delay is much higher than in proactive approaches. The route must be discovered before any data packet can be sent. Piggy-backing solves this issue to a certain extent. If route caches are used, this delay can be further reduced to the cost of possible stale route information, which may result in inconsistencies. DSR has no built-in mechanism to locally repair broken links. If a route is no longer available, the source must discover a new one by sending a new RREQ message. Thus, the performance degrades with increasing mobility. Finally, the routing overhead of DSR is directly proportional to the path length, which has implications on the supported network topologies and application scenarios. This also includes the storage requirements for the complete route in each data packet, which is directly proportional to the path length as well.

10.2.3   Ad Hoc on Demand Distance Vector

Another very popular routing protocol is AODV (Perkins and Royer 1999; Perkins *et al.* 2003). AODV is a reactive routing protocol, searching paths between the source and the destination of a particular message on demand. Basically, the same route discovery procedure is used as in DSR. New routes are searched by flooding the network with discovery messages and, finally, replies are used to set up the requested routes. In AODV, all nodes remember from where a packet came and populate their routing tables with that information. Thus, all involved nodes, i.e. intermediate nodes in a given path, maintain routing tables instead of using source routing. The AODV routing protocol maintains the established routes as long as they are needed by the source. Sequence numbers are used to ensure

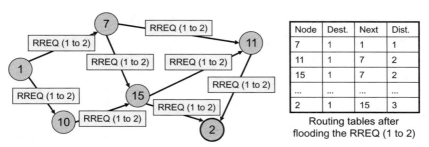

Node	Dest.	Next	Dist.
7	1	1	1
11	1	7	2
15	1	7	2
...	...	...	...
2	1	15	3

Routing tables after
flooding the RREQ (1 to 2)

Figure 10.6 Route discovery process in AODV. Route Request messages are flooded through the entire network. Intermediate nodes update their routing tables to include or update the path towards the source of the RREQ in this first phase of the route discovery process

the freshness of routes and to handle stale routes. Essentially, AODV has been developed to be loop-free, self-starting and scale to large numbers of mobile nodes.

The route setup procedure of AODV works as follows. The routing protocol uses Route Request (RREQ)/Route Reply (RREP) cycles to set up routes from a source node to a destination. When a source node desires a route to a destination for which it does not already have a route, it broadcasts a RREQ packet across the network. The diameter of the flooded area is only limited by an appropriate Time To Live (TTL). All nodes that receiving this RREQ packet update their routing information for the source node and set up backwards pointers to the source node in the route tables. The principles of this route setup are shown in Figure 10.6. Besides the destination address, the RREQ contains the current sequence number, which is used to prevent stale routes and routing loops. If an intermediate node receives a RREQ which it has already processed, it discards the RREQ and does not forward it. Flooding is stopped at the destination or if a node has a valid route to the destination. If a RREQ is received multiple times, the duplicate copies are discarded.

In order to reduce the overhead created by flooding the RREQ, AODV uses an expanding ring search strategy for the route setup. Actually, the first RREQ will be sent with a small TTL. Thus, only a small portion of the entire network will be flooded. If the search was not successful, subsequent tries are made with increasing TTL values.

Figure 10.6 depicts an example of the route-discovery process. Node 1 is searching a route towards node 2. The RREQ is flooded through the entire network until either a loop has been detected or the final destination has been reached. During this discovery process, all intermediate nodes update their routing table accordingly, i.e. the routing entry pointing to node 1 is either updated or created.

If the RREQ reaches the final destination, a Route Reply (RREP) message is unicasted back to the source. With the help of the routing information previously obtained while forwarding the corresponding RREQ, the intermediate nodes are able to send the RREP further back to the start of the chain, until it eventually reaches the originating node. As the RREP is propagated back to the source, all intermediate nodes set up forward pointers to the destination. This way, the routing tables in all involved nodes are updated. Once the source node receives the RREP, it may begin to forward data packets to the destination. If

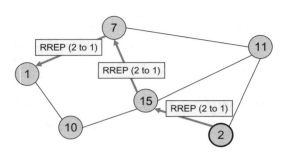

Node	Dest.	Next	Dist.
7	1	1	1
7	2	15	2
11	1	7	2
15	1	7	2
15	2	2	1
...	...	...	...
2	1	15	3

Routing tables after
sending the RREP (2 to 1)

Figure 10.7   Route discovery process in AODV. The destination sends a Route Reply message back to the source to announce the shortest path. The RREP can be unicasted as all intermediate nodes updates their routing tables to include the address of source during the first phase of the route discovery (RREQ)

the source later receives another RREP containing a greater sequence number or containing the same sequence number with a smaller hop count, it may update its routing information for that destination and begin using the better route.

This second phase of the route setup is depicted in Figure 10.7. Node 2 issues a RREP towards node 1. This RREP is forwarded according to the routing tables, which have been created during the first phase. In our example, the RREP is forwarded by nodes 15 and 7 towards source node 1. All these nodes update their local routing tables to include a path towards node 2. After receiving the RREP, node 1 may start to transmit data messages to node 2.

In order to prevent unnecessary overhead in the network due to route maintenance, intermediate nodes are encouraged to answer a RREQ on behalf of the destination node. Such a shortcut is possible if an intermediate node already has routing information toward the searched destination. Such behavior is depicted in Figure 10.8 (left). In this example, node 1 searches a path to node 15. Node 7 already has an active entry to this destination in its local routing table. Therefore, it will send a RREP back to node 1 on behalf of the searched destination. This way, the number of RREQs in the network can be dramatically reduced, especially if particular destinations are addressed by multiple nodes in the network.

As long as the particular route remains active, it will continue to be maintained. A route is considered active as long as there are data packets periodically traveling towards the destination along that particular path. For this purpose, each routing table entry contains a timeout. Once the source stops sending data packets, the links will time out and eventually be deleted from the individual routing tables of the intermediate nodes.

If a link break occurs while the route is active, e.g. due to mobility or failures of intermediate nodes, the node upstream of the break is requested to propagate a RERR message to the source node. This RERR is used to inform the source about a problem on the path and, therefore, about the unreachability of the destination. After receiving the RERR, the source node can re-initiate the route discovery if it still needs this particular route. This failure detection process is depicted in Figure 10.8 (right). Node 15 detects a

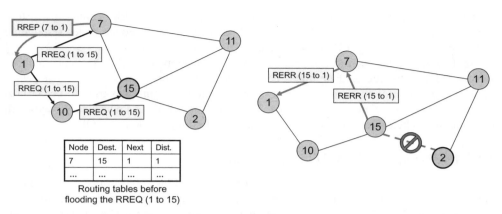

Figure 10.8 Route maintenance in AODV. If an intermediate node already has an active routing table entry pointing towards the searched destination, it will answer the RREQ on behalf of this destination (left). Routes are maintained active as long as data are forwarded to the particular destination. If such an active route breaks, e.g. due to node mobility, a RERR is propagated to the source (right)

problem with its direct link to the destination node 2. Therefore, it creates a RERR, which is sent upstream to source node 1.

Advantages Similarly to DSR, AODV reactively searches routes on demand. Therefore, AODV does not periodically maintain globally valid routing tables. All nodes only care for local information that includes only active routes. As all nodes in the network are required to maintain active routes, i.e. routes to destinations for which active data traffic can be observed, AODV usually needs less connection setup delay compared with DSR (which only encourages nodes to cache routing information). In typical ad hoc networks, AODV performs well, even in networks with limited mobility. Unsolicited RERR messages can improve the performance by announcing broken links even if no data packet needs to be forwarded. Destination sequence numbers are used to find the most up-to-date route to a particular destination. Thus, they provide a measure to detect routing loops and stale routes.

Compared with DSR, there is no overhead related to the network diameter, i.e. the path length, but to the route setup delay. As all intermediate nodes maintain a local routing table entry pointing towards the destination, the complete path information must not be stored in each data packet.

Disadvantages Especially in low-traffic networks with arbitrary connections, i.e. with a low possibility of searching for a destination for which a path is already actively maintained in some intermediate nodes, the route discovery process may consume an essential amount of the overall bandwidth requirements. Thus, the application in networks with special energy constraints requires essential changes in the protocol behavior of AODV. Further control overhead is induced due to multiple RREP packets in response to a single RREQ. This problem may happen as AODV circularly searches for routes to a particular destination by flooding the RREQ messages through the network. If many nodes maintain paths to a common sink node, the probability is high to find multiple intermediate nodes responding to the request by sending a RREP on behalf of the searched destination.

Intermediate nodes can also lead to inconsistent routes if the source sequence number is very old and the intermediate nodes have higher but not the latest destination sequence number. Finally, in first versions of AODV, a periodic beaconing was used to detect broken links to directly connected neighbors. This beaconing leads to unnecessary bandwidth consumption. Therefore, it has been removed from the final version of AODV.

10.2.4   Dynamic MANET on Demand

Dynamic MANET on Demand (DYMO) (Chakeres and Perkins 2007) is a new reactive (on-demand) routing protocol, which is currently developed in the scope of the IETF MANET working group. DYMO builds upon experience with previous approaches to reactive routing, especially with the routing protocol AODV. It aims at a somewhat simpler design, helping to reduce the system requirements of participating nodes, and simplifying the protocol implementation. DYMO retains established mechanisms of previously explored routing protocols such as the use of sequence numbers to enforce loop freedom. At the same time, DYMO provides enhanced features, such as covering possible MANET–Internet gateway scenarios.

Protocol behavior

The basic protocol behavior is quite similar to the principles discussed for AODV (see Section 10.2.3). Additionally, DYMO introduces and implements the concept of path accumulation, as depicted in Figure 10.9. Besides route information about a requested target, a node will also receive information about all intermediate nodes of a newly discovered path. Similarly to source routing as used by DSR, the complete path is stored in the RREQ and RREP messages. This information can be evaluated at each intermediate node to update its local routing tables accordingly. This is in contrast to AODV, in which the route discovery messages carry the source and destination addresses only. Thus, intermediate AODV nodes can only generate routing table entries for both the source and the destination node.

An example is illustrated in Figure 10.9. When using AODV, the RREQ from source node A to destination node D enables the intermediate nodes (including the final destination) to learn paths towards the transmitting node A. Similarly, the RREP from node D to node A provides routing information to the sender of the message, i.e. to node D. In DYMO, the route-discovery message is extended to cover routing information to all nodes which processed this RREQ or RREP, respectively. Thus, with a single route discovery, all nodes in our example learn complete topology information. In more complex ad hoc scenarios, an essential part of the network can be discovered using only a few routing messages.

In order to efficiently deal with highly dynamic network scenarios, links on known routes may be actively monitored, e.g. by using the MANET Neighborhood Discovery Protocol (NHDP) (Clausen et al. 2006) or by examining feedback obtained from the data-link layer. Detected link failures are announced to selected nodes in the MANET using RERR messages. These error messages are sent to all nodes in the range, informing them of all routes that have now become unavailable. If the received RERR message in turn invalidates any routes in the local routing table of the node receiving the RERR, it will

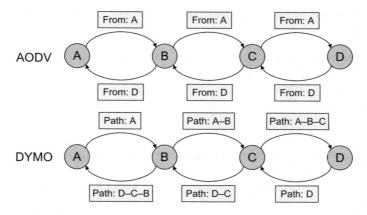

Figure 10.9 Comparison of the routing information dissemination in AODV and DYMO; DYMO transports complete path information in the route discovery messages to enable intermediate nodes to learn as much as possible about the network topology used

again inform all its neighbors by multicasting a RERR containing the routes concerned. This allows effective flooding of information about a link breakage through the MANET.

Advantages   Compared with AODV, DYMO primarily benefits from the possibility of passively learning topology information of intermediate nodes. Then, if such a node needs to send a message to an already known destination if yet another RREQ for this destination arrives, the intermediate node can use the stored topology information. Thus, the network load due to protocol overhead can be essentially reduced in most scenarios.

Disadvantages   The same drawbacks as discussed for AODV still hold for DYMO. In low-traffic scenarios, the protocol–internal message exchange represents a remarkable amount of the overall traffic. Additionally, the route-discovery messages are larger for DYMO compared with AODV, as complete path information needs to be stored. Therefore, the packet size for RREQs and RREPs can grow to significant sizes in large networks, i.e. in networks with a huge diameter.

Performance aspects

Dietrich *et al.* (2007) performed a number of simulation experiments in order to evaluate the protocol in different MANET scenarios and under different traffic conditions. For the performance analysis, a number of different setups have been simulated, each consisting of 100 nodes running our implementation of DYMO, 99 of which continuously generated packets addressed to one node acting as a packet sink located in a corner of the playground. Combinations of the following scenarios have been evaluated:

  1. Nodes were arranged either to form a 10×10 grid or in a completely random manner.

2. The playground size was adjusted so that the average distance between neighboring nodes corresponded to either one hop or three hops (according to the grid scenario).

3. Nodes sent a new packet either following an exponential distribution with a mean value of 1 second or a mean value of 10 seconds, or following a random, bursty pattern, which consisted of waiting between 0 and 5 minutes, then sending ten packets spaced 0.49 seconds apart.

Frequency of route setups The traffic overhead induced by DYMO was gaged by relating the number of route request messages to the number of application-layer messages sent by DYMO via established routes. As depicted in Figure 10.10 (left), the number of route request messages exchanged per application-layer message sent decreased significantly if packets were sent at an interval that could keep established routes from expiring. In static scenarios, node density only played a minor role insofar as it improved node connectivity, thus shortening routes, while, at the same time, it increased the number of collisions, thus slightly raising the median number of route request messages that needed to be resent. Similar results were obtained if nodes were arranged in a grid.

However, in the dynamic scenarios shown in Figure 10.10 (right), in which 10 % of all nodes moved according to a random waypoint model, the higher node density played a key role in reducing the number of route requests. Here, a larger number of potential routes to the sink meant a higher probability that the chosen route included more static nodes, thus reducing the probability of this route's breaking if one of the involved nodes moved out of communication range.

Data packets dropped by DYMO As a measure of DYMO's aptitude for finding routes, we compared the number of data packets dropped by the network layer with the number

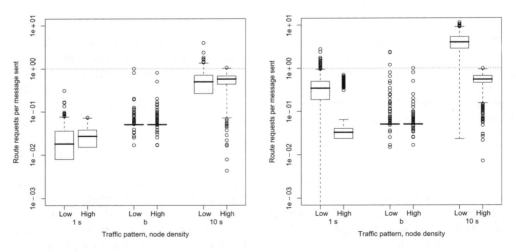

Figure 10.10 Frequency of route setups. Scenario: random deployment, static nodes (left); random deployment, 10 % mobile (right)

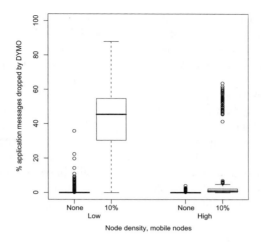

Figure 10.11 Data packets dropped by DYMO. Scenario: random deployment, 1 s mean intra-packet spacing

of packets requested to be sent. Figure 10.11 displays the results of this comparison. In the stationary scenario, almost no packets got lost due to problems at the network layer. The few outliers result from particular nodes' being, in principle, unable to establish a path towards the sink in the random deployment. In the mobile scenarios, the probability to successfully set up a path and to transmit messages essentially relies on the probability of route failures. Obviously, the low-density scenario tends to show a higher number of route failures compared with the high-density example. Similar results were obtained for other intra-packet spacings and/or if nodes were arranged in a grid.

10.3   Optimized route stability

In our discussion on ad hoc routing approaches, we have seen a wide variety of techniques and algorithms to find shorter paths between two arbitrary nodes in the network. The metric for path selection is usually the length of the path in terms of hops. This metric can also be translated to shortest end-to-end delay or similar measures. Nevertheless, in mobile networks, none of the discussed approaches provides measures for optimized path selection with respect to node mobility.

Agarwal *et al.* (2000) studied possible optimizations of route stability in mobile environments in the context of their Route-Lifetime Assessment Based Routing (RABR) protocol. They employ a new measure for the stability of single routes – the link affinity, as first proposed by Paul *et al.* (1999). This link affinity basically characterizes the strength of relationship between two nodes. The RABR protocol is a distributed routing scheme that exploits this measure to provide paths that are more stable and less congested in a specific context. The overall objective is to enable ad hoc networks to become more effective by making the nodes in the system communication-aware in the sense that each node should know its affinity with its neighbors and should be aware of the impact of its movement on the communication structure of the underlying network.

In order to show the capabilities of exploiting the link stability, i.e. the affinity between two nodes, to improve the overall network behavior, we briefly discuss the main ideas of link affinity. In general, the stability of a route depends on all the links in that route and is determined by the strength of the weakest link. The link affinity can be described as the time for which the link is working (Paul *et al.* 1999).

Basically, the affinity between two nodes n and m, a_{nm}, is a prediction of the lifetime of the link l_{nm}. The affinity a_{nm} – please note again that a_{nm} describes a time interval – is the time needed by node n to move out of range of node m. The movement of node n can be estimated by periodically sampling the signal strength between the two nodes. Based on these measurements, the affinity a_{nm} can calculated as shown in Equation 10.1. In this equation, S_{thresh} denotes the threshold signal strength below which a link l_{nm} can be assumed to be disconnected and $\Delta S_{nm(avg)}$ is the rate of change of signal strength averaged over the last few samples:

$$a_{nm} = \begin{cases} high & \text{if } \Delta S_{nm(avg)} > 0 \\ \dfrac{S_{thresh} - S_{nm(current)}}{\Delta S_{nm(avg)}} & \text{otherwise} \end{cases} . \tag{10.1}$$

The calculated link affinities, which can actually be updated using scenario-dependent sampling rates, build the basis for link-stability-based routing. In principle, the link metric in number of hops can be simply replaced by accumulated link affinities. This accumulation is done by minimizing overall link affinities along the path, i.e. by searching for the weakest link. Considering a path of length l from node x_0 to node x_l using intermediate nodes $(x_1, x_2, \ldots, x_{l-1})$, the path affinity $p_{x_0,x_1,\ldots,x_l}$ can be expressed as given in Equation 10.2:

$$p_{x_0,x_1,\ldots,x_l} = \min_{0 \leq i < l} (a_{x_i,x_{i+1}}). \tag{10.2}$$

Based on this, the routing protocol has to select a path between two arbitrary nodes using the maximum path affinity. Interestingly, the path affinity can also be exploited to adaptively modify the timeouts for the routing table entries. After expiring the path affinity, i.e. after the estimated time interval for breaking the weakest link (the minimum link affinity), the route can no longer be considered stable.

Using this affinity definition, Agarwal *et al.* (2000) additionally proposed an affinity-based dynamic power-adjustment scheme to optimize the overall power consumption in the entire wireless network. The basic idea is to estimate the necessary transmission power of a node in the path to reach its next hop neighbor. This way, the transmission energy can be minimized for individual transmissions and, thus, for the overall end-to-end communication. For a large number of packets being forwarded by the nodes along a route, the power conservation effected by a dynamic variation of transmission power can be quite significant.

The quality of the estimation of the transmission energy can be adapted by modifying the sampling intervals for measuring the link affinities. In the following, we denote the sampling interval by τ. The objective for the optimization is the required power for a transmission to a specific neighboring node in the time interval $[t, t + \tau]$. This transmission power denoted as $P_{t,t+\tau}$ can be calculated as given in Equation 10.3, where P_T represents the normal (maximum) transmission power and $0 < \frac{S_{thresh}}{S_{t,t+\tau}} \leq 1$ is a factor representing the estimated 'distance' of the neighboring node in the given time interval:

$$P_{t,t+\tau} = P_T \times \frac{S_{thresh}}{S_{t,t+\tau}}. \tag{10.3}$$

In the factor $\frac{S_{thresh}}{S_{t,t+\tau}}$, S_{thresh} represents the minimum signal strength necessary for success-fully receiving and decoding a message. $S_{t,t+\tau} \geq S_{thresh}$ is the approximate signal strength to be expected for transmissions in the next time interval τ according to the link affinity. If $S_{t,t+\tau}$ is close to S_{thresh}, the correction factor for transmission energy estimation becomes about 1; thus, the maximum transmission energy will be used for the communication. Otherwise, adapted values $P_{t,t+\tau} < P_T$ will be used.

We finally need to discuss the estimation of the signal strength $S_{t,t+\tau}$ according to two measurements or approximations, respectively. First, the received signal strength S_H of exchanged probe messages is recorded. Secondly, the link affinity a is approximated, as shown in Equation 10.1. Based on these values, $S_{t,t+\tau}$ can be calculated as shown in the following equation:

$$S_{t,t+\tau)} = \begin{cases} S_H - (S_H - S_{thresh}) \times \frac{\tau}{a} & \text{if moving further and } \tau < a \\ S_H & \text{if moving closer and } \tau < a \\ S_{thresh} & \text{otherwise} \end{cases} \tag{10.4}$$

Let us discuss the three cases given in Equation 10.4 individually. If $\tau \geq a$, then the link can be assumed to be stable for the complete time interval τ. Thus, the power estimation can help to improve energy savings. If the nodes are moving away from each other, i.e. the distance is increasing, the speed of this change can be estimated by $\frac{\tau}{a}$. Thus, the necessary signal strength for communications in the next time interval will follow the expression $S_H - (S_H - S_{thresh}) \times \frac{\tau}{a}$. If the distance between the nodes is getting smaller, the node movement cannot easily be predicted. Thus, S_H is used as an approximation. Finally, if $\tau \geq a$, the link is assumed to break within τ. Thus, we need to transmit with the maximum power in order to increase the probability of a successful transmission.

10.4   Dynamic address assignment

All the ad hoc routing techniques and protocols discussed in this chapter rely on the avail-ability of globally unique address identifiers, which are assigned to all nodes prior to any communication. Especially in typical Internet-oriented MANET applications, this require-ment can easily be fulfilled by using Ethernet addresses in WLANs. In other application domains, the assignment of unique addresses to all nodes in the network can become diffi-cult in terms of manageability and scalability. Interestingly, many scenarios in the domain of WSNs do not depend on fixed node addresses due to various reasons:

- The application itself does not require globally unique addresses, e.g. if measurement results need to be transmitted and analyzed. The application only needs to know about the region or position of an identified event.

- The deployment and maintenance becomes much easier if address-less or data-centric operations (which will be discussed in Chapter 11) are enabled. Nodes can be de-ployed (or replaced) without changing the node programming.

In this section, we briefly outline possible strategies to encounter address assignments in self-organizing ad hoc networks. Besides a global overview, including some information about typical Internet-based centralized approaches for address allocation, we outline two techniques for distributed address assignment, which can be used in general ad hoc networks as well as in WSNs.

10.4.1   Overview and centralized assignment

In order to prevent static address assignments using pre-programmed node addresses, which need to be enforced and maintained during the overall lifetime of the network, dynamic address allocation or assignment strategies have been developed. Basically, these mechanisms can be grouped into two classes: centralized address management and self-organized address assignment based on Duplicate Address Detection (DAD) techniques. Both kinds of mechanisms ensure the uniqueness of assigned address identifiers, either in a certain part of the network or even in the entire network.

The best known example of centralized address management is Dynamic Host Configuration Protocol (DHCP) (Droms 1997, 2004), which has been developed for use in Internet scenarios using IPv4 or IPv6, respectively. DHCP is a client–server-based network protocol. Each DHCP server maintains one or more address pools. Such a pool describes available and currently used addresses, respectively. This working principle ensures the uniqueness of assigned IP addresses. Unfortunately, all the assignments are based on the Ethernet address of each client, which is unique worldwide. Therefore, DHCP only provides an IP address to Ethernet address binding.

Similarly, techniques for operation in much more dynamic environments such as mobile ad hoc networks have been proposed by Bernardos and Calderon (2005). In the context of ad hoc and sensor networks, Passive Auto-Configuration for Mobile Ad Hoc Networks (PACMAN) needs to be mentioned as an optimized solution for efficient dynamic address allocation (Weniger 2005). This approach is directly based on the lessons learned from the DHCP development.

The advantage of DHCP-based schemes is the provision of unique addresses based on a central server or, at least, on a common allocation mechanism. In consequence, no detection scheme for duplicate addresses is necessary. In contrast, DAD mechanisms have been proposed for higher scalability in ad hoc networks. Basically, nodes select addresses by local means and the DAD algorithm ensures the uniqueness of the choice. In the following sections, two techniques are described which provide address allocation features for ad hoc and sensor networks, either on a global scale or for use in a limited part of the network.

Weniger (2003) considers a number of reasons which motivate duplicate address detection algorithms. Besides manufacturing problems and malicious or selfish address modifications, the following reasons have been identified:

- The node address has been chosen randomly either due to the unavailability of unique hardware addresses or because it was infeasible to use the pre-programmed address.

- Privacy reasons demanded randomization of parts of the node address (Narten and Draves 2001).

- The heterogeneity of the network devices prevents the selection of a pre-programmed unique hardware address.

Based on the detailed summary of dynamic addressing techniques in MANETs by Sun and Belding-Royer (2004) and a paper by Weniger and Zitterbart (2004), which provides an overview and future directions for such address-allocation solutions, the following categories of address-allocation techniques can be distinguished. Most of the named approaches focus on IP networks, while some are also applicable in general-purpose MANETs:

- Decentralized approaches – Basically, two categories of decentralized solutions can be distinguished. First, methods like ad hoc address autoconfiguration (Perkins *et al.* 2001) focus on random address allocation and subsequent distributed DAD. Secondly, full-state techniques such as MANETconf (Nesargi and Prakash 2002) force the maintenance of globally valid address lists at each node in the network; thus, a new node can obtain a new address from any of the previously available nodes followed by a state update. Another self-organized address-allocation scheme was proposed by Toner and O'Mahony (2003), which is based on the original DAD algorithm.

- Leader-based approaches – Besides centralized approaches such as DHCP, dynamically chosen leaders help to reduce the overhead of dynamic address allocations. Best known examples include Dynamic Address Configuration Protocol (DACP) (Sun and Belding-Royer 2003), which basically relies on a number of elected address authorities for address allocation, and Dynamic Address Allocation Protocol (DAAP) (Patchipulusu 2001), in which the role of the leader is migrated to the node that last joined the network.

- Best-effort approaches – As the problem of unique address assignment is known to be hard in large scale networks, best-effort approaches such as weak DAD (Vaidya 2002) explicitly allow address duplicates while enforcing the routing towards the 'right' destination by including additional keying information in the routing protocol. Thus, such techniques are only feasible if specific routing protocols are employed.

10.4.2   Passive Duplicate Address Detection

Previous DAD techniques rely on the active distribution of address information through the network. The original DAD (Perkins *et al.* 2001) enforces a node which lacks an address to determine whether a candidate address that it has selected is still available. For this, address request and reply messages are exchanged between all nodes in the MANET. Similarly, the already mentioned leader-based approaches actively query elected leaders to allocate new addresses (Weniger and Zitterbart 2002). This can be named active DAD. In contrast, Passive Duplicate Address Detection (PDAD) tries to detect duplicates without disseminating additional control information.

PDAD (Weniger 2003) was specifically developed for application in MANETs using Internet protocols and addressing schemes. Basically, PDAD is not an algorithm for choosing and allocating addresses, but for detecting duplicate addresses. The primary objective during the development of PDAD was its application in networks with high node mobility. Therefore, the result was a light-weight scheme for address allocation with passive duplicate detection.

The behavior of the algorithm is as follows. Each node randomly chooses an address by itself. Thus, there is no need for a particular server or any other required network infrastructure. In a second step, the node performs the DAD algorithm to verify the uniqueness of

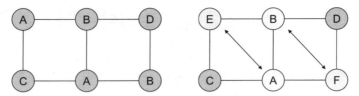

Figure 10.12 Duplicate address detection in PDAD using two-hop neighborship information; starting with a random address allocation (left), PDAD updates the addresses of nodes
A and B to remove the conflicting address selections (right)

the selected address by passively observing the network traffic. PDAD continuously checks
routing information for bandwidth-efficient DAD. In particular, information about the local
neighborhood, routing information and sequence numbers is used to detect multiple sources
of traffic which are sharing the same address identifier.

An example is shown in Figure 10.12. Starting with random address allocations by
all the nodes in the network, PDAD can detect duplicates, for example, by using two-
hop neighborhood information. In the given example, nodes A and B (these addresses
are allocated to multiple nodes) can detect the conflicts and resolve them by reallocation
address identifiers.

Essentially, PDAD is focusing on the detection of duplicate addresses throughout an
entire MANET. Thus, the scalability of this approach is at least bounded by the scalability
of the employed routing protocols, as PDAD strongly depends on particular ad hoc routing
information.

While PDAD provides an efficient address-allocation algorithm for use in small-sized
networks, in the case of data-centric data communication, additional messages must be
created to enable PDAD in such networks. Obviously, this results in unnecessarily high
overhead. Additionally, PDAD's objective is to maintain globally unique addresses. In
many scenarios, there is no such requirement.

10.4.3 Dynamic Address Allocation

In many cases, address-based communication is only needed in a very localized environment, e.g. for maintenance issues, while, generally, data-centric communication mechanisms
are employed. In such scenarios, the allocation of network-wide unique addresses to all
nodes would lead to unacceptable overhead (or it may even be totally infeasible). In order
to address this challenge, Dynamic Address Allocation (DAA) (Dressler and Chen 2007;
Yao and Dressler 2007) has been developed, which works in a localized environment.
Therefore, the overhead due to management of topology and uniqueness of the addresses
becomes very low. In its first version, DAA specifically focused on single-hop communication that is usually more reliable compared with a multi-hop approach. Basically, DAA
employs selected solutions from PDAD and DHCP to create an efficient and robust dynamic
address-allocation scheme for management and control especially in WSNs.

The DAA algorithm benefits from many ideas learned from DHCP and PDAD. A selected device, e.g. a server performing management and control operations, initiates and
controls the address-assignment process. While this server maintains a list of previously

allocated addresses, each address is randomly chosen by all neighboring devices. A DAD algorithm is employed to ensure the uniqueness of the independently selected addresses. After finishing the address-allocation step, the server can continue in maintaining the surrounding nodes by contacting each of them individually. As multiple servers may initiate address allocations simultaneously (while independently), DAA employs a round-based allocation and assignment scheme, which also simplifies the handling of unreliable communication and failing nodes. Below, the basic working principles of DAA are depicted.

Basic DAA algorithm

The basic DAA algorithm (Yao and Dressler 2007) assumes that only one server exists and no collisions occur throughout the complete DAA process. The DAA process is always started by a dedicated node. We call this node 'server' because we assume that this node will provide management and control operations for the neighboring sensor nodes. The basic algorithm is illustrated using the example depicted in Figure 10.13.

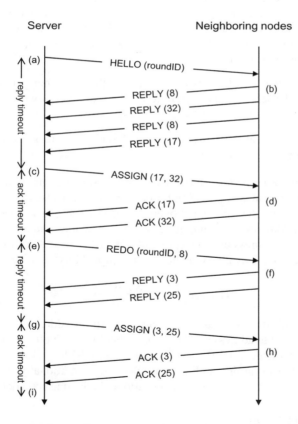

Figure 10.13 The basic DAA algorithm for single server without collisions; in this example, unique addresses are assigned to four neighboring nodes

(a) The server initiates a new round of the DAA process by broadcasting a 'HELLO' message with a new `roundID`.

(b) Upon reception of the 'HELLO', all neighboring nodes store the new `roundID` and the associated server address. Then, each of them randomly chooses an address and submits it to the server in a 'REPLY' message.

(c) The server stores the received addresses in a local list. After the expiration of the reply timeout, the server verifies the uniqueness among all received addresses (DAD). All unique addresses are acknowledged by broadcasting an 'ASSIGN' message (in our example, addresses 17 and 32 are assigned in the first step).

(d) After receiving an 'ASSIGN' for the self-selected address, the node assigns this address to its radio interface for further unicast communication. The success of the operation is acknowledged by an 'ACK' message. Furthermore, the node will ignore all future messages received from the same server with the current `roundID`.

(e) All acknowledge assignments are stored by the server. Additionally, after a specific 'ACK' timeout, all nodes with non-unique address selections or unacknowledged unique ones are requested to re-allocate an address by sending a 'REDO' message containing all the mentioned address identifiers (in our example, address 8 is non-unique and must be re-allocated).

(f) In response to the 'REDO' message, all nodes that have not yet obtained a unique address from the server in the current round have to randomly choose an address and send it back to the server in a 'REPLY' message.

(g-i) The server will repeat the same operations described at steps (c)–(e) until all nodes are allocated unique address identifiers.

Additional neighboring nodes can be discovered by sending one more 'REDO' message containing only the current `roundID` and processing the responses.

The applicability and the introduced overhead of DAA strongly depends on the possibility of duplicate addresses. Such duplicate address selections can appear if two or more nodes choose the same address upon reception of a 'HELLO' or 'REDO' message. Assuming n surrounding nodes and a available addresses, the probability of an address collision is

$$P_{duplicate} \sim 1 - \frac{a!/(a-n)!}{a^n}. \tag{10.5}$$

Typical address identifiers as used in WSNs use 16-bit addresses, i.e. a total number $a = 2^{16} = 65536$ possible addresses. Then, the probability of duplicates in a round will be about 0.00068 for 10 nodes, 0.0028 for 20 nodes and 0.072 for 100 nodes. Considering typical sensor networks, the probabilities are quite low. Additionally, the probability can be controlled by verifying the application scenario and adapting the address space.

Improved DAA algorithm

The original DAA algorithm was proposed under the assumption of reliable communication, i.e. the absence of collisions in the wireless radio channel. The improved DAA algorithm

(Dressler and Chen 2007) includes some special rules to care for possible packet losses due to collisions. We reuse the example shown in Figure 10.13 to outline possible problems if collision occurs during the DAA process.

The probability of collisions for server-side broadcasts ('HELLO', 'ASSIGN', 'REDO') can be well controlled by adapting the corresponding timeouts (replyTimeout, ack-Timeout). On the other hand, all messages sent by the neighboring nodes ('REPLY', 'ACK') are prone to collisions because there is no timing control in the original DAA algorithm. A complete analysis of possible packet losses is given in Dressler and Chen (2007). Here, we only focus on two selected problems and the corresponding solutions:

- Lost 'REPLY' messages – If 'REPLY' messages get lost, the server is not aware of a problem, and therefore of the nodes in its neighborhood, and will terminate the current DAA round. In order to prevent those nodes whose 'REPLY' messages get lost from being abandoned by the server, a new rule is added to the server – the concept of null sub-rounds. In such null sub-rounds, the server receives no responses for a 'HELLO' or 'REDO' message. Then, the server is allowed to terminate the current DAA round only if it has experienced a certain number of null sub-rounds consecutively.

- Lost 'REPLY' and 'ACK' messages – Consider the case that two nodes selected the same address, i.e. duplicate address selections exist that must be detected and corrected. If one 'REPLY' gets lost, the server is not aware of the duplicate address. This could be detected by analyzing the final 'ACK' messages. Now, consider the possible case that the 'ACK' from the same node will be discarded due to a collision as well. In consequence, the address will be registered by the server and assigned to the radio interface of both nodes. The improved DAA algorithm focuses on preventing collisions during the 'ACK' phase by specifically scheduling these messages. For this, the server arranges the addresses in the 'ASSIGN' message in a specific order and assigns a time schedule to each address slot. Upon reception of the 'ASSIGN' message, each node listed in the message has to send its 'ACK' at the specified time point. Thus, either two similar 'ACK' messages will collide or none at all. From a collision, the server will learn that – with a high probability – the particular address has been selected multiple times and initiate a re-allocation.

The improved DAA algorithm is depicted in Figure 10.14. The same example as depicted and discussed for the basic DAA algorithm is used again. In this case, two messages get lost and, therefore, the duplicate address assignment cannot be detected in the first sub-round. As the acknowledgments are well scheduled, the duplicate can be detected either due to a collision (as depicted in Figure 10.14) or due to multiple received 'ACK' messages containing the same address. The DAA algorithm is terminated by a number of null sub-rounds.

10.5 Conclusion

In this chapter, we discussed the basic properties of address-based ad hoc routing techniques. All the discussed routing protocols aim at finding an adequate path towards a well specified destination and to use the discovered path to forward data packets along this path. In the

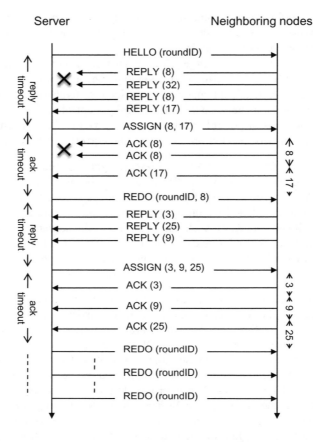

Figure 10.14 The improved DAA algorithm provides mechanisms for handling colliding radio packets; the same four nodes are considered as in the given example for the basic DAA algorithm

context of MANETs, two basic approaches can be distinguished: proactive and reactive (on-demand) routing strategies.

Following a discussion of the general characteristics of address-based routing techniques, their requirements and possible constraints, we outlined a number of specific examples of ad hoc routing protocols. Basically, these examples were selected for two reasons. First, all these protocols represent the state of the art in mobile ad hoc networking and, secondly, very different strategies can be observed by examining these protocols.

Starting with the proactive approach of DSDV, we saw that this Internet-style routing has advantages in terms of route discovery time but we also recognized several problems in unstable, e.g. mobile, environments. Additionally, energy savings due to sleep states are hard to achieve, as periodic routing updates require all nodes to actively participate in the routing process all the time. On the other hand, the reactive approaches DSR, AODV and DYMO directly focus on the specific requirements in MANETs. As no globally synchronized state has to be maintained, such protocols are usually said to be self-organizing in terms of local interactions to discover a (partial) view of the global system. DSR moves all the complexity into a transmitted message in order to prevent any state information in intermediate nodes. In contrast, AODV and DYMO encourage small data messages for efficient use of the wireless medium. The quality and applicability of each of these two groups strongly depend on the particular application characteristics (data rate, node capabilities, energy budget and others).

Independently of the particular solution, all routing techniques can be evaluated according to the following two main characteristics. Depending on the application scenario, different routing approaches may be 'optimal' for the current situation. Unfortunately, it is a hard problem to adaptively exchange the routing protocol 'on demand'.

- Scalability – Similarly to the discussion on MAC protocols, the scalability is the primary limiting factor for routing protocols in MANETs and especially in WSNs. Usually, the scalability can be described by observing the particular resource requirements, e.g. needed memory for routing tables, bandwidth for routing updates, processing power for path calculation, and energy for periodic updates.

- Performance – The performance of ad hoc routing techniques can be evaluated in two dimensions. First, the route discovery time, i.e. the elapsing time before the first data packet can be transmitted, describes the startup delay. Secondly, the quality of the routes, i.e. the path length in terms of hops or other measures, specifies the network performance that an application will perceive during an ongoing transmission. In general, all the routing techniques are limited by the upper bound of the wireless network capacity as formulated by Gupta and Kumar (2000).

The primary prerequisite for all the discussed routing protocols is the availability of unique address identifiers associated to all nodes in the overall network. As this may be difficult to achieve in some wireless networks, dynamic address-allocation techniques have been developed. We briefly introduced the key ideas and characteristics of dynamic address-allocation techniques. In the context of WSNs, two approaches have been selected for extended discussion – PDAD enables the address allocation in a global context, while DAA is focused on localized environment.

11

Data-Centric Networking

Having discussed general ad hoc routing issues in Chapter 10, we learned about the main problems of address-based routing, i.e. its limited scalability in large-scale ad hoc networks and its overhead in terms of route maintenance and route setup delay. These limitations explicitly hold in the case of temporal or spatial mobility. Besides these algorithmic issues, another issue has been identified which correlates to the primary pre-requisite of address-based routing – the availability of network-wide unique address identifiers assigned to all participating nodes.

In this chapter, we will explore the capabilities, algorithms, and limitations of data-centric networking alternatives. All these techniques address the typical scalability problems of classical address-oriented routing strategies (Iwata *et al.* 1999). In contrast to address-based routing, data-centric networking approaches aim at managing data instead of trans-mitting data to a well defined destination node. While the final objectives are similar – in general, networking is always about transmitting data from a producing source to a consum-ing destination – in some networks such as WSNs, the relevance is not about the particular source node but about the measured data. Additionally, data-centric approaches aim at re-ducing the necessary state information in all networked nodes to an absolute minimum in order to increase the global scalability.

Several approaches have been proposed in the literature in the context of data-centric networking. The brief introduction to the main principles in Section 10.1.1 only scratched the surface of the many ideas and objectives behind this technology. Basically, the following concepts and goals are supported by data-centric networking approaches:

- Data dissemination – The objective of data dissemination is to forward data messages towards a (or perhaps multiple) destination is comparable to the primary goal of address-based routing approaches. Actually, specific routing techniques perform the selection of appropriate next-hop nodes to forward the data to. These techniques rely either on probabilistic schemes or on domain- or application-specific measures.

- Network-centric pre-processing – Based on the principles of data-centric networking, all nodes receiving a message may analyze and evaluate the content of received messages before taking forwarding decisions. This way, for example, data aggregation

Self-Organization in Sensor and Actor Networks   Falko Dressler
© 2007 John Wiley & Sons, Ltd

techniques can be employed to reduce the number and size of the messages in transit and, thus, to reduce the overall communication load in the network. Additionally, the quality of the data can be improved by adding additional information about the current environment using data fusion principles.

- In-network operation and control – Finally, data-centric networking is not only about transporting data but it can include complete data-management functionalities. Considering the relocation of the classical sink node into the network itself, all data processing as well as operation and control must be moved into the network. Such in-network operation and control techniques try to reduce the communication load by adding processing tasks to (a set of) the nodes in the network.

Specific examples for data-centric networking are flooding-based techniques, which usually employ probabilistic schemes to reduce the typical overhead of pure flooding. Additional improvements try to reduce the covered area in the network based on spatial or temporal information known from neighboring nodes or previous transmissions. Diffusion techniques basically represent inverse routing techniques. Based on the transmission of interest on the reception of particular data messages, nodes will set up reverse path information towards this destination. Agent-based approaches try to migrate topology information in a probabilistic scheme to all nodes in the network. Additionally, aggregation and other pre-processing techniques can be employed to reduce the overall network load.

The evaluation of data-centric networking approaches is basically similar to address-based routing techniques. In some data-centric approaches, there are state maintenance tasks necessary which induce some initial startup delay comparable to the route setup time and the convergence time. Also, the general protocol overhead can be evaluated. Nevertheless, there are a number of additional measures which describe the feasibility of a particular approach to a much better degree. The following evaluation criteria must be especially considered for data-centric networking solutions:

- Probability to reach the destination – Most data-centric approaches employ probabilistic schemes for their internal operation. Therefore, there is a non-zero application-dependent probability of missing the appropriate destination for a particular data message.

- Number of duplicates – In all solutions based on flooding techniques, there is a high probability of observing duplicated messages within the network. Appropriate counter-measures need an essential amount of storage on intermediate nodes; thus, applications of such forwarding techniques must consider a non-zero probability of receiving duplicated messages.

- General network load – As there is no 'shortest path' known from source to destination, data messages will be forwarded either on unnecessarily long paths or even to nodes that have no interest in the particular data message. Thus, all data-centric approaches will contribute to the overall network load.

In the following, we will first provide a more detailed overview to data-centric networking, resulting in a general classification of appropriate algorithms and techniques. Afterwards, specific solutions will be discussed, starting with probabilistic flooding and agent-based techniques and continuing with diffusion and aggregation approaches. Further in-network processing alternatives will be discussed in the context of Sensor and Actor Networks (SANETs) in Section 14.4.

11.1   Overview and classification

The main goal of this section is to develop an understanding on the various groups and classes of data-centric networking techniques. We start with the differentiation that we used in the introduction to this chapter. According to this, we can distinguish between the following categories. Figure 11.1 depicts the general classification according to the mentioned scheme. The classes and sub-classes will be described in the following sections:

- Data dissemination – Basically, data dissemination is about forwarding data through the network. We can distinguish flooding techniques, probabilistic data forwarding, also known as gossiping, agent-based approaches and reverse path techniques.

- Network-centric operation – Besides data-forwarding issues, data manipulation and control tasks are primary objectives of data-centric networking. The following two classes need to be considered in particular:

  - Network-centric pre-processing – Pre-processing activities include data aggregation, e.g. using statistical measures such as average, mean and standard deviation to describe multiple data records in a single one, and data fusion, i.e. the combination of multiple types of data or the attachment of meta information.

  - In-network operation and control – In-network operation usually means to work without a central base station. Thus, all data-processing activities, including the final decision processes, are moved into the network. Examples are rule-based sensor network approaches (see Section 14.4) or the application of Grid computing to WSNs (which is not discussed in this book).

Additionally, we will see that, depending on the methods under consideration, a number of independent classifications are possible. All classification schemes have in common that particular approaches and algorithms as proposed in the literature typically do not fit a single category but incorporate mechanisms from other categories as well. Thus, multiple dimensions and criteria need to be investigated to classify data-centric networking approaches.

In addition to this classification, the following cross-cutting categorizations need to be considered. In the presentation of selected examples, we will see that these additional issues have a high impact regarding the efficiency of a particular algorithm:

- Probabilistic scheme – Probabilistic schemes are employed in most data-centric networking techniques. The broad spectrum of algorithms ranges from pure probabilistic data forwarding, i.e. random selection of adequate next hop nodes, to randomized distribution of processing capabilities among the available nodes.

- Spatial information – Location information can be exploited in several ways. In the data-dissemination domain in particular, several approaches have been proposed

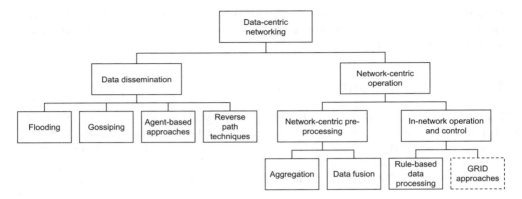

Figure 11.1 Classification scheme for data-centric networking

to use neighborhood information (one- or two-hop information). Additionally, path information can be obtained by propagating interest information through the entire network.

• Temporal information – Historical information can be cached and reused for subsequent data transmissions. For example, nodes can learn about the 'usual' structure and behavior in the network in an adaptive way.

We will discuss the particular aspects of the classification scheme shown in Figure 11.1 below and provide selected examples in the next sections. In general, we will mainly focus on data-centric routing techniques.

11.1.1   Data dissemination

Data-centric routing and data dissemination can be analyzed in two dimensions. First, methods and algorithms can be distinguished according to the used forwarding or path discovery techniques used, respectively. Secondly, it is necessary to differentiate the utilization of spatial and temporal information to improve data-dissemination efficiency. Both dimensions will then be separately described below.

Forwarding and path discovery

In order to transmit data from a source to a destination, basically, two principles can be employed in the context of data-centric networking. First, data can be sent using the push principle, i.e. data will be transmitted without the knowledge of possible receivers. Flooding (Liu *et al.* 2006) and gossiping (Haas *et al.* 2002a) are the best known approaches in this category. Actually, flooding, probabilistic flooding and gossiping, and optimized techniques have been intensively studied in the last few years for their applicability in WSNs. While gossiping has been investigated for many years in the networking community (Hedetniemi *et al.* 1988), its benefits and advantages have just been recognized in the context of data dissemination in sensor networks. We will intensively study available approaches and optimizations in Section 11.2.

Secondly, data can be requested using the pull principle. This way, a receiver requests specific data (in other words, it expresses its interest to receive this kind of data). The nodes in the network are informed about this receiver and its interest and the request messages can be exploited to obtain information about the location of the receiver. This information allows reverse path forwarding of data messages towards the receiver. The best known examples are Reverse Path Forwarding (RPF) and directed diffusion. The concept of RPF to broadcast link-state updates in the reverse direction along the spanning tree formed by the minimum-hop paths from all nodes to the source of the update is, for example, used by Topology Broadcast based on Reverse Path Forwarding (TBRPF) (Bellur and Ogier 1999). TBRPF uses the topology information received along the broadcast trees to compute the minimum-hop paths that form the trees themselves. Actually, TBRPF is not a data-centric routing solution, as particular node addresses are needed. Conversely, directed diffusion (Intanagonwiwat *et al.* 2000) is using reverse path determination as well, while it solely relies on meta information in the data messages to build and to manage necessary state information. Directed diffusion will be further described in Section 11.4.

Finally, combinations can be thought of using flooding or gossiping techniques if no path to a particular destination is known and reverse path forwarding in the other case. Agent-based approaches rely on this working behavior. In the scope of this book, we will briefly discuss rumor routing (Braginsky and Estrin 2002) in Section 11.3. Being a combination of flooding and probabilistic path determination, rumor routing can be adapted to various network scenarios.

Use of spatial and temporal information

In many cases, the exploitation of meta information allows improvement in the behavior of routing and data-dissemination techniques. Basically, meta information includes spatial and temporal information. There is a large amount of such meta information available – and it is differently used in many approaches. So, it is not our objective to provide a broad review on this topic, but to give a rough idea on how to exploit spatial and temporal knowledge to improve data-centric networking approaches. Selected principles will be further discussed in Sections 11.2–11.4.

Available spatial information basically refers to varying degrees of topology information. While most address-based routing protocols employ complete path information from a source node to a sink node (thus, they are at least aware of a part of the complete network topology), data-centric approaches usually exploit the one- or two-hop neighborhood. For example, Cai *et al.* (2005) and Liu *et al.* (2006) use one-hop neighborhood knowledge for efficient flooding. Spatial gossip (Kempe *et al.* 2001) is an example of exploiting distance-based propagation bounds for resource-efficient gossiping. All reverse-path techniques inherently rely on spatial information for message forwarding and state maintenance.

In many cases, temporal information is much easier to exploit, as no further communication is required. Nevertheless, this advantage comes with significant storage requirements, to observe a particular behavior or environments over the time. Temporal information is, for example, exploited to express and to expire interest in specified sensor data using directed diffusion (Intanagonwiwat *et al.* 2000). Flooding and gossiping strategies can also be improved by using histories of transmitted packets or by coping with wireless losses and unpredictable node failures that affect the network connectivity over time (Kyasanur *et al*. 2006).

11.1.2   Network-centric operation

Data-centric networking specifically enables in-network data processing. In this way, the networking algorithms can care for data handling and manipulation, depending on the current objectives and environmental conditions. Additionally, control tasks can be derived from received data packets by means of local rules.

Network-centric pre-processing

In energy-constrained networks such as WSNs, independent transmissions from all sensors to a central base station are quite inefficient. Thus, we need methods for combining data into fewer compact information particles at the sensor nodes or at intermediate nodes, which can reduce the number of packets transmitted to the base station, resulting in conservation of energy and bandwidth (Krishnamachari *et al*. 2002). Data-aggregation techniques have been proposed to address this issue (Rajagopalan and Varshney 2006). Basically, data aggregation is defined as the process of aggregating the data from multiple sensors to eliminate redundant transmission and provide more compact information to the base station. We will discuss specific aggregation techniques later, in Section 11.5.

Multi-sensor data fusion is an evolving technology, concerning the problem of how to fuse data from multiple sensors in order to make a more accurate estimation of the environment (Qi *et al*. 2001). Basically, data fusion also supports the combination of heterogeneous types of data or the attachment of meta information (Brooks and Iyengar 1997). Important technical issues include the degree of information sharing between nodes and how nodes fuse the information from other nodes (Chong and Kumar 2003). Processing data from more sensors generally results in better performance, but also requires more communication resources and, thus, energy.

Data-centric processing has been investigated, for example, by Chu *et al*. (2002). They analyzed information-driven sensor querying and routing for ad hoc heterogeneous sensor networks. Essentially, they started with the following problem statement: How to dynamically query sensors and route data in a network so that information gain is maximized while latency and bandwidth consumption is minimized. The developed Information Driven Sensor Querying (IDSQ) algorithm allows optimized sensor selection using a hierarchical cluster-based scheme.

In-network operation and control

In-network operation usually means to work without a central base station. Thus, all data-processing activities, including the final decision processes, are moved into the network. A good example is the rule-based approach for data-centric networking, further described in Section 14.4 (Dressler 2006a, 2007). This system employs a simple set of local rules that enable the sensor node to process data messages according to their internal meanings. Supported reactions include data aggregation, local actuation control, modification of messages and selective discarding.

11.1.3   Related approaches

A series of data-centric routing approaches that needs to be named at this place is ant-like routing. As learned from the domain of Swarm Intelligence (SI), this bio-inspired approach for routing has already demonstrated its capabilities in the context of MANETs and WSNs. We choose not to focus on these algorithms in this section, as we will study ant-like routing approaches in detail in Part V, in which we will study a number of related bio-inspired approaches. Nevertheless, a short introduction to ant-like routing seems necessary and sufficient at this point.

AntNet is one of the first approaches in the context of ant-like routing. Di Caro and Dorgio (1998) describe it as a distributed, mobile agents-based Monte Carlo system that was inspired by investigations into the ant colony metaphor for solving optimization problems. Multiple agents (ants) are used to explore the network. The information exchange between the agents is indirect and asynchronous, using the network itself.

Based on the investigations in AntNet, Di Caro et al. (2005) developed a new routing approach for use in MANETs–AntHocNet. Similar to AntNet, AntHocNet's design is based on the specific self-organizing behavior of ant colonies and the capabilities of ants to find the shortest paths through the network. AntHocNet is based on the framework of Ant Colony Optimization (ACO) (Dorigo et al. 1999). Based on the natural principles of ants, nodes in AntHocNet forward data stochastically. When a node needs to forward a message towards a destination, it randomly selects one of the possible next hop nodes with a given probability. This probability is adaptively modified with respect to the better paths. The probabilistic routing strategy leads to distribution of the data load according to the estimated quality of the paths. Automatic load balancing and congestion avoidance are supported inherently.

Labella and Dressler (2006) reused the AntHocNet algorithm to support objective-driven path selection. As an objective, for example, the distribution of multiple tasks, which need to be fulfilled in a sensor network, can be used. All these ant-like routing approaches will be described in more detail in Part V.

The use of emergent routing techniques has also been proposed by Brooks et al. (2003a,b). In this approach, adaptive methods for routing in WSNs are investigated. All these techniques represent emergent behaviors that, first, perform global adaptation using only locally available information. Secondly, they have strong stochastic components and, thirdly, they use both positive and negative feedback to steer themselves.

11.2   Flooding, gossiping and optimizations

As we have seen in Chapter 10, most address-based routing approaches exhibit essential scalability problems. Node mobility further complicates the environment. Thus, adequate communication protocols must not only take into account the restrictions on resources imposed by the nodes in ad hoc networks, but must also be robust enough to deal with the dynamic network topology (Scott and Yasinsac 2004). Flooding and probabilistic flooding, i.e. gossiping, are such approaches, in which, basically, no state information is required, except for some optimizations. Their reliability depends on the network topology and the principles of the used probabilistic scheme. Optimized versions exploit spatial and temporal meta information for more efficient data forwarding. Below, we discuss the basic principles of flooding and gossiping, as well as possible optimizations.

11.2.1   Flooding

Principles of flooding

Flooding is a very simple communication principle. Basically, it relies on nodes' receiving a message to rebroadcast this message to their direct neighbors. Flooding usually covers all the nodes in a network. Nevertheless, it can also be limited to a set of nodes that is defined by a geographical area or by a Time To Live (TTL) parameter. Geographically restricted flooding is also called geocast flooding (Ko and Vaidya 2002). We will outline the use of spatial information in the discussion of advanced flooding and gossiping strategies. The limitation according to a TTL refers to the maximum network diameter. Each node that is receiving a flooded message only rebroadcasts it if the TTL of the message is greater than 0. The TTL value is decremented at each hop before rebroadcasting.

The basic principles of flooding are depicted in Figure 11.2 and Algorithm 11.1. In the given example, the node in the top-left corner sends a single message in step 1. This message is received by its direct neighbors that, in step 2, rebroadcast the message to all their neighboring nodes. This process is repeated until the TTL value is decremented to 0. As shown in Figure 11.2, the source node receives a copy of the message in step 2 and participates again in step 3 by rebroadcasting this copy.

Discussing flooding and broadcasting, these two mechanisms are usually distinguished in the following way. Broadcasting refers to a transmission that is received by all nodes within the transmission range of the broadcasting node. In wireless networks, the broadcasting range is equivalent to the radio transmission range. Conversely, flooding uses broadcasting in the local neighborhood and requires the receiving nodes to rebroadcast the message.

Additionally, in the literature, flooding is often referred to as being directed, i.e. to allow intermediate nodes to rebroadcast a received message exactly once (Pleisch *et al.* 2006;

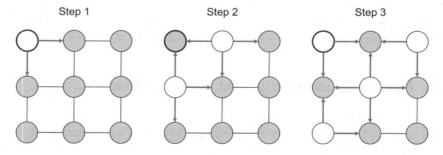

Figure 11.2   Principles of flooding through the network. Shown are the first three steps of a single message that originated in the top-left corner being flooded through the network. This process continues until nodes decide to discard the message, e.g. using a time-to-live field to limit the maximum number of hops

Algorithm 11.1 Flooding

Require: Maximum network diameter MAXTTL.
Ensure: Flooding of a message through the network.

1: **while** Receive a new flooding packet P **do**
2:     $TTL(P) \leftarrow TTL(P) + 1$
3:     **if** $TTL(P) < MAXTTL$ **then**
4:         Rebroadcast packet P
5:     **end if**
6: **end while**

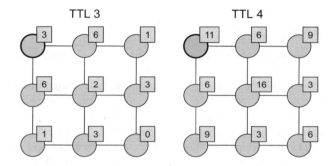

Figure 11.3 Broadcast storm as initiated by transmitting a single message from the top-left node; shown are the numbers of received messages duplicated at each node for two different maximum hop counts

Scott and Yasinsac 2004). Thus, in such references, the flooding protocol is meant to work as follows: upon receiving the message for the first time, each node retransmits the message to all neighbors. This behavior would require two measures that cannot be assumed to be available in all scenarios. First, all nodes will be required to store state information about already observed messages and, secondly, the communication protocol must ensure that all messages can be uniquely identified. In autonomously working self-organizing networks such as WSNs, both requirements cannot be fulfilled.

Although flooding is a very simplistic protocol, flooding has the virtue of being reliable, while requiring minimal state retention (Scott and Yasinsac 2004). Unfortunately, flooding often results in redundant messages, consuming valuable bandwidth and power as well as causing contention, collision and packet loss. This problem is generally referred to as the broadcast storm problem in dense networks (Ni *et al.* 1999). Figure 11.3 shows the message implosion problem. As all nodes rebroadcast the received message as long as the TTL is greater than 0, depending on the network topology, a huge number of copies can be generated. Figure 11.3 depicts some numbers for a relatively small network.

Algorithm 11.2 Edge Forwarding

Ensure: Reliable rebroadcasting by a minimum number of nodes in the broadcast range.

1: **while** Receive a new flooding packet P **do**
2:     Start a process on packet P
3:     Wait for T time units – *overhearing period*
4:     **if** Each one-hop neighbor is already covered by at least one broadcast of P **then**
5:         Terminate process on packet P
6:     **else**
7:         Rebroadcast packet P
8:     **end if**
9: **end while**

Topology-assisted flooding

Because traditional (simple) flooding approaches suffer from the huge number of duplicated messages, causing resource contention and collisions in the wireless radio channel, more advanced approaches have been proposed that exploit spatial information. Basically, we can summarize these solutions as topology-assisted flooding. Below, we will briefly discuss a number of selected approaches for topology-aware flooding to reduce the network overhead.

Information about the one-hop neighborhood has been exploited by Cai *et al.* (2005). The developed technique was named Edge Forwarding. In this approach, each node is required to track neighboring nodes within the one-hop distance. Rebroadcasting of a received message is then accomplished only by hosts near the perimeter of the broadcast coverage. The algorithm assumes that each node knows the exact position information of all neighboring nodes. The basic algorithm works as follows (see Algorithm 11.2). After receiving a message, the node determines its distance towards the sender. Then, an overhearing period is initiated. The length of is inversely proportional to the estimated distance. Thus, distant nodes will finish this period first and initiate a rebroadcast. All other nodes will receive this broadcast and terminate their overhearing period.

Additionally, the approach as presented in Cai *et al.* (2005) includes a forwarding rule checking the necessity for rebroadcasting the message by detecting loops (we have already discussed this issue) and a confirmation period to face possible uncertainties in the location estimation due to node mobility.

Another approach relying on knowledge about the one-hop neighborhood has been studied by Liu *et al.* (2006). This algorithm relies on the idea of computing an optimal subset of neighboring nodes as forwarding nodes instead of performing the decision on whether to rebroadcast the message or not at the receiver. The computed list of forwarders is attached to the transmitted message. The complete procedure to calculate the minimal forwarding set is presented in Liu *et al.* (2006).

Clustering techniques can also be used to facilitate efficient flooding in ad hoc networks. Kwon and Gerla (2002) demonstrated the feasibility of this approach – which relies on the same basic principle as the solution by Liu *et al.* (2006), i.e. to calculate a minimum subset of forwarding nodes – by dynamically structuring the network into clusters. The purpose is to group all nodes into manageable sets and the basic idea is to distribute the work load to the clusterheads and the gateway nodes. We will discuss different clustering methods in detail in Chapter 12. Basically, any of them can be used for clustering the network. The improved efficiency of cluster-based flooding has been successfully demonstrated (Kwon and Gerla 2002). Other approaches to reduce the overhead caused by flooding are addressed by probabilistic schemes, as discussed below.

11.2.2   Pure gossiping

While flooding is a reliable transmission protocol, it is prone to the broadcast storm problem in dense networks (Ni *et al.* 1999). A simple solution to this packet-implosion problem is to reduce the number of redundant messages. However, reducing the number of redundant broadcasts leads to a lower degree of reliability. Hence, the challenge that we face is to strike a balance between message overhead, i.e. the level of redundancy, and reliability.

The difficulty is always to know which messages are necessary and which are redundant. A number of flooding-based approaches have been proposed to address the broadcast storm problem (Williams and Camp 2002). In the context of this chapter, we want to focus on one particular solution–probabilistic flooding, which is also known as gossiping. In the absence of better information, gossip assumes that all messages are equally important.

Pure gossiping tries to retain the advantages of flooding while reducing the overhead due to duplicated messages. The basic idea is to decide at each node whether to forward the packet to all neighboring nodes according to a given probability (Barrett *et al.* 2003). The behavior of gossiping has been studied in detail by Haas *et al.* (2002a). In their paper, the authors have shown that gossiping, i.e. essentially, tossing a coin to decide whether or not to forward a message, can be used to significantly reduce the number of messages sent through the network. We will discuss the limitations and possible optimizations later.

In Figure 11.4, the principles of gossiping are depicted in direct comparison to the already shown example for flooding. All nodes that receive a message first flip a coin to determine whether they should rebroadcast the message. According to a given probability p, nodes either forward or discard the message. In the example shown, two marked nodes decide to drop the message to reduce the network overhead. The principles are also shown in Algorithm 11.3.

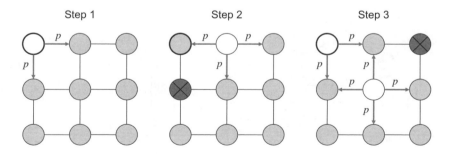

Figure 11.4   Principles of gossiping. We reused the flooding example from Figure 11.2, although, in this example, each transmission is initiated with a certain probability p. In the given example, two nodes decided not to forward the received message according to the probabilistic scheme

Algorithm 11.3 Gossiping
Require: Maximum network diameter MAXTTL, gossiping probability p.
Ensure: Simple gossiping with probability p.

1: **while** Receive a new flooding packet P **do**
2:     $TTL(P) \leftarrow TTL(P) + 1$
3:     $p_h \leftarrow \mathrm{random}(0, 1)$
4:     **if** $TTL(P) < \mathrm{MAXTTL} \wedge p_h < p$ **then**
5:         Rebroadcast packet P
6:     **end if**
7: **end while**

Results from investigations in the percolation theory (Broadbent and Hammersley 1957; Grimmett 1989) show that such probabilistic flooding approaches will result in either almost all nodes' receiving the packet or almost no nodes, receiving the packet. Haas *et al.* (2002a) discovered that gossiping exhibits a certain type of bimodal behavior (Birman *et al.* 1999): Assuming a gossiping probability p, then, in sufficiently large 'nice' graphs, there are execution patterns depending on p in which the gossip quickly dies out, while, in almost all of the executions in which the gossip does not die out, almost all of the nodes receive the message.

Thus, in almost all executions of the algorithm, either hardly any nodes receive the message or most of them do. The open issue is to make the fraction of executions in which the gossip dies out relatively low while also keeping the gossip probability p low in order to reduce the message overhead.

In addition to the general search for the optimal gossiping probability p, special network topologies must be considered for evaluating the applicability of gossiping. Haas *et al.* (2002a) named it 'nice graphs', in which the gossiping probability can be well chosen. There are a number of specific network topologies in which gossiping will usually fail. The most extreme example is a linear network, as depicted in Figure 11.5. As each node rebroadcasts a received message with a static probability p, the overall probability of a

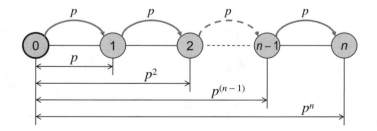

Figure 11.5 Limitations of gossiping become obvious in special network topologies, particularly linear networks; as pure gossiping forwards messages with a certain probability p, in this special case, the probability of a message to reach the end of the linear network is p^n with n being the network diameter, i.e. the number of hops to be traversed

message to reach the end of the linear network is $P(p) = p \times p \times \ldots \times p = p^n$, with n being the network diameter, i.e. the number of intermediate nodes that will flip the coin to check for possible rebroadcasting. Obviously, the resulting probability p^n will be quite small (actually, it tends to become close to 0) for increasing network diameters and low gossiping probabilities.

As a concluding remark, it should be noted that there is an upper bound of the gossiping probability p above which no further improvement in the gossiping range can be achieved. Sasson et al. (2002) studied this so-called phase transition phenomenon in MANET topologies. The threshold $p_c < 1$ at which the flooded message will almost surely reach all nodes within the multi-hop broadcast range can be explained using the percolation theory and the phase transitions in random graphs. This phase-transition property has been well studied in the context of random graphs; for example, Erdös and Rényi (1960) have shown that a random graph will be connected with a probability close to 1 if the number of edges is greater than $p_c(E) = \frac{N \log N}{2}$.

11.2.3 Optimized gossiping

As discussed before, probabilistic flooding approaches, i.e. gossiping, offer a simple alternative to deterministic approaches. With gossiping, nodes in the network are required to forward packets with a pre-specified probability $0 < p < 1$ (for $p = 1$, we speak about flooding). If the probability is well chosen, the percolation theory says that all nodes in the entire network will receive the transmitted message with very high probability – even though only a non-deterministic subset of nodes have forwarded the message. Gossiping is a very simple data-dissemination solution, while it is able to achieve better reliability and load balancing than most deterministic solutions (Haas et al. 2002a; Kyasanur et al. 2006).

Obviously, the problem is to choose the gossiping probability p as optimal, as the best value for p is closely associated with the topology of the network. In the absence of topology information, estimating the optimal gossiping probability p is difficult and error-prone. Additionally, spatial and temporal mobility in WSNs further contribute to the complexity of the challenging task of carefully choosing the gossiping probability. Thus, the gossiping quality will become sub-optimal over the network lifetime. In the following, we discuss a number of approaches for optimizing the gossiping probability based on locally available information in a self-organizing adaptive manner.

A two-threshold scheme

Haas et al. (2002a) first studied the behavior of gossiping in detail. Especially, they investigated the problem of gossiping in linear networks as depicted in Figure 11.5. They proposed a gossiping technique based on a two-threshold scheme. The first approach considers typical network topologies and the problem of quickly dying-out gossips. Following the notion used by the authors, we use GOSSIP(p) to refer to pure gossiping. Instead, GOSSIP(p, k) describes a modified gossiping protocol in which pure flooding is used for the first k hops. Then, the standard gossiping technique with probability p is employed.

Basically, GOSSIP($1, k$) refers to flooding and GOSSIP($p, 0$) stands for pure gossiping, Between these two extremes, a number of possible solutions for optimal values for p and k can be determined. Using a regular network topology and an ideal MAC protocol, i.e. one that does not lose any packets, the authors discovered that a gossiping probability $p = 0.72$

Algorithm 11.4 Gossiping using the two-threshold scheme

Require: Maximum network diameter MAXTTL, initial flooding range k, gossiping probability p, minimum neighbor count n, and gossiping probability p_{sparse} for sparse
networks.
Ensure: Optimized gossiping compared to the 'pure' standard version.

  1: **while** Receive a new flooding packet P **do**
  2:     $TTL(P) \leftarrow TTL(P) + 1$
  3:     $p_h \leftarrow$ random$(0, 1)$
  4:     $N \leftarrow$ neighbor count
  5:     **if** $TTL(P) <$ MAXTTL **then**
  6:       **if** $k < TTL(P) \vee (N \geq n \wedge p_h < p) \vee (N < n \wedge p_{sparse} < p)$ **then**
  7:         Rebroadcast packet P
  8:       **end if**
  9:     **end if**
10: **end while**

---

and an initial flooding range $k = 4$ is sufficient so that almost all nodes in a medium-sized
network (consisting of about 1000 nodes) will receive the message. Similar values can be
obtained for random topology networks.

This gossiping scheme can be further improved using simple optimizations. In irregular
random networks, the node density will be different in the various parts of the network. The
idea is to make the gossiping probability a function of the node density, i.e. on the degree
of neighboring nodes. The two-threshold scheme introduces two new parameters $-p_{sparse}$
and n −with the final idea of using the second probability $p_{sparse} > p$, which is a node that
has fewer neighbors than n.

The resulting gossiping function is GOSSIP(p, k, p_{sparse}, n). The need to increase the
gossiping probability in sparse networks can be illustrated by analyzing the original gossiping behavior. If a node broadcasts a message, the further forwarding depends on the
'willingness' of the neighbors to rebroadcast the message. If only a few neighbors exist, the
accumulated forwarding probability is quite low. Thus, increasing the gossiping probability,
especially in these cases, greatly improves the overall gossiping efficiency. The resulting
procedure is depicted in Algorithm 11.4. In their investigations, Haas *et al.* (2002a) discovered that for $p \geq 0.8$, GOSSIP$(p, 4)$ has the same performance as GOSSIP$(0.6, 4, 1, 6)$.
However, GOSSIP$(p, 4)$ used 13% more messages compared with the improved solution.

Node density and node distribution

Further improvements in gossiping have been proposed by Scott and Yasinsac (2004). Similarly to the previous scheme, the authors exploit the node density to determine the gossiping
probability p. In contrast, network dynamics is explicitly allowed and incorporated into the
communication technique. As the network topology can change over the time due to spatial
and temporal mobility, the node density also changes over the time. Basically, the algorithm assumes that the node density can be approximated by the overlapping broadcast
area of two nodes (that can easily be estimated based on the knowledge of the broadcast

range and the distance between the two nodes). Similarly, Zhang and Agrawal (2005) propose dynamic probabilistic broadcasting in MANETs exploiting node distribution and node movement.

Spatial gossip as proposed by Kempe *et al.* (2001) exploits the idea that in many settings – consider a network of sensors or a cluster of distributed computing hosts – new information generated at individual nodes is most 'interesting' to nodes that are nearby. Based on this observation, distance-based propagation bounds can be formulated as a performance measure for gossiping.

Smart gossip using overhearing

Approaches like edge routing (Cai *et al.* 2005) can be adapted to gossiping as well. Smart gossip represents a class of data-dissemination techniques that relies on observations of the behavior of neighboring systems to estimate an optimal gossiping probability (Kyasanur *et al.* 2006). Thus, smart gossip automatically and dynamically adapts transmission probabilities based on the underlying network topology. The main advantage is that the resulting protocol, which is completely decentralized, is capable of coping with wireless losses and unpredictable node failures that affect network connectivity over time. In smart gossip, nodes extract information from overheard gossip messages (which contain a parent node identifier and a required gossiping probability $p_{required}$) and apply simple rules to attempt to deduce whether the sender of the message is a parent, a child or a sibling.

The application using smart gossip needs to specify its reliability requirements in the form of an average reception percentage τ_{arp}. As the application requires that even the farthest node will receive the message with the given probability, τ_{arp} can be translated into a relative per-hop probability τ_{rel}, which can be estimated by solving $(\tau_{rel})^{\text{MAXTTL}} = \tau_{arp}$.

Exploiting the knowledge about the parent nodes, each node can dynamically adapt its local forwarding probability. Assuming that a particular node has K parent nodes, then it is sufficient for each parent to use a gossiping probability p_{parent} which ensures that the probability that at least one of them transmits is greater than τ_{rel}. Thus, $p_{required}$, which is announced by a node with K parents, can be estimated using Equation 11.1:

$$(1 - p_{required})^K < (1 - \tau_{rel}). \tag{11.1}$$

Destination attractors

Another step for exploiting spatial information has been proposed in the context of parametric probabilistic sensor network routing (Barrett *et al.* 2003). Again, the goal is to optimize the gossiping probability p according to specific meta information. The concept of this approach was named 'destination attractor'. This approach follows a quite simple and sensible algorithm. If a packet is getting closer to the destination, then its rebroadcasting probability is increased.

Based on these ideas, the gossiping probability P_{R_i} for a packet at the ith node R_i in its path from source to destination can be calculated as depicted in Equation 11.2, where k is a factor describing the slope of the increment or decrement of the gossiping probability, respectively:

$$P_{R_i} = \begin{cases} (1+k)P_{R_{i-1}} & \text{closer to destination} \\ (1-k)P_{R_{i-1}} & \text{further to destination} \\ P_{R_{i-1}} & \text{same or indeterminate.} \end{cases} \tag{11.2}$$

The rebroadcasting probability can also be expressed using the change in the distance to the destination $d(R_{i-1}, D) - d(R_i, D)$. Using the distance information, the formula can be written as shown in Equation 11.3. A non-recursive description can be easily derived, as given in Equation 11.4.

$$P_{R_i} = \{1 + k[d(R_{i-1}, D) - d(R_i, D)]\}P_{R_{i-1}} \tag{11.3}$$

$$= \{1 + k[d(S, D) - d(R_i, D)]\}. \tag{11.4}$$

Obviously, the retransmission probability functions for destination attractors depend on the locally available knowledge of the distance $d(R_i, D)$ from node R_i to destination D, the distance $d(S, D)$ from source S to destination D, and time step i. Several methods can be thought of for estimating $d(R_i, D)$. Barrett *et al.* (2003) used the following light-weight method distance estimation. Each successfully received broadcast message will be acknowledged by sending out a packet to the originator of the broadcast containing the currently known distance towards the destination. This way, the correct distance estimation will propagate slowly from the destination towards the source.

Weighted Probabilistic Data Dissemination

Flooding and all previously discussed gossiping techniques in are based on the same assumptions. First, all nodes are considered to be identical and, secondly, each node transmits its messages with the same (usually low) data rate. Regarding the behavior of a typical sensor network – WSNs were primarily developed and deployed for monitoring environmental properties – these assumptions hold in this application scenario as well. Nevertheless, if we allow network-centric pre-processing and measurement-dependent reactions, the basic conditions are changing rapidly. For example, in a monitoring scenario, each sensor node will send its information on a fairly regular basis, while, if dramatic changes are recognized, all the sensors discovering this anomaly will start sending traces about this event in a burst transmission. Congestion will obviously occur without implemented countermeasures (Dressler 2005).

Besides uniform congestion control, the need for QoS measures is rising in the context of WSNs. To give an example, we quickly analyze data aggregation in sensor networks (a more detailed study is provided in Section 11.5). Consider the following two cases. First, all messages are forwarded individually using gossiping to a common destination. According to the gossiping probability p, each message will be transmitted independently and reach the destination with a cumulated probability P. The overall probability $P_a \gg P$ that at least one transmission is successful is much larger than P for $P \neq 0$ and a large number of messages. If data aggregation is used, multiple messages are combined into a single packet. Thus, with increasing aggregation ratios, i.e. large numbers of messages' being combined into a single message, the overall probability to reach the destination P_A will be almost equal to P. Thus, aggregation techniques are conflicting with the gossiping strategies.

The addressed problem is depicted in Figure 11.6. Obviously, high-priority emergency messages (in the given example, specific shock information) must contend with regular measurement data for the communication resources. Thus, the probability of such emergency messages' reaching the base station needs to be increased.

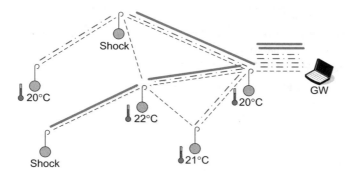

Figure 11.6   Motivation for developing priority-based gossiping techniques such as the WPDD algorithm; in emergency situations, the delivery of high-priority alarm messages needs to be ensured by superseding normal measurement information

The WPDD algorithm has been developed to tackle these problems (Dressler 2006c). The main objectives can be summarized as follows: data-centric data dissemination, inherent congestion control, and QoS features, i.e. overload detection and prevention and priority-oriented data transmission. The basic algorithm is outlined below.

Each message is given a priority $I(msg)$. This priority depicts the importance of the particular operation. Based on this priority, the message is forwarded to a percentage of the direct neighbors and to an even lower percentage of remotely accessible nodes. The forwarding probability can now be calculated using two measures. First, the message-specific probability $G(I(msg))$, which is basically a translation of the message priority to an upper bound for the gossiping calculation, represents the message dependent part. Secondly, $W(S_i)$ depicts the node weighting for node S_i, i.e. the local gossiping probability of node S_i. Now, the forwarding condition can be expressed as follows.

$$G(I(msg)) > W(S_i). \tag{11.5}$$

The complete procedure is shown in Algorithm 11.5. Basically, WPDD represents an enhanced and freely configurable version of gossiping. All discussed optimizations relying on spatial and temporal meta information can be applied as well. Congestion control can be achieved by counting the received packets in pre-specified time intervals (Dressler 2005).

This forwarding strategy can be directly applied to various application scenarios in WSNs, especially if aggregation techniques need to be employed. The algorithm ensures that high-priority messages arrive at the destination with an increased probability.

Forward Error Correction

Finally, methodology needs to be named in the context of optimized probabilistic flooding, which has been introduced in the Mistral algorithm by Pleisch *et al.* (2006). We know that selective rebroadcasting of flooded messages is a way to limit the number of redundant transmissions. The downside of gossiping is that a message may no longer reach all intended nodes. In particular, if a node has only a few neighbors, none of these neighbors

---

Algorithm 11.5 Weighted Probabilistic Data Dissemination

Require: Maximum network diameter MAXTTL, message specific priority $I(msg)$, node
    specific gossiping probability $W(S_i)$.
Ensure: Gossiping considering message priorities.

1: **while** Receive a new flooding packet P **do**
2:     $TTL(P) \leftarrow TTL(P) + 1$
3:     **if** $TTL(P) < MAXTTL$ **then**
4:         **if** $G(I(P)) > W(S_i)$ **then**
5:             Rebroadcast packet P
6:         **end if**
7:     **end if**
8: **end while**

---

may rebroadcast the message. Selective flooding thus balances message overhead against
reliability. The key idea of Mistral is to introduce a new mechanism that allows fine-tuning
the balance between message overhead and reliability by compensating for messages that
are not being rebroadcasted. This compensation is based on a technique borrowed from
Forward Error Correction (FEC) (Huitema 1996). Every incoming data packet is either
rebroadcasted or added to a compensation packet. The compensation packet is broadcasted
at regular intervals and allows the receivers to recover one missing data packet. One of the
fundamental advantages of FEC is that it imposes a constant overhead.

11.3   Agent-based techniques

Motivated by the observation that highly efficient data-centric routing mechanisms can offer
significant power cost reductions (Xu *et al*. 2001) and improve network lifetime (Dietrich
and Dressler 2006), Braginsky and Estrin (2002) proposed a routing scheme that they
called 'rumor routing'. This technique focused on the forwarding of queries to events in
the network. The authors used the same key assumption that, because of the large amount
of possible system and data redundancy in WSNs, data become disassociated from specific
nodes and resides in regions of the network.

    Rumor routing considers an event as an abstraction, identifying anything from a set
of sensor readings to the node's processing capabilities. Nevertheless, events are assumed
to be localized phenomena, thus occurring in a fixed region of space. Queries represent
requests for particular events. Depending on the amount of data to be transmitted from
the event to the originator of the query, it may make sense to invest in discovering short
paths from the source to the sink. However, in many applications, the quality of the path
may not be very important, since the application may only request a small amount of
data. In such cases, flooding of neither queries nor the final events represents an optimal
data-dissemination strategy.

    As outlined in Braginsky and Estrin (2002), the rumor routing algorithm is intended
to fill the region between query flooding and event flooding. If too many queries are
generated in the network or if a huge number of data transmissions are to be expected
from the event sources, query or event may achieve better results compared with rumor

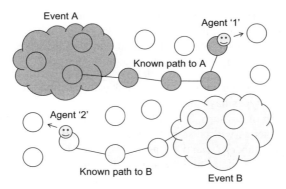

Figure 11.7 Prerequisites for rumor routing. Events represent geographically distributed regions of space. Some nodes already 'know' a path towards specific events. Two agents are crawling the network and distributing information about the known events

routing. Nevertheless, an application aware of this ratio can also use hybrid communication techniques, exploiting the best results from both rumor routing and flooding.

The proposed rumor-routing scheme is a logical compromise between flooding queries and flooding event notifications (Braginsky and Estrin 2002). The algorithm exploits ideas from multi-agent systems by employing agents to create paths leading to each event. Basically, each generated query can be sent on a random walk until it finds the event path. As soon as the query discovers the event path, it can be routed directly to the event. Rumor routing proposes to use flooding as a last resort if the random search fails. In their original paper, the authors demonstrated that the number of paths and the number of query attempts increase the likelihood of delivery exponentially, making rumor routing tunable to a wide variety of application requirements.

Figure 11.7 depicts the main ideas of rumor routing. A given number of agents is crawling the network. They carry information about already visited events and distribute this information on their random way through the network. Thus, paths towards the particular events are generated. In general, agents may propagate path information for multiple events as well.

The complete algorithm for rumor routing works as follows. We distinguish between node and agent activities and between query and event messages:

- Node behavior – Each node needs to maintain information about its local neighborhood and an event table holding forwarding information to known events. Rumor routing does not explicitly propose a method for maintaining the neighborhood information. Periodic 'HELLO' messages can be used or other communications can be exploited for this issue. The event table is being updated by arriving agents. In the original algorithm, such event information is considered to live unrestricted in time. Nevertheless, an appropriate timeout seems to be necessary to support spatial and temporal mobility in the network.

- Agent activities – An agent is created whenever a node witnesses a new event. Rumor routing proposes to handle this creation probabilistically in order to limit the number of agents for particular events. After its creation, the agent travels in the form of a long-living packet through the network. Its attached information is used for updating the event tables at visited nodes. The agent can also learn about other events from these nodes and carry this information though the network.

All queries and events (agents) are forwarded using a random path way through the network. Both message types have an upper bound for their travel. The agents continue for a maximum path length of L_a hops. Similarly, queries are limited by a maximum hop count of L_q. Additionally, queries will be stopped at nodes that already know a path towards the searched event. Both numbers L_a and L_q are strongly application- and network-topology-dependent. If a query is unable to find the searched event, the originating node may decide to retransmit the query or to flood the query through the network. Retransmission is a risk, but the chance of delivery is exponential with the number of transmissions. Hopefully, only a very small percentage of queries would have to be flooded.

An example for rumor routing is depicted in Figure 11.8. In the first step (left), node Z knows paths to both events A and B. With the arrival of the agent, node Z updates its local event tables according to the knowledge carried by the agent. In particular, node Z updates its path towards event A. Additionally, the agent's table is updated and node Z knows a path to event B which the agent has not yet included in its local table. On the way through the rest of the network, the agent will propagate routing information towards both events.

Braginsky and Estrin (2002) propose several possible enhancements to rumor routing. Two of them should be mentioned in the conclusion of this section. First, non-random next-hop selection can improve the path-finding process. Especially, if limited location information is available, this may be exploited to direct agents to less explored regions in the network. Secondly, constrained flooding can be used to find a path to the searched destination. Limited flooding may provide a more efficient path-finding method compared with the random search. In this case, a cost-efficient balance between limited flooding and random search needs to be implemented, e.g. using gossiping approaches.

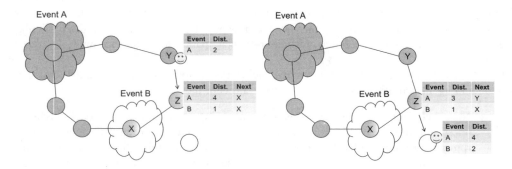

Figure 11.8 Propagation of paths towards events using the rumor-routing agent system. In the shown example, an agent arrives at node Z and updates the locally stored event table according to its carried knowledge. Additionally, the agent's table is updated using information from node Z

11.4   Directed diffusion

Motivated by robustness, scaling and energy-efficiency requirements, Intanagonwiwat *et al.* (2000) proposed a different data-centric message-forwarding technique for application in WSNs, explicitly focusing on the operation paradigm of such sensor networks. They named this technique 'directed diffusion' according to the two primary principles of the algorithm: data generated by sensor nodes are named by attribute–value pairs. A node can express its 'interest' for named data by sending (diffuse) appropriate requests into the network. Data matching the interest is then 'drawn' down (directly) towards that node. For optimization issues, intermediate nodes can cache or transform data, and may direct interests based on previously cached data. In this section, we describe the basic directed-diffusion algorithm along with its key techniques and a number of selected optimizations.

11.4.1   Basic algorithm

The basic directed-diffusion algorithm has been developed to cover the typical communication requirements in WSNs. Sensor nodes measure particular physical phenomena and distribute this information to dedicated sink nodes. The main issue in data-centric networking is to identify the location of the sink nodes and – this issue has not been targeted by the previously discussed data-dissemination techniques – the current demand at these sinks, i.e. their willingness to receive and to process the sensor data (Intanagonwiwat *et al.* 2000). Thus, the mechanism was developed for use in WSNs in which nodes can coordinate to perform distributed sensing of environmental phenomena. All nodes in a directed-diffusion-based network are application-aware. This enables diffusion to achieve energy savings by selecting empirically good paths and by caching and processing data within the network.

Before outlining the core components of directed diffusion, its basic principles should be explained using a simple example depicted in Figure 11.9. Generally, directed diffusion consists of two basic mechanisms: interest distribution and data propagation. Interest is diffused (basically flooded) through the entire network, as shown in Figure 11.9(a). Intuitively, this interest may be thought of as exploratory; it tries to determine whether there are any sensor nodes that detect the requested measures. Such interest messages are renewed periodically to keep them up to date. Each node in the network that forwards this interest message sets up a gradient towards the source of the interest, i.e. the sink node (Figure 11.9(b)). A special reinforcement process is employed to weight the gradients based on their qualities, e.g. loss ratio or hop count. If a sensor node receives the message that actually can provide the requested data, it will finally start to produce measurement data and to transmit the results along the chosen gradient towards the sink, as depicted in Figure 11.9(c).

After this brief introduction to directed diffusion, we now go into the details of the four key techniques used by the basic algorithm, following the original paper by Intanagonwiwat *et al.* (2000).

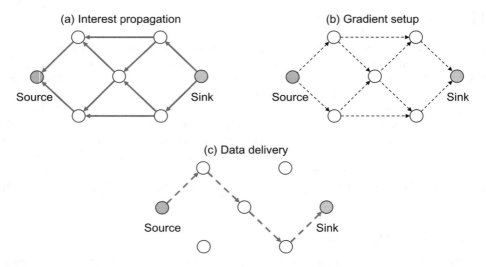

Figure 11.9   Interest and data-propagation mechanisms used in directed diffusion. First, interest is flooded through the entire network (a), the intermediate nodes store this interest information for setting up initial gradients towards the sink (b), and, finally, the data are delivered along the (reinforced) data path (c)

Naming

All actions in directed diffusion rely on well described, i.e. named, interests and data. One possible naming scheme is a list of attribute–value pairs. A typical example to be used in WSNs can be described as follows:

type = temperature	// collect temperature values
interval = 20 s	// send one sample every 20 s
duration = 1 h	// send item for the next hour
location = first floor	// interest is limited to the first flood
precision = 0.5	// minimum sensor precision

This interest description specifies exactly which data should be transmitted to the sink node which is expressing this interest. The data sent in response to received interests are also named using a similar naming scheme. In our example, the following data may be generated and transmitted to the sink node:

type = temperature	// this message carries temperature values
location = first floor, room 01.134	// specific sensor location
value = 20°C	// measured value
precision = 0.7	// quality of the sensor readings
timestamp = 01:40:40	// time of measurement

Interests and gradients

Each sink in the network broadcasts its interest through the entire network. Intuitively, the initial interest may be thought of as exploratory. It tries to determine whether there are any sensor nodes that can deliver the requested information. Thus, the initial interest should specify a low data rate. Later, we describe how the desired data rate is achieved by reinforcement. The interest is periodically refreshed by the sink by re-sending the same interest with a monotonically increasing timestamp attribute. This is necessary for two reasons. First, the interest propagation is not reliable and, secondly, the periodic refreshment defines an inherent method for timing out unnecessary data transmissions due to failing or leaving sink nodes.

Every node in the network maintains an interest cache, which can be compared to a data-centric routing table. Each item in the cache corresponds to a distinct interest. As interest entries in the cache do not contain information about the sink, interest state scales with the number of distinct active interests. The interest cache is maintained by processing received interest messages and by expiring unrefreshed entries.

Interest processing also includes the re-broadcasting of the received interest message. Directed diffusion proposes to forward the interest based on a small set of rules, thus 'diffusing' the interest through the network. For example, interest forwarding is suppressed if similar messages have been sent recently. In summary, interest propagation sets up a state in the network to facilitate 'pulling down' data towards the sink.

Data propagation

After receiving the interest message, a node matching the interest will become a source node and starts measuring and forwarding the results along the previously created gradients towards the sink. Intermediate nodes that receive a data message from its neighbors attempt to find a matching interest entry in their cache to forward the message. If no match exists, the data message is silently dropped. This data-propagation scheme allows the support of multiple sources (as gradients are source-independent) as well as multiple sinks (as gradients may point to multiple neighbors).

Reinforcement

For searching available sources and for setting up the gradients, the sink initially diffuses an interest for a low-event-rate notification. Once sources detect a matching interest, they start sending low-rate events, possibly along multiple paths, towards the sink. The reception of such messages informs the sink about active sources. Now, the sink may reinforce selected sources to collect sensor readings at higher rates. This reinforcement is initiated by re-sending the original interest message but with a smaller interval.

In summary, directed diffusion operates on local requirements in form of interests, their diffuse distribution through the network and temporary state maintenance in the form of gradients. Depending on the number of nodes and the number of active interests, the utilization of the network is very low or comparable to other routing approaches. Directed diffusion allows switching between different design choices for the implementation or even their runtime-change to adapt to changing environments. For example, the interest propagation can employ pure flooding, directed flooding based on the location or directional propagation using previously cached data. Data propagation can be implemented in the form of single-path delivery of probabilistic multipath forwarding. Additionally, data-caching or aggregation algorithms can be employed for robustness and data reduction. This example shows again that multiple self-organization mechanisms can be successfully coupled to build a communication protocol that makes efficient use of the available resources. Because there is no addressing scheme needed and the interest information is periodically updated, directed diffusion supports mobility as well as additional or failing nodes.

11.4.2 Mobility support

In its basic form, directed diffusion does not provide protocol-inherent measures to support node mobility. Instead, the timeout schemes for interest propagation and gradients are used to detect node failures (and node movements). A number of enhancements for specific mobility support for diffusion-based WSNs have been proposed by Choksi *et al.* (2002).

First, so-called aggressive diffusion is added to directed diffusion for timeout handling. Instead of relying on static timeouts to reconstruct paths in the routing tree, nodes use a simple heuristic to initiate path reconstruction. If a node does not receive any packet for a given time interval (the original paper suggests 2.5 s), the sink restarts the path-creation process by diffusing interest messages through the network. The additional cost for frequent interest propagations accumulates for potential node mobility.

Secondly, mechanisms known from cellular networks, namely handoff techniques, have been added to the basic directed diffusion algorithm. Actually, the sink continuously observes the signal strength of neighboring nodes. If a given threshold triggers an alarm, a special reinforcement process is initiated. For slowly moving nodes, handoffs are ideal because the sink always joins the optimal point in the existing tree. Nevertheless, because the handoff only selects between immediate neighbors, it is limited to altering the last hop of a potentially long routing path.

For faster-moving nodes, the depicted handoff mechanism will lead to a rapidly repeated path-discovery process. To cope with this problem, a third approach was introduced, which is called proxy. Basically, the proxy scheme introduces special intermediate nodes that serve as anchors for established routing paths. Thus, the path-discovery process can be stopped at these proxies, leading to a more balanced and stable network. Proxies introduce more complexity in to the network, as these systems must keep state information that is increasing with the number of nodes and data types.

Finally, a method called anticipatory diffusion was introduced. This method is based on the observation that in many cases, the relative position of the source to the sink is essential to evaluate the particular interest. This can be called relative interest. If nodes are moving, the costs for setting up this relative interest may become too high to fulfill all system requirements. To alleviate these costs and thereby improve performance for such queries, the notion of anticipatory interest was introduced. Basically, the idea is to set up paths to the sources in advance to a move of the sink. Thus, predictive mobility models are needed to estimate the relative motion in the network.

Choksi *et al.* (2002) discovered that – depending on the mobility model – the enhanced diffusion algorithm outperforms the original directed diffusion. One particular exception is mobility according to the random waypoint model. The main reason is the insufficient prediction quality in random movements.

11.4.3 Energy efficiency

Energy is perhaps the most important resource in WSNs. We outline this importance in our discussion of the network lifetime as the main characteristic of sensor network algorithms in Section 17.3. In the original directed-diffusion algorithm, node energy has not been considered as an optimization parameter. Conversely, Handziski *et al*. (2004) studied possible improvements of the energy efficiency of directed diffusion in particular using passive clustering. The envisioned advantage is the drastically reduced communication overhead compared with diffusion in unstructured networks.

We have already discussed the applicability of clustering mechanisms in flooding approaches. Passive clustering (Kwon and Gerla 2002) has been proposed for on-demand creation and maintenance of clustered structures in WSNs. The applicability of such clustering techniques in the context of directed diffusion has been investigated to analyze possible energy savings.

The basic idea of using passive clustering for directed diffusion is to first build an 'overlay' structure of the network in the form of clusters. Only clusterheads and border nodes are then required for diffusing messages through the network. This assumption holds for interest messages as well as for data transmissions along selected gradients. In short, the basic principle of the proposed optimization to directed diffusion can be described as follows: ordinary (non-clusterhead, non-gateway) nodes do not forward the interest and exploratory data messages that they receive.

Figure 11.10 depicts the working principle of directed diffusion in a clustered environment. As can be seen, the entire network is first clustered into a number of groups of nodes. Each group is coordinated by a clusterhead and the communication between clusters is maintained by gateway nodes. Now, interest flooding and gradient setup can be restricted to intra-cluster communication organized by the clusterheads and inter-cluster communication provided by the gateway nodes.

Based on the described ideas, the following observations have been made by analyzing directed diffusion in networks structured by passive clustering (Handziski *et al*. 2004):

- The number of redundant transmissions during the broadcasting period is significantly reduced (basically shown for interest diffusion).

- The size of all directed-diffusion messages that are broadcasted is increased by the size of the passive clustering header.

- Passive clustering reduces the level of connectivity available to directed diffusion.

In conclusion, it can be shown that the use of clustering techniques to provide structures for interest propagation throughout the entire network can enable large energy savings. The main reason is the reduced flooding among all the available nodes. This behavior finally also leads to improved delivery ratios and lower delay, as fewer messages will collide. An open issue in the proposed algorithm is the tendency of passive clustering to build hot spots, i.e. clusterheads and gateway nodes, which are more frequently used for energy-intensive communication. We will discuss optimized clustering methods in Chapter 12.

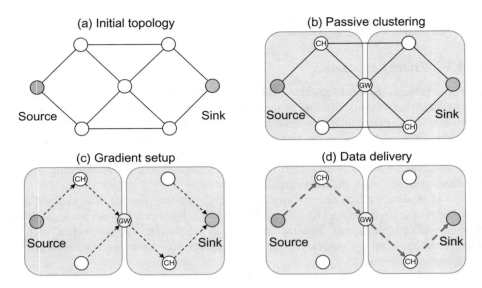

Figure 11.10 Directed diffusion using passive clustering. First, the initial topology is clustered into a number of groups of nodes each coordinated by a clusterhead (CH). Connections between clusters are maintained by gateway nodes (GW)

11.5   Data aggregation

In our discussion of networking aspects in WSNs, we outlined that sensor networks are primarily being developed for collecting information from sensor nodes distributed over a specific geographical region. The nodes participate in obtaining sensor readings and transmit the results eventually to a particular base station for further processing. We have also seen that sensor nodes are strongly resource-restricted – especially in terms of energy and communication bandwidth. Therefore, it might be inefficient to let all the sensor nodes transmit the obtained measurement results directly to the base station over a multi-hop sensor network.

Thus, the following two main reasons motivate the development of methods for combining data particles into higher-quality information, i.e. data-aggregation and data-fusion techniques (Akyildiz *et al.* 2002b; Rajagopalan and Varshney 2006):

- Energy constraints and network congestion – The transmission of data in a WSN is much more energy-expensive compared with local computation efforts. Thus, employing aggregation techniques will help a lot in saving energy at individual sensor nodes. Additionally, as a kind of side effect, the reduced number of transmitted messages towards the base station will also help in reducing the congestion in the network (especially near the base station).

- Redundancy and correlation – As measured sensor data are often generated by nearby nodes, a certain degree of overlap and redundancy is created. Obviously, the efficiency of the network can be improved by removing this redundancy as early as possible in the communication path towards the base station. Additionally, measured data can be expected to be highly correlated, allowing further improvements in information quality by using data-fusion approaches (possibly exploiting further available meta information).

Additionally, the amount of data generated in large sensor networks might become enormous and even too huge for a base station to process. Obviously, we need methods and techniques to address this issue to counteract the resulting scalability problems. Data aggregation is one particular approach to reduce the amount of data being forwarded through a sensor network. In this section, we discuss the principles and objectives of data aggregation and data fusion. Additionally, possible aggregation topologies as proposed in the literature are depicted, focusing on their advantages and limitations.

11.5.1 Principles and objectives

Before introducing the primary techniques and methods of data aggregation techniques, we need to clarify two generally distinct terms–data aggregation and data fusion.

Principles

According to the following definitions, both approaches aim to pre-process data on their way through the network. Nevertheless, the main objectives are quite different – while both approaches may lead to a reduction of the number of transmitted messages.

Definition 11.5.1 Data aggregation – Data aggregation is the process of combining multiple information particles (in our scenario, multiple sensor messages) into a single message that is representing all the original information. Examples of aggregation methods are statistical operations like the mean or the median.

According to this definition, we can write $m_A = aggregate(m_0, m_1, \ldots, m_n)$, with m_i being the input messages, $aggregate()$ being a particular aggregation function, and m_A being the aggregated outcome. Obviously, the quality of the information is reduced by the aggregation, as only statistical correctness can be achieved. This is a main difference from compression techniques.

Definition 11.5.2 Data fusion – Data fusion is the process of annotating received information particles with meta information. Thus, data from different sources are combined to produce higher-quality information, e.g. by adding a timestamp or location information to received sensor readings.

Thus, the process of data fusion can be expressed (using, again, a number of available information fragments m_i) as $m_F = fusion(m_0, m_1, \ldots, m_n, a_0, a_1, \ldots, a_p)$, with a_j being additional meta information locally available to the fusion function, $fusion()$ being a particular processing element and m_F being the resulting fused outcome. Thus, data fusion can be thought of as an extension of data aggregation (assuming no available meta information). Nevertheless, data fusion can work also on a single input message and, therefore, result in increased message sizes or numbers.

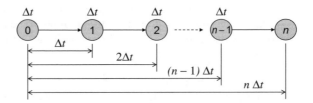

Figure 11.11 Aggregation delay introduced by a number of linearly ordered aggregation processes; in the worst case, aggregation will introduce an artificial network delay of $n \times \Delta t$

In the rest of this section, we will use the term 'aggregation' for both processed data aggregation and data fusion. As we will not go into detail on which aggregation or fusion functions can be used – the development of which is inherently challenging, as they are highly application-dependent – we will concentrate on general objectives, limitations and topological constraints of aggregation in WSNs. Nevertheless, it should be mentioned that the data accuracy strongly depends on the employed aggregation function. For example, filtering allows the removal of spikes and obvious errors in the measurement process.

Objectives and limitations

In order to aggregate a given number of messages, these messages must be available to a sensor node. While this observation sounds simple, this requirement introduces the most challenging issues for data aggregation. Figure 11.11 depicts this problem. Each aggregating node will introduce an artificial delay, as messages need to be queued up for aggregation. In a network with n aggregating nodes to be traversed by a particular message towards the sink, the worst-case delay introduced by queuing and aggregation processes, i.e. an artificial delay needed only for aggregation, is $n \times \Delta t$.

As Δt defines both the artificial delay as well as the possible aggregation ratio, this value must be carefully tuned according to current the application and scenario demands (if Δt increases, the number of messages to be aggregated will increase while the delay is increasing as well and vice versa). The primary objective is to optimally balance between the introduced end-to-end latency and aggregation ratio.

This objective can be translated to the energy efficiency of individual nodes and the network lifetime in general (as discussed in Chapter 17). A data-aggregation scheme is considered energy-efficient if it maximizes the functionality of the network. Issues to be regarded are energy constraints, QoS awareness, reliability and security (Rajagopalan and Varshney 2006).

A number of trade-offs needs to be looked at in the context of data aggregation. In particular, the trade-offs related to energy, accuracy and latency need to be mentioned.

Solis and Obraczka (2006) explored in-network aggregation as a power-efficient mechanism for collecting data in sensor networks. In this context, they also investigated in-network aggregation trade-offs for data aggregation in sensor networks; the focus was on trade-offs between energy efficiency, data accuracy and freshness in particular. Using analytical and simulation models, it was shown that timing models play a significant role in the accuracy and freshness of data aggregation. By carefully selecting when to aggregate and forward data, cascading timers (a model introduced by the authors of the mentioned article) achieves considerable energy savings while maintaining data freshness and accuracy.

Similarly, Boulis *et al.* (2003a,b) studied the energy–accuracy trade-offs for periodic data-aggregation problems in sensor networks. This work was motivated by the finding that aggregation techniques need to be evaluated focusing on a new dimension–accuracy (which is usually not the first parameter to be optimized in WSNs). This dimension can be exploited to produce more energy-efficient algorithms. The less accuracy required, the more energy can be saved using proper algorithmic techniques.

11.5.2 Aggregation topologies

Besides the specifically employed aggregation functions, data aggregation in WSNs mainly differs according to the underlying aggregation topology. While, in general, arbitrary topologies can be thought of (we will discuss this aspect later on, in Section 14.4), the aggregation accuracy can be strongly increased if the aggregation topology – which actually represents an overlay structure to the underlying network topology – is known in advance and explicitly exploited for efficient data aggregation.

In the literature, multiple aggregation principles are discussed, which can basically be reduced to the following three primary aggregation topologies: chain-based aggregation, tree-based aggregation and grid-based (or cluster-based) aggregation.

Chain-based aggregation

The main idea of chain-based data aggregation is to reduce the communication effort by transmitting the measured data only to the closest neighbor, which will then collect and aggregate the sensor information. In order to achieve this method effectively, the network needs to be arranged in the form of (possibly multiple) single chains towards the sink node. An example is depicted in Figure 11.12. As can be seen, four aggregation chains are created and data are delivered along these chains towards the sink node.

Lindsey *et al.* (2001, 2002) presented a chain-based data-aggregation protocol to improve the energy × delay product, which they named Power-Efficient Data-Gathering

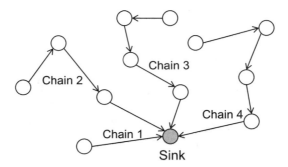

Figure 11.12 Chain-based aggregation in a sensor network. The overall topology is organized in the form of chains towards the sink node. Junctions are not allowed in this technique. All nodes are allowed to aggregate data flowing towards the sink node

Protocol for Sensor Information Systems (PEGASIS). In PEGASIS, nodes are organized into a linear chain for data aggregation. The nodes can either form a chain by employing a greedy algorithm or the sink can determine the chain in a centralized manner. A prerequisite for greedy chain formation is the (partial) knowledge of the network topology available at all participating nodes. Based on this information, the farthest node from the sink will initiate the chain formation. In each step, the closest node towards the sink is selected as the next chain member. Lindsey *et al.* (2002) called this method the chain-based binary scheme.

Obviously, the effectiveness of chain-based data-aggregation techniques strongly depends on the construction process. First, the construction process itself needs to be energy-efficient and scalable, e.g. in terms of necessary global topology information. Secondly, the outcome of the construction process, i.e. the aggregation chain, needs to be optimal in terms of path length and resource utilization. Such chain-construction schemes have been analyzed and a number of approaches can be found in the literature. For example, Du *et al.* (2003) developed an energy-efficient chain-construction algorithm which uses a sequence of insertions to add the least amount of energy consumption to the whole chain. It consumes less transmission power compared with the closest-neighbor algorithm.

Tree-based aggregation

Tree-based data aggregation follows a different approach compared with chain-based aggregation. Instead of building non-overlapping chains of nodes, a tree is being created in which data aggregation is performed at junctions of the logical tree structure. An example of such a tree structure is shown in Figure 11.13. Using the same network topology as presented for chain-based aggregation, a highly different logical tree-based aggregation network is created. Aggregation is no longer performed by all nodes in the network, but by dedicated aggregating nodes (junctions in the tree).

Such tree-based aggregation has several advantages compared with chain-based aggregation. First, a number of routing protocols already construct spanning trees in the network, which can easily be exploited for the aggregation functions. Additionally, the overall end-to-end latency can be reduced, as the path length is usually reduced in a tree compared with

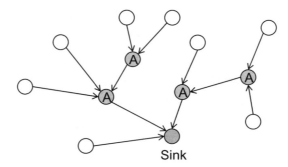

Figure 11.13 Tree-based aggregation in a sensor network. The topology is organized in the form of a loop-free spanning tree with the sink node at its root. Data aggregation takes place at all junctions (indicated by an A)

chains. Again, the main criterion for evaluating the efficiency of tree-based aggregation is the construction process.

Depending on the available topology information, well known algorithms for calculating the minimum spanning tree can be used. Conversely, Ding *et al.* (2003) proposed an energy-aware distributed heuristic to construct and maintain a data-aggregation tree in sensor networks. The algorithm is initiated by the sink, which broadcasts a control message. Thus, the sink performs the role of the root node in the aggregation tree.

Grid-based aggregation

A third structure for aggregation in WSNs is grid-based aggregation. Essentially, this technique can also be named cluster-based aggregation, as a number of dedicated nodes are selected to perform aggregation and all the other nodes maintain pointers to one of these aggregation nodes. The basic principles are shown in Figure 11.14. In this example, three nodes have been selected to perform data aggregation, while all the other nodes just send their sensor readings to one of them. The main advantage of grid-based aggregation is its simplicity. The selection of a given number of dedicated nodes can be performed in a highly distributed way (please refer to Chapter 12 for further information on clustering). Nevertheless, this advantage comes with some increased costs for communication, as, usually, a cluster structure cannot be as efficient as a minimal spanning tree.

Vaidyanathan *et al.* (2004) proposed two data-aggregation schemes which are based on dividing the region monitored by a sensor network into several grids. In grid-based data aggregation, a set of sensors is assigned as data-aggregation nodes in fixed regions of the sensor network. The sensors in a particular grid transmit the data directly to the dedicated data-aggregation nodes of that grid. Additionally, a second scheme similar to the tree-based approach is used within a grid. Vaidyanathan *et al.* (2004) called the resulting two-stage aggregation technique 'hybrid data aggregation', which, according to the presented measurements, increases throughput, decreases congestion and saves energy.

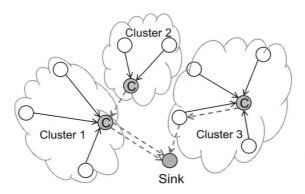

Figure 11.14 Grid-based aggregation in a sensor network. A number of grid nodes organize the sensor network in the form of multiple clusters. The grid nodes (indicated by a C) are responsible for data aggregation. Aggregated data forwarded by a grid node cannot be further aggregated

11.6   Conclusion

Data-centric networking has a number of advantages compared with address-based routing. Basically, the objectives behind the development of data-centric techniques are motivated by scalability issues in typical WSNs. Additionally, these sensor networks have a number of specific characteristics that cannot be found in other ad hoc networks. In particular, sensor nodes usually collect redundant sensor information, as sensor ranges are overlapping, e.g. for higher redundancy. Thus, the specific sensor does not need to be identified as long as the received sensor information can be associated to a given region in space and time. Additionally, sensor information typically describes itself, e.g. as a temperature reading from a particular geographic area.

In this chapter, we outlined a number of data-centric networking approaches, starting with (probabilistic) flooding techniques. Even though such mechanisms are hard to control and they tend to produce duplicated messages, their scalability is fascinating, as only a little or even no state information needs to be maintained. Similarly, agent-based techniques and directed-diffusion approaches follow the same characteristics while reducing the overhead (and unnecessary redundancy) by employing simple approximations and distributed path-discovery mechanisms.

The properties of data-centric networking approaches can be studied best by summarizing their main advantages and drawbacks:

- Advantages – The most obvious advantage of most data-centric communication techniques is their simplicity. The necessary algorithms are straightforward and easy to implement, even on low-resource sensor nodes. Additionally, all these algorithms inherently allow distributed operation without complex coordination frameworks. Some of the presented approaches do not require any coordination at all. The main result is a great reduction in necessary state maintenance compared with address-based routing techniques.

- Disadvantages – The mentioned advantages come with some associated costs. For example, there is a non-zero probability of losing messages during their transmission. While such behavior is common in wireless networks, e.g. collisions are dominant factors for unreliable transmissions, probabilistic schemes inherently increase this loss probability. Assuming a given degree of redundancy, this problem may be alleviated by the sensor network itself. Additionally, the non-zero probability to observe duplicated messages may contribute to network congestion.

Depending on the scenario and the employed algorithms, data-centric networking is a potential source of unnecessary network load essentially if a high degree of redundancy is needed. Therefore, a careful implementation and optimized configuration of system parameters is necessary. Furthermore, data-aggregation techniques can be exploited to improve the behavior of the communication system. Nevertheless, aggregation introduces additional complexity into the network. Again, the system needs to adapt its parameters to find an optimized balance between resource consumption (e.g. bandwidth and energy) and performance measures (e.g. latency and redundancy).

12

Clustering

In the previous chapters, we often mentioned clustering methods as powerful building blocks for distributed operations. Basically, clustering refers to a structuring process applied to unstructured data (Hartigan 1975). Thus, a loose definition of clustering could be 'the process of organizing objects into groups whose members are similar in some way'. In effect, clustering techniques are able to group items according to specific measures and algorithms.

While commonly used for data management and pattern recognition, clustering techniques are employed in all domains of computer science today. Focusing on communication systems, clustering methods can be exploited for greatly improved scalability of communication methods, especially in ad hoc and sensor networks. The best known domains for efficient usage of clustering techniques in SANETs are the following two:

- Optimized resource utilization – Clustering techniques have been successfully used for time and energy savings. These optimizations essentially reflect the usage of clustering algorithms for task and resource allocation.

- Improved scalability – As clustering helps to organize large-scale unstructured ad hoc networks in well defined groups according to application-specific requirements, tasks and necessary resources can be distributed in this network in an optimized way.

Finally, clustering provides techniques and measures to reduce the size of the network to be controlled by grouping nodes into clusters which are of manageable size. This structuring leads obviously to reduced state information needed for coordination among the nodes in the network, e.g. reduced routing tables or synchronization demands in MAC protocols. In turn, this reduced state information ensures greatly improved scalability of any state-based algorithm in MANETs and WSNs.

At the same time, the efficiency of the optimization techniques, e.g. routing algorithms, will be similar to the application in unstructured networks, if and only if 'relevant' groups of nodes can be found. Therefore, the primary difficulty is how to find these groups. In our brief discussion of clustering in general and particular clustering techniques, we will see that this search for optimizing the group forming is highly application-dependent. Usually,

Self-Organization in Sensor and Actor Networks Falko Dressler
© 2007 John Wiley & Sons, Ltd

the measure used for associating nodes to a cluster is more important than the clustering algorithm itself. In the literature, a number of basic clustering algorithms are described which can easily be used for varying problems (Berkhin 2002).

In this chapter, we provide a brief introduction to clustering. The objective is to introduce the basic mechanisms and algorithms. As will become obvious, the main criterion for clustering schemes is the similarity measure used to merge nodes or objects to separate clusters. Additionally, we discuss some selected examples of clustering algorithms in ad hoc and sensor networks to show the applicability for improving the scalability. Hereby, we focus on possible energy savings in wireless communication.

12.1   Principles of clustering

Clustering deals with grouping objects according to a given measure. Thus, it is about bringing a higher degree of structure to an unstructured set of objects. Formally, clustering can therefore be considered as unsupervised learning. The final objective of clustering objects into groups is to sort them in such a way that 'similar' objects are put into the same group while 'dissimilar' objects go to different collections. The objective of clustering is therefore the maximized similarity within a cluster and the minimized similarity between different clusters.

The definition of this similarity measure is itself a complex process. It is rather obvious that no absolute best criterion, i.e. similarity measure, will be optimal for any structuring problem. Thus, the definition of this measure is application dependent and may easily lead to suboptimal solutions if not carefully defined.

Figure 12.1 depicts the formation of groups of objects, In this example, 12 objects, e.g. network nodes, are shown that are grouped into three clusters. Obviously, multiple solutions can be found, depending on the clustering algorithm and the similarity measure.

In the last few decades, clustering has been successfully applied to a huge number of problems. Examples are marketing, e.g. finding groups of customers with similar behavior

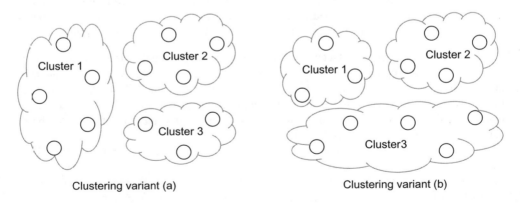

Clustering variant (a)   Clustering variant (b)

Figure 12.1 Clustering of objects into groups; obviously, different solutions can be created, depending on the algorithm and, most importantly, on the similarity measure

given a large database of customer data containing their properties and past buying records, biology, e.g. classification of plants and animals given their features, city planning, e.g. identifying groups of houses according to their house type, value and geographical location, and the world wide web, e.g. document classification – clustering weblog data to discover groups of similar access patterns. In sensor networks, routing optimization, resource and task allocation and energy-efficient operation need to be mentioned in particular.

In the chapters on networking aspects of self-organizing systems, we have already discussed clustering in the context of data-centric networking. Handziski *et al.* (2004) employed passive clustering techniques to improve the energy efficiency of directed diffusion. Similar passive clustering schemes have been used by Kwon and Gerla (2002) for efficient flooding in ad hoc networks. The PEGASIS system (Lindsey and Raghavendra 2002) deals with power efficient data gathering in sensor networks using clustering techniques.

12.1.1 Requirements and classification

Before starting with a brief classification of clustering methods, we need to outline the basic requirements and demands on clustering techniques. As discussed before, clustering is regarded to optimize the resource utilization and to improve the scalability of algorithms in distributed and self-organizing systems such as ad hoc and sensor networks. These advantages can only be achieved if the clustering itself is not too expensive in terms of computational costs and communication requirements. Therefore, we can summarize the following key requirements on the clustering algorithms.

Requirements

First of all, the clustering technique must be scalable. While this might be an obvious observation, not all clustering algorithms fulfill this requirement if employed in sensor networks. The scalability greatly depends on the clustering process and the required context information, i.e. state or topology. Based on the following discussion of particular clustering algorithms, this scalability issue will become clearer.

Additionally, clustering techniques are required to deal with different types of attributes; thus, the more generic an algorithm is, the more flexible is its application in a particular scenario. The generic application for different clustering objectives includes the minimization of necessary domain knowledge in order to determine input parameters. Often, single objects cannot easily be grouped in to a specific cluster, e.g. because the similarity measure cannot provide any relationship. Thus, a further requirement is the ability of the algorithm to deal with noise and outliers. For improved scalability, the clustering technique must be insensitive to the order of input records and be able to generate clusters with arbitrary shape.

Classification

In general, clustering techniques can be classified in to two dimensions according to the used similarity measure and to the basic behavior of the algorithm. First, the similarity measure needs to be addressed. In this dimension, we can distinguish between distance-based clustering and conceptual clustering.

- Distance-based clustering – In distance-based clustering, the similarity between two objects is represented by a distance measure. Thus, a number of objects belongs to the same group, i.e. it is similar according to the measure, if they are close in terms of the distance between any pair of those objects.

- Conceptual clustering – In contrast, conceptual clustering refers to particular characteristics that must be fulfilled by objects of the same group. Thus, in this case, similarity defines a concept common to all objects belonging to the same cluster.

In the second dimension, the organization of the clustering process itself needs to be analyzed. We can distinguish between centralized algorithms and distributed self-organized variants that basically differ in the necessary state or topology information.

- Centralized – Based on global state information, i.e. the knowledge about all objects in the field and their characteristics, a centralized optimization process can be initiated which will always lead to a best solution for the clustering. In the case of multi-objective optimizations, perhaps an approximation is needed, which is feasible according to the available global knowledge.

- Distributed or self-organized – In distributed clustering, which can also be regarded as self-organized clustering, clusters are usually formed dynamically. First, a (potential) clusterhead is selected, e.g. based on some election algorithm known from distributed systems. Afterwards, the group membership and resource management are maintained by the clusterhead. This process may include a dynamic reallocation of the clusterhead functionality. In summary, such clustering represents a distributed (multi-dimensional) optimization process.

Roughly, clustering techniques can be classified into the following four groups. First, exclusive clustering aims at grouping nodes into exclusive domains such that every node belongs to exactly one cluster. The best known example of exclusive clustering is k-means (Hartigan and Wong 1979; MacQueen 1967). Secondly, overlapping clustering allows nodes to simultaneously belong to multiple clusters. Usually, conceptual clustering methods belong to this category. Additionally, fuzzy objective function algorithms lead to overlapping clusters, e.g. Fuzzy C-Means (FCM) (Bezdek 1981). Thirdly, hierarchical clustering follows the idea of identifying each object initially as a single cluster. Then, in multiple iterations, the two nearest clusters are merged into a bigger one. After a few iterations, the algorithm reaches the final cluster structure (Johnson 1967). Fourth, and finally, probabilistic clustering is used on a completely probabilistic approach. This allows a simple (while effective) load balancing as used, for example, in LEACH (Heinzelman et al. 2000). More detailed classifications of clustering schemes can be found in the literature (Berkhin 2002; Hartigan 1975).

Typical problems

Every application using clustering techniques must be aware of a number of typical problems and limitations of clustering. Amongst others, the most challenging issue is the definition of similarity measures. For distance-based clustering techniques, which are widely used in the context of ad hoc and sensor networks, the effectiveness of the clustering method primarily depends on the definition of the distance measure. For example, the used 'distance' can be the approximated physical distance between nodes in a geographical area, the difference in the available energy resources, or the communication range of neighboring sensor nodes. If an obvious distance measure does not exist, we must 'define' it, which is not always easy, especially in multi-dimensional spaces.

Furthermore, current clustering techniques do not address all possible application requirements adequately (and concurrently) and dealing with a large number of dimensions and a large number of data items can be problematic because of the time complexity. Finally, clustering algorithms will not always be able to find the optimal configuration of clusters. In many cases, configurations can be invented that represent a converged state for the clustering algorithm but do not have the minimum distortion.

12.1.2   k-means

k-means is one of the simplest clustering distributed algorithms (Hartigan and Wong 1979; MacQueen 1967). k-means can be thought of as self-organizing, as it prevents the storage of any state information. It represents a quite simple unsupervised learning algorithm. The objective of k-means is to cluster a given number of objects to k disjunct groups.

The main idea is to define k centroids–one for each cluster. Because the initial selection of the centroids strongly affects the final clustering, the centroids should be placed carefully. Usually, the best choice is to place the centroids as far away as possible from each other. Alternatively, multiple runs of k-means with random start configurations can be used to improve the clustering quality.

Based on the initial configuration with k centroids, k-means iteratively updates the clustering until the algorithm converges, i.e. no further changes can be determined. Each iteration includes the following algorithm steps. Each data point (object) is associated with the nearest centroid. When no point is pending, a first clustering is finished with k disjunct clusters. Based on this configuration, the k centroids are recalculated as the barycenter of each cluster. Then, the next iteration can be started to improve the clustering.

This iterative process will update the centroids and the associated clusters in each step until no further improvement can be achieved. The k-means algorithm aims at minimizing an objective function–in this case, a squared error function, depicted in Equation 12.1. In this formula, $\left\| x_i^{(j)} - c_j \right\|^2$ is a chosen distance measure between a data point $x_i^{(j)}$ and the cluster center c_j. This objective function is an indicator of the distance of the n data points from their respective cluster centers.

$$J = \sum_{j=1}^{k} \sum_{i=1}^{n} \left\| x_i^{(j)} - c_j \right\|^2 . \tag{12.1}$$

The complete procedure of k-means is depicted in Algorithm 12.1. k-means has already been successfully used in the context of wireless ad hoc networks by Fernandess and Malkhi (2002) in order to limit the amount of routing information stored and maintained at individual hosts. The k-clustering framework allows division of the wireless network into non-overlapping sub -networks in which every two wireless nodes in a sub-network are, at most, k hops from each other.

Algorithm 12.1 k-means clustering algorithm

Require: Number of centroids k.
Ensure: Exclusive clustering of n objects into k disjunct clusters.

1: Initialize centroids c_j $(j = 1, 2, \ldots, k)$, e.g. by randomly choosing the initial positions c_j or by randomly grouping the nodes and calculating the barycenters
2: **repeat**
3:    Assign each object x_i to the nearest centroid c_j such that $\left\| x_i^{(j)} - c_j \right\|^2$ is minimized
4:    Recalculate the centroids c_j as the barycenter of all $x_i^{(j)}$
5: **until** Centroids c_j have not moved in this iteration

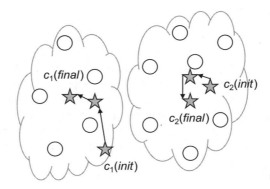

Figure 12.2 Demonstration of the k-means algorithm. Starting with two centroids $c_j(init)$, the final (converged) clustering can be found in two iterations with the new centroids $c_j(final)$

To conclude the discussion of k-means, a brief example should help in understanding the iteratively converging nature of k-means. Figure 12.2 depicts a number of nodes, two moving centroids c_j and the final cluster structure.

12.1.3 Hierarchical clustering

Hierarchical clustering aims at iteratively merging two disjunct clusters into a single larger one. Initially, this method assumes that each object represents a cluster itself. Thus, in each iteration, the number of clusters is reduced by one. This kind of hierarchical clustering is called agglomerative because it merges clusters iteratively. In theory, there is also a divisive hierarchical clustering, which starts from a single cluster containing all objects and then iteratively divides this set into smaller pieces. Nevertheless, divisive hierarchical clustering is rarely used in practice.

According to Johnson (1967), hierarchical clustering starts with a set of N items to be clustered and an $N \times N$ distance (or similarity) matrix. In an iterative procedure, as depicted in Algorithm 12.2, it searches for the closest (most similar) pair of clusters and

---

Algorithm 12.2 Hierarchical clustering algorithm

Require: N items to be clustered and a $N \times N$ distance matrix.
Ensure: Complete hierarchical tree (for k clusters, the $k - 1$ longest links have to be cut).

1: Assign each item to a cluster (N items result in N clusters each containing one item)
2: Let the distances between the clusters be the same as the distances between the items
    they contain
3: **repeat**
4:    Find the closest pair of clusters and merge them into a single cluster
5:    Compute distances between the new cluster and each of the old clusters
6: **until** All items are clustered into a single cluster of size N

---

merges them into a single cluster. Then, it computes distances (similarities) between the new cluster and each of the old clusters to update the similarity matrix. These two steps are repeated until all items are clustered into a single group of size N. In order to generate k clusters, the algorithm can be stopped after $N - k$ iterations.

Again, the clustering behavior primarily depends on the computation of the distances (similarities). Hierarchical clustering depends on the distance estimation between multiple clusters. Three possible methods have been proposed for this computation. In single-linkage clustering, the distance between two clusters is approximated by the shortest distance between any member of one cluster to any member of the other cluster. Therefore, this is also called the minimum method. In contrast, complete-linkage clustering the maximum, i.e. the greatest distance between members of both clusters, is considered. Finally, average-linkage clustering uses the average distance from any member of one cluster to any member of the other cluster.

The working principle of single-linkage clustering is depicted in Figure 12.3. Six nodes are iteratively merged into disjunct clusters. In each iteration, the minimum distances between members of all established clusters are computed and the two closest clusters are merged accordingly. The algorithm terminates if the whole data set is merged into a single cluster.

The main weakness of agglomerative clustering methods is the low scalability. The computational complexity is at least $O(n^2)$, where n is the number of objects.

12.2 Clustering for efficient routing

A wide range of methods have been developed that employ clustering techniques to improve the scalability and the energy efficiency in ad hoc and sensor networks. First, a number of general-purpose clustering algorithms for use in wireless ad hoc networks have been proposed recently (Banerjee and Khuller 2001; Chen *et al.* 2004; Fernandess and Malkhi 2002). Some of those specifically focus on MANETs, such as the adaptive clustering approach by Lin and Gerla (1997) or the stability-aware cluster-routing protocol by Chiu *et al.* (2003). Additionally, optimized task allocation in sensor networks has been investigated (Younis *et al.* 2003). In the context of this section, two approaches for energy-efficient communication in WSNs are discussed in more detail–Low-Energy Adaptive Clustering

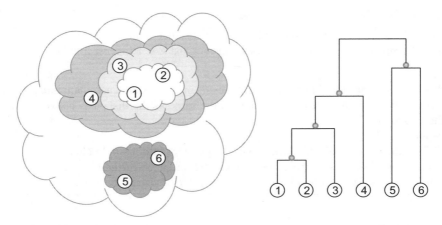

Figure 12.3 Single-linkage clustering iteratively merges the two clusters with the minimum distance between any cluster items; on the left-hand side, the cluster structure is depicted, while on the right-hand side, the emerging spanning tree is shown

Hierarchy (LEACH) (Heinzelman *et al*. 2000) and Hybrid Energy-Efficient Distributed Clustering Approach (HEED) (Younis and Fahmy 2004). Both aim at reducing the energy load in sensor networks in order to increase the overall network lifetime.

12.2.1 Low-Energy Adaptive Clustering Hierarchy

In the context of WSNs, Low-Energy Adaptive Clustering Hierarchy (LEACH) (Heinzelman *et al*. 2000) is a well known clustering-based protocol that utilizes randomized rotation of local cluster base stations (clusterheads) to evenly distribute the energy load among the sensors in the network. In this context, energy load is defined as the utilization in terms of wakefulness plus needed transmission energy. LEACH uses localized coordination to enable scalability and robustness for dynamic networks, and incorporates data aggregation into the routing protocol to reduce the amount of information that must be transmitted to the base station. The primary goals are to equally distribute the energy load to all available nodes and to enhance the lifetime of the entire network. The main capabilities of LEACH can be summarized as follows:

- Self-organization – LEACH is a self-organizing, adaptive clustering protocol that uses randomization to distribute the energy load evenly among the sensors in the network. All nodes organize themselves into local clusters, with one node acting as the local base station or clusterhead.

- Energy distribution – Randomized rotation of the high-energy clusterhead position among the various available sensor nodes is included in order to not drain the battery of a single-sensor node too much.

- Data aggregation – The protocol performs local data aggregation to 'compress' the amount of data being sent from the clusters to the base station, thus further reducing energy dissipation and enhancing system lifetime.

Operating principles

In LEACH, the cluster maintenance and all communications are achieved and coordinated by the clusterheads. Thus, these nodes must perform essentially more computational and communication efforts compared with the other nodes, and, in effect, draw more energy from their batteries. In order to spread this energy usage over multiple nodes, the clusterhead nodes are not fixed, but relocated over the time. More precisely, the clusterheads are self-elected at different time intervals. The decision to become a clusterhead also depends on the amount of energy left at the node. In this way, nodes with more energy remaining will perform the energy-intensive functions of the network.

Basically, the algorithm works as follows. A given number of sensor nodes elect themselves to become local clusterheads in a predefined time schedule with a certain probability. Afterwards, the clusterhead nodes broadcast their status to the other sensor nodes in the network. Each sensor node determines to which cluster it wants to belong by choosing the clusterhead that requires the minimum communication energy, i.e. selecting the closest clusterhead. Figure 12.4 depicts the resulting clustering at two different points in time. At time t_1, three nodes elect themselves to become clusterhead. The resulting cluster forms according to the proximity of the surrounding nodes to the closest clusterhead. A quite different clustering becomes visible at time $t_1 + d$.

The operation of LEACH is broken up into rounds, as depicted in Figure 12.5, in which each round begins with an advertisement phase in which the clusterheads are self-elected. Then, in the cluster setup phase, the clusters are formed. Finally, a steady-state phase follows in which data transfer can occur. The individual phases will be discussed below.

Advertisement phase   In the advertisement phase, the current clusterhead assignment is decided. This self-election process works as follows. At the beginning of a round, each node

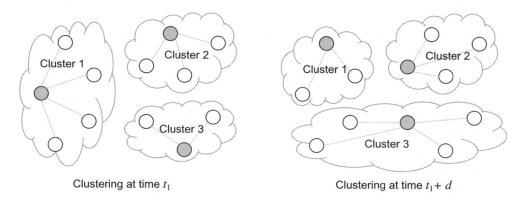

Clustering at time t_1 Clustering at time $t_1 + d$

Figure 12.4   Clustering using LEACH; shown are the clusters at two different time points

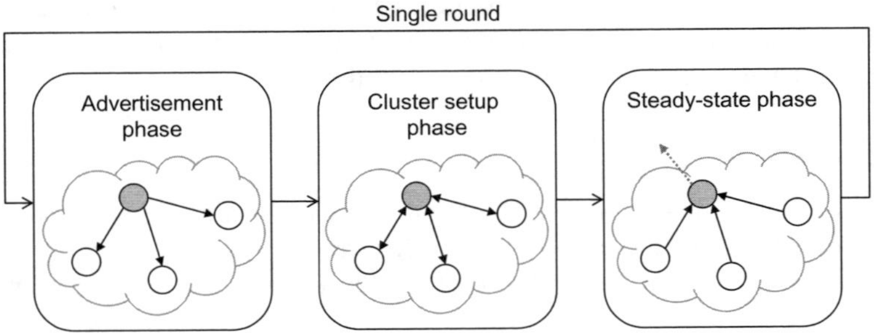

Single round

Figure 12.5 Operation principle of LEACH; the algorithm works in rounds, each divided into three phases: advertisement, setup and steady-state

decides whether or not to become a clusterhead. This decision process depends on only a single a priori determined value–the desired percentage P of clusterheads in the network. Thus, P represents a measure of the number of clusters to be created. Each node elects itself to be a clusterhead with a certain probability. For this, each node chooses a random number n in the interval $[0, 1]$. If this number is less than a threshold $T(n)$, the node will become a clusterhead for the current round r. This threshold is calculated according to Equation 12.2, where G denotes the set of nodes that have not been clusterhead in the last $\frac{1}{P}$ rounds.

$$T(n) = \begin{cases} \dfrac{P}{1 - P \times (r \bmod \frac{1}{P})} & \text{if } n \in G \\ 0 & \text{otherwise.} \end{cases} \qquad (12.2)$$

Therefore, in the first round, i.e. round 0, each node has the same probability P for becoming a clusterhead. In the following rounds, the probability must be increased for the remaining nodes because there are fewer nodes left for becoming a clusterhead in round $r + 1$.

After deciding on the clusterhead position, each node that has elected itself a clusterhead for the current round broadcasts an advertisement message to all other nodes. All clusterheads transmit their advertisement using the same (maximum) transmission energy – the non-clusterhead nodes must keep their receivers on during this phase in order to receive these advertisements. Finally, each non-clusterhead node decides about the cluster to which it will belong for the current round based on the measured received signal strength of the advertisement. If two or more clusterheads seem to be the same distance from a particular node, this node randomly chooses between these clusterheads. Therefore, the communication energy towards the currently selected clusterhead is minimized.

Cluster setup phase During the cluster setup phase, each node must inform the chosen clusterhead node that it will be a member of the cluster by transmitting this information back to the clusterhead. According to the received messages, the clusterhead nodes establish the necessary state information about the created cluster. Additionally, the clusterheads might create a Time Division Multiple Access (TDMA) schedule for efficient communication within the cluster in terms of collision avoidance and QoS support. This TDMA schedule is broadcasted back to all nodes in the local cluster.

Steady-state phase   Finally, nodes can use the created-cluster structure for energy-efficient communication. Assuming the worst case that nodes always have data to send, they send it during their allocated transmission time slots to the clusterhead. This transmission uses a minimal amount of energy, as chosen, based on the received strength of the clusterhead advertisement. If no data are available for transmission, non-clusterhead nodes can skip their transmission window. Additionally, these nodes can completely turn off their radio until the node's allocated transmission time. Both mechanisms further foster the minimized energy dissipation of these nodes.

On the other hand, all clusterhead nodes must keep their receiver turned on to receive all the data from the nodes in their local cluster. The clusterhead is also responsible for forwarding messages through the entire network, e.g. using a general purpose routing protocol among the clusterheads. After a certain time, which is a priori determined, the next round begins.

Discussion

Heinzelman *et al.* (2000) have shown in simulations that LEACH can achieve a reduction by as much as a factor of eight in energy dissipation compared with conventional routing protocols. In addition, LEACH is able to distribute energy dissipation evenly throughout the sensors, doubling the useful system lifetime of the networks that were simulated.

In summary, it can be said that LEACH operates according to locally taken decisions that are broadcasted to all neighboring nodes. Based on the local decision and some local communication, the nodes organize themselves into a larger compound for energy-aware operation. Therefore, LEACH combines multiple mechanisms for self-organization: probabilistic algorithms (choice of becoming clusterhead) and neighborhood information (setup of the clusters). This combination, together with location information (the clusters are used for efficient data communication), LEACH provides an optimized behavior for communication in ad hoc networks based on self-organization methods. Obviously, in order to minimize the overhead, the steady state phase should be carefully compared to the setup phase. Mobility is supported by LEACH, whereas new nodes have to be synchronized to the current round. Node failures may lead to fewer clusterheads' being elected than desired.

12.2.2   Hybrid Energy-Efficient Distributed Clustering Approach

Hybrid Energy-Efficient Distributed Clustering Approach (HEED) (Younis and Fahmy 2004) represents a direct successor of LEACH. HEED also aims at using clustering techniques for energy-efficient communication in WSNs and SANETs.

According to Younis and Fahmy (2004), the clustering problem for the HEED algorithm can be described as follows. Assume that n nodes are dispersed in a field. The goal is to identify a set of clusterheads which cover the entire field. Each node v_i, where $1 \leq i \leq n$, must be mapped to exactly one cluster c_j, where $1 \leq j \leq n_c$, and n_c is the number of clusters ($n_c \leq n$). A node must be able to directly communicate with its clusterhead without necessary multi-hop routing, i.e. in a single hop.

Clusterheads can use a common routing protocol to compute inter-cluster paths for multi-hop communication between clusters or to a central base station.

The following requirements have been observed during the development of HEED:

- Clustering is completely distributed. Each node independently makes its decisions based only on local information.

- Clustering terminates within a fixed number of iterations (regardless of the network diameter).

- At the end of each clustering process, each node is either a clusterhead or not a clusterhead (which is referred to as a regular node) that belongs to exactly one cluster.

- Clustering is efficient in terms of processing complexity and message exchange.

- Clusterheads are well-distributed over the sensor field, and have relatively high average residual energy compared with regular nodes.

HEED uses three protocol phases to set up the cluster structure: initialize, cluster setup and finalize. In comparison to LEACH, the most important difference is the exploitation of the currently available remaining energy at each node for the (still probabilistic) self-election of clusterheads. Before a node starts executing HEED, it sets its probability of becoming a clusterhead, CH_{prob}, according to Equation 12.3, where C_{prob} denotes the initial amount of clusterheads among all n nodes. It is only used to limit the initial clusterhead announcements. $E_{residual}$ is the estimated current residual energy in the node, and E_{max} is a reference maximum energy (corresponding to a fully charged battery), which is assumed to be identical for all nodes.

$$CH_{prob} = C_{prob} \times \frac{E_{residual}}{E_{max}}. \tag{12.3}$$

The HEED approach is called hybrid, as clusterheads are probabilistically selected based on their residual energy, and nodes join clusters such that communication cost is minimized. The objective function for the reduced communication cost is the average minimum reachability power, defined as the mean of the minimum power levels required by all nodes within the cluster range to reach the clusterhead.

Similarly to LEACH, HEED exploits the availability of multiple transmission power levels at typical sensor nodes. HEED terminates in a constant number of iterations (like LEACH), independently of the network diameter. Additionally, it guarantees good clusterhead distribution and does not assume uniform energy consumption for clusterheads.

Simulation results demonstrate that HEED prolongs network lifetime (Younis and Fahmy 2004). The operating parameters, such as the minimum selection probability and network operation interval, can be easily tuned to optimize resource usage according to the network density and application requirements.

12.3 Conclusion

In this chapter on clustering techniques, we discussed the main principles and objectives of clustering algorithms. Basically, clustering can be used in various domains of computer science. As it generally represents an unsupervised learning process, it is specifically applicable to coordination and communication problems in autonomous SANETs, as discussed in Part III.

We selected general-purpose clustering algorithms like k-means or hierarchical single-linkage clustering for studying the main principles of clustering algorithms. As can be seen, distributed operation is possible for structuring large numbers of objects, e.g. nodes in a WSN or a SANET. Nevertheless, all distributed clustering algorithms, i.e. such techniques that rely only on locally available information in order to improve the scalability, do not necessarily always lead to optimal solutions.

Clustering can be regarded as a generic helper function in self-organizing systems. Therefore, it was not easy to prepare this chapter. For example, we already discussed selected clustering techniques in the context of data-centric networking in Chapter 11. In general, there is a wide range of algorithms available in the context of WSNs and SANETs which relies on efficient clustering mechanisms. The primary goal is always to reduce the number of objects in a given network to improve the scalability of a particular networking algorithm or a coordination procedure.

We discussed such advantages specifically for energy-efficient communication in ad hoc and sensor networks using two prominent examples–LEACH and HEED. Both algorithms exploit a number of techniques for self-organization for improved scalability. Basically, both approaches rely on probabilistic techniques for clusterhead self-election and a subsequent local optimization process.

Appendix II

Networking Aspects – Further Reading

In the following, a selection of some interesting textbooks, papers and articles is provided that is recommended for further studies of the concepts of self-organization. Additionally, a list of major journals and conferences is provided that are either directly focused on self-organization and related aspects or broadly interested in publishing special issues or organizing dedicated workshops in this domain. Obviously, this list cannot be complete or comprehensive. It is intended to provide a starting point for further research.

Textbooks

- M. Barbeau and E. Kranakis, *Principles of Ad-hoc Networking*, John Wiley & Sons Ltd, 2007.

- N. Bulusu and S. Jha, *Wireless Sensor Networks: A Systems Perspective*, Artech House Publishers, 2005.

- C. De Morais Condeiro and D. P. Agrawal, *Ad Hoc & Sensor Networks: Theory and Applications*, World Scientific Publishing Company, 2006.

- S. S. Iyengar and R. R. Brooks, *Distributed Sensor Networks*, Chapman & Hall/CRC, 2004.

- H. Karl and A. Willig, *Protocols and Architectures for Wireless Sensor Networks*, John Wiley & Sons Ltd, 2005.

- B. Krishnamachari, *Networking Wireless Sensors*, Cambridge University Press, 2006.

- J. F. Kurose and K. W. Ross, *Computer Networking: A Top-down Approach Featuring the Internet*, 3rd edn, Pearson/Addison Wesley, 2005.

- A. Leon-Garcia and I. Widjaja, *Communication Networks: Fundamental Concepts and Key Architectures*, 2nd edn, McGraw-Hill, 2004.

Self-Organization in Sensor and Actor Networks   Falko Dressler
© 2007 John Wiley & Sons, Ltd

- C. S. R. Murthy and B. S. Manoj, *Ad Hoc Wireless Networks*, Prentice Hall PTR, 2004.

- C. E. Perkins, *Ad Hoc Networking*, 1st edn, Addison-Wesley Professional, 2000.

- C. S. Raghavendra, K. M. Sivalingam and T. Znati, *Wireless Sensor Networks*, Kluwer Academic Publishers, 2004.

- B. Tavli and W. Heinzelman, *Mobile Ad Hoc Networks: Energy-Efficient Real-Time Data Communications*, Springer, 2006.

Papers and articles

- J. Ahn and B. Krishnamachari, 'Fundamental Scaling Laws for Energy-Efficient Storage and Querying in Wireless Sensor Networks', Proceedings of 7th ACM International Symposium on Mobile Ad Hoc Networking and Computing (ACM Mobihoc 2006), Florence, Italy, May 2006.

- Ö. B. Akan and I. F. Akyildiz, 'Event-to-Sink Reliable Transport in Wireless Sensor Networks', *IEEE/ACM Transactions on Networking (TON)*, vol. 13(5), pp. 1003–1016, October 2005.

- K. Akkaya and M. Younis, 'A Survey of Routing Protocols in Wireless Sensor Networks', *Elsevier Ad Hoc Networks*, vol. 3(3), pp. 325–349, 2005.

- I. F. Akyildiz, W. Su, Y. Sankarasubramaniam and E. Cayirci, 'Wireless Sensor Networks: A Survey', *Elsevier Computer Networks*, vol. 38, pp. 393–422, 2002.

- C. L. Barrett, S. J. Eidenbenz and L. Kroc, 'Parametric Probabilistic Sensor Network Routing', Proceedings of 9th ACM International Conference on Mobile Computing and Networking (ACM MobiCom 2003), San Diego, CA, USA, 2003.

- D. M. Blough and P. Santi, 'Investigating Upper Bounds on Network Lifetime Extension for Cell-Based Energy Conservation Techniques in Stationary Ad Hoc Networks', Proceedings of 8th ACM International Conference on Mobile Computing and Networking (ACM MobiCom 2002), Atlanta, Georgia, USA, September 2002, pp. 183–192.

- A. Boulis, S. Ganeriwal and M. B. Srivastava, 'Aggregation in Sensor Networks: An Energy–Accuracy Trade-off', *Elsevier Ad Hoc Networks*, vol. 1, pp. 317–331, 2003.

- T. Camp, J. Boleng and V. Davies, 'A Survey of Mobility Models for Ad Hoc Network Research', *Wireless Communications and Mobile Computing*, Special Issue on Mobile Ad Hoc Networking: Research, Trends and Applications, vol. 2(5), pp. 483–502, 2002.

- J.-H. Chang and L. Tassiulas, 'Maximum Lifetime Routing in Wireless Sensor Networks', *IEEE/ACM Transactions on Networking (TON)*, vol. 12(4), pp. 609–619, August 2004.

- I. Dietrich and F. Dressler, 'On the Lifetime of Wireless Sensor Networks', University of Erlangen, Dept of Computer Science 7, Technical Report 04/06, December 2006.

- P. Gupta and P. R. Kumar, 'The Capacity of Wireless Networks', *IEEE Transactions on Information Theory*, vol. 46(2), pp. 388–404, March 2000.

- Z. J. Haas, J. Y. Halpern and L. Li, 'Gossip-Based Ad Hoc Routing', Proceedings of 21st IEEE Conference on Computer Communications (IEEE INFOCOM 2002), June 2002, pp. 1707–1716.

- E.-S. Jung and N. H. Vaidya, 'A Power Control MAC Protocol for Ad Hoc Networks', *ACM/Kluwer Wireless Networks (WINET)*, vol. 11(1–2), pp. 55–66, 2005.

- M. Mauve and J. Widmer, 'A Survey on Position-Based Routing in Mobile Ad-Hoc Networks', *IEEE Network*, vol. 15(6), pp. 30–39, 2001.

- O. Younis and S. Fahmy, 'HEED: A Hybrid, Energy-Efficient, Distributed Clustering Approach for Ad-hoc Sensor Networks', IEEE *Transactions on Mobile Computing*, vol. 3(4), pp. 366–379, October–December 2004.

Journals

- *Ad Hoc Networks* (Elsevier)

- *Communications Magazine* (IEEE)

- *Computer* (IEEE)

- *Computer Communication Review* (ACM)

- *Computer Networks* (Elsevier)

- *Network* (IEEE)

- *Journal of Selected Areas in Communications* (IEEE)

- *Transactions on Mobile Computing* (IEEE)

- *Transactions on Networking* (IEEE/ACM)

- *Wireless Communications* (IEEE)

- *Wireless Communications and Mobile Computing* (Wiley Interscience)

- *Wireless Networks* (ACM/Springer)

Conferences

- EmNetS – IEEE Workshop on Embedded Networked Sensors

- EWSN – European Workshop on Wireless Sensor Networks

- ICC – IEEE International Conference on Communications

- INFOCOM – IEEE Conference on Computer Communications

- MASS – IEEE International Conference on Mobile Ad Hoc and Sensor Systems

- MobiCom – ACM International Conference on Mobile Computing and Networking

- Mobihoc – ACM International Symposium on Mobile Ad Hoc Networking and Computing

- MSWiM – ACM International Symposium on Modeling, Analysis and Simulation of Wireless and Mobile Systems

- Networking – IFIP International Conference on Networking

- SECON – IEEE International Conference on Sensor and Ad hoc Communications and Networks

- SenSys – ACM Conference on Embedded Networked Sensor Systems

Part III

Coordination and Control: Sensor and Actor Networks

The growing interest in wireless sensor networks and the developed communication tech-niques and algorithms lead to another research challenge–the need for self-organized coordination and control. Self-organization as a control paradigm for massively distributed systems offers new opportunities in system design and applications. On the other hand, further constraints need to be considered. The concept of actors was introduced in the context of sensor networks to enable network-centric actuation control in conjunction with the ability of the system to sense the environment. In this part, coordination and control techniques are investigated, focusing on concepts needed for sensor and actor networks. The main parameters of interest are again scalability and energy consumption. In addi-tion, real-time constraints have to be regarded because most actuation tasks are strongly time-bounded.

Outline

- Sensor and actor networks

  Sensor and actor networks can be seen as a new class of sensor networks with added functionality. In particular, actuation devices enable the sensor network to interact with the environment in two ways: the environment can be sensed and, after analyzing this information, actuators can manipulate the environment accordingly. We discuss mobile robot systems as typical actors and outline the emerging research challenges.

- Communication and coordination

  Coordination among a autonomously acting nodes is strongly needed for any col-laborative task execution. In the best case, all participating systems can perfectly synchronize their state and time. Unfortunately, this is often not the case due to limitations in inter-node communication. Lamport introduced the concept of logical clocks for coordination based on weak synchronization. Based on selected algorithms, the principles and limitations of time synchronization in sensor networks are outlined. Based on this information, distributed coordination is discussed based on case studies of promising algorithms. Finally, in-network operation and control are investigated as the main paradigm to build self-organizing sensor and actor networks.

- Collaboration and task allocation

Task-allocation algorithms build the basis for efficient collaboration. In the past, such algorithms have usually been addressed and developed in the context of multi-robot task allocation. This field is briefly introduced, focusing on the demands applicable in sensor and actor networks. Afterwards, examples of task-allocation techniques are investigated regarding the differentiation between intentional cooperation and emergent cooperation.

- Coordination and control – further reading

13

Sensor and Actor Networks

Wireless Sensor Networks (WSNs) have been developed for the distributed sensing of physical phenomena. In Part II, we discussed several communication aspects relevant to WSNs. This sensor data are usually delivered to an (external) base station, which processes the measurement results and performs control actions within the supervised region, e.g. by initiating actuation tasks according to the sensor data.

The main idea of Sensor and Actor Networks (SANETs) is to complement sensor networks with network-inherent actuation facilities. In addition to sensing and processing tasks, network nodes may perform actuation tasks as well. The benefit of such hybrid sensor/actuator networks is the integrated communication and control plane supporting all nodes in the entire network. The term 'SANET' was introduced by Akyildiz and Kasimoglu (2004). We will use this term throughout the entire book. We present a detailed discussion of SANETs in Section 13.1. In the following, we mostly refer to a form of SANETs that is usually cited in the literature: the combination of mobile robot systems with WSNs (Batalin and Sukhatme 2003a; Melodia *et al.* 2007). This is obviously not the only form of actors used in SANETs. Nevertheless, we can use this example without loss of generality for most algorithms and approaches discussed in this book.

In this chapter, we discuss the main ideas and objectives of SANETs. This includes the basic requirements on self-organization in terms of scalability and responsiveness as well as communication and coordination issues. Current research objectives are presented and explained in detail in the following chapters. Finally, inherent limitations are discussed, which need to be addressed by future research initiatives in the context of SANETs.

13.1 Introduction

According to Akyildiz and Kasimoglu (2004), Sensor and Actor Networks are distributed wireless systems of heterogeneous devices referred to as sensors and actors. While sensors are supposed to observe the environment and to measure physical phenomena, actors collect and process sensor data and consequently perform actions on the environment. The term 'actor' has been introduced to differentiate complex actuation systems from single actuators

Self-Organization in Sensor and Actor Networks   Falko Dressler
© 2007 John Wiley & Sons, Ltd

(Akyildiz and Kasimoglu 2004; Melodia *et al.* 2007). Actors represent a single network entity that performs network-related functionalities systems. At the same time, they are able to act on the environment using several actuators. The most commonly used examples of actors are mobile robot systems or groups of mobile robots as described, for example, by Batalin and Sukhatme (2003a) and by Johnson *et al.* (2006). Such robots represent resource-rich devices able to move around while communicating and coordinating among themselves and with distributed sensor systems. Therefore, SANETs are considered heterogeneous by default – providing improved capabilities and features to solve more sophisticated problems while introducing new challenges in terms of coordination and communication aspects.

13.1.1 Composition of SANETs – an example

Obviously, SANETs can be seen as an extension of WSNs by embedded actuation facilities. Figure 13.1 shows the main ideas of composing SANETs based on components of sensor networks and mobile robots.

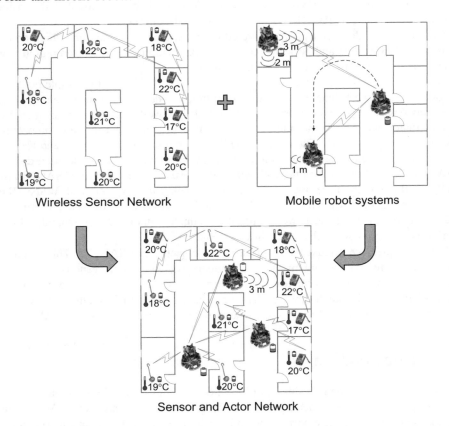

Figure 13.1 SANETs can be seen as an extension of WSNs by embedded actuation facilities; in particular, mobile robots are often used to represent actors in Sensor and Actor Networks because they represent resource-rich devices able to move around and to communicate and coordinate with distributed sensor systems

In particular, functionalities from WSNs are enriched by adding mobile robot systems to the scenario. Our example depicts a floor-monitoring scenario. Several tasks have to be achieved by the sensor nodes and the robots. Among others, the following tasks are listed to demonstrate the requirements of the scenario:

- Sensor nodes – Sensor nodes are employed to measure the temperature at dedicated places and to establish an ad hoc network infrastructure able to carry exchanged sensor messages. Sensor nodes need to be deployed with a high degree of redundancy in order to prevent system outtakes due to single node failures.

- Mobile robots – The mobile-robot systems are used for floor monitoring as well. They are responsible for observations in the building, e.g. by taking pictures at pre-defined places. If only a few robots will be used, they have to operate either fully autonomously (resulting in less accurate coordination) or a separate network infras-tructure, e.g. a WLAN network, must be installed for inter-robot communication.

The combination of WSNs and mobile robots to SANETs is depicted in Figure 13.1 as well. The available resources of the sensor and robot systems are utilized by both of them in order to enhance the capabilities of the entire SANET. Obviously, the sensor network can be used as a communication network between distant robot systems. Additionally, the robots can exploit the available sensor data to optimize their local behavioral strategies. On the other hand, mobile robots can perform operations on behalf of the sensor network, e.g. the deployment of new sensor nodes or the carriage of data between disrupted network parts.

13.1.2   Properties and capabilities

Discussing the properties and capabilities of SANETs, we first outline possible benefits arising from the coupling of sensor-network and robot technology. This list is concluded by some high-level application descriptions outlining the capabilities of SANETs. Afterwards, we discuss requirements of SANETs, concluded by a summarizing definition.

Benefits of coupled sensor networks and mobile robots

Several applications for SANETs are concerned with enhancing and complementing existing sensor network applications (Melodia *et al.* 2007). Depending on the available actuation facilities, a wide range of possible improvements can be envisioned. Staying with our example–the combination of WSNs with mobile robot systems–the SANET can benefit from interactions in two directions, namely sensor network-assisted teams of mobile robots and robot-assisted sensor networks. Actually, a wide range of capabilities of sensor networks and mobile robot systems can be exploited for such assistance relationships, either to optimize the network for existing applications or to act as enablers for future scenarios requiring support from sensors and actors simultaneously.

Sensor-network-assisted teams of mobile robots   For example, the navigation of mobile robots in known environments such as office or lab environments but also in unknown ter-ritories can be optimized using stationary sensor networks. In this case, the sensor network provides functions for on-demand localization and it can operate in the form of intelligent landmarks. Additionally, the sensor network can provide a communication infrastructure between unconnected robots for coordination and control among those robot systems. Some selected support functions are summarized in the following:

- Localization, based on well-known geographic positions of sensor nodes and distance estimations, e.g. based on the measured radio signal strength (Priyantha *et al.* 2005).

- Intelligent landmarks, providing storage and computational facilities to build an intelligent environment in which robots can coordinate among each other (Pei *et al.* 2000).

- Cooperative tracking relies on the intensive collaboration of robots with stationary sensor nodes that are used to observe well defined regions (Jung and Sukhatme 2001).

- Communication infrastructure, based on established ad hoc routing principles to enable communication and coordination between distant robots (Batalin and Sukhatme 2003c).

Robot-assisted sensor networks   Mobile robots can improve the behavior of WSNs in multiple ways. For example, robot systems can be used to deploy further sensor nodes or to repair failed ones in order to maintain disruption-free networking with minimized redundancy. This includes possible strategies to relocate sensor nodes or for energy harvesting by robot systems. Finally, mobile robots can be used as transporters for data between disrupted parts of the network or to provide a communication relay into other networks such as the Internet. A brief summary of selected helpers is provided in the following:

- WSN deployment, optimized by laying out new sensor nodes in uncovered regions or in the geographical proximity of nodes that are estimated to fail early due to battery outages (Howard *et al.* 2002).

- Energy harvesting, supported by mobile robot systems, e.g. inductive energy transmission between resource-rich robots and distributed sensor nodes (Rahimi *et al.* 2003).

- Software management, based on on-demand composition of software modules according to a global objective, with subsequent node reprogramming accomplished by mobile robots (Fuchs *et al.* 2006).

- Communication relay, provided by mobile robots to bridge communication holes in WSNs and to connect different network types (Shah *et al.* 2003).

Based on the mentioned capabilities, a huge number of applications can be thought of that will benefit from the employment of SANETs. Such networks will become an integral part of surveillance applications, home and factory automation, and environmental monitoring. Sensors can be used to detect changes in the environment and emergency situations. In addition, simple actuators such sprinklers or heaters can be used to act in fire emergencies and to control the climate in buildings, respectively. Mobile robot systems provide more powerful solutions for intruder detection or observations in inaccessible environments. Further examples are discussed in Section 13.1.4.

SANET properties and definition

According to the composition of a SANET and its typical application range as discussed so far, we can extract a number of basic properties and requirements for SANETs that finally lead to a common definition of the term 'SANET'. In principle, SANETs show properties similar to WSNs. In addition, more sophisticated management and coordination mechanisms are needed to control the entire network, consisting of sensor nodes and actors such as mobile robots. The following three requirements must be considered for the development of algorithms and solutions in the context of SANETs.

Broad heterogeneity   By definition, SANETs are composed of heterogeneous subsystems. This heterogeneity is reflected in multiple dimensions (Dressler *et al.* 2007b). First, different hardware components will be used to accomplish the needs of the application scenarios. Secondly, installed software modules may differ between multiple nodes according to application requirements and available resources, such as attached sensors and actuators, or processing and storage capabilities. Finally, the deployed and working nodes may differ in terms of current system parameter settings, as these will be adapted according to observed environmental conditions.

Concurrent objectives   In SANETs, two concurrent objectives – coordination and communication – need to be considered simultaneously (Melodia *et al.* 2005). The main issue is that coordination essentially relies on communication and, at the same time, SANET algorithms focus on energy-efficient operation and the ability to work in delay-and loss-tolerant networks. In general, this property can be found in other networks as well, especially in WSNs. Nevertheless, it became especially challenging in the context of sensor–actor and actor–actor coordination.

Self-organization and emergence   A main characteristic of autonomously working systems is their need for self-organization techniques for management and control operations. Many algorithms used for self-organized coordination and communication show non-linear behavioral properties if used in multiple interacting subsystems. Therefore, the emergent behavior of the collaborating autonomous systems needs to be considered (Dorigo *et al.* 2004; Mataric 1995a).

Additionally, it needs to be mentioned that node mobility is an inherent feature of most SANETs (we already discussed the use of mobile-robot systems as a typical scenario to study algorithms and techniques for coordination and communication among sensors and actors). In the context of SANETs, spatial mobility is perhaps the most cited feature, while temporal mobility can be found as well (similarly to WSNs). Both types of mobility enforce the requirement for adaptive and efficient coordination techniques.

Based on our first discussions about SANETs, we can now derive a proper definition of the term 'Sensor and Actor Network'. Actually, this definition is based on the findings on Mobile Ad Hoc Networks (MANETs) and WSNs, as discussed in Part II.

Definition 13.1.1 Sensor and Actor Network – A SANET typically consists of heterogeneous and mobile nodes able to sense their environment (sensor) and to act on it (actor). The most prominent challenges of SANETs are communication and cooperation issues.

Similar to WSNs, SANETs are assumed to be strongly resource-restricted in terms of communication, processing and storage capabilities, and in terms of available energy.

13.1.3 Components of SANET nodes

Before starting the discussion of specific research objectives in SANETs, a brief overview to the architecture of SANET nodes, including the description of a typical robot system, is provided below.

Figure 13.2 shows the architecture of typical SANET nodes. The main composition is similar to a sensor node. The embedded system is operated by a micro controller that also provides access to memory and external storage options. A radio transceiver is employed for wireless radio communication. As we speak about Sensor and Actor Networks, nodes may have attached sensor subsystems to measure environmental conditions. In addition, a number of actuators can be used to interact with this environment. All parts of a SANET node are powered by a battery. We will summarize typical systems as used for SANETs in the following and depict some example systems.

- Main processing and storage system – Similar to small sensor nodes, micro controllers are often used as low-power processing units. Nevertheless, the application range for actuators often requires more powerful processing units, e.g. for picture or even video compression. In such cases, often, embedded PC systems based on industry PC-104 boards are employed, providing much higher computational power.

- Sensors – In addition to the spectrum of sensing devices available of sensor nodes as used in typical sensor networks, mobile robots are able to carry and to operate

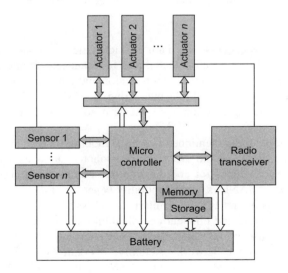

Figure 13.2 Composition of a single SANET node; in addition to communication, processing and sensing devices, a number of actuators are attached for performing action tasks, such as moving around, turning switches on and off, and others

more resource-intensive devices. Examples of such sensors are video cameras and 3D laser sensors. The idea of employing mobile robot systems to assist stationary sensor networks also encouraged the use of costly high-quality sensors to calibrate low-cost devices attached to sensor nodes.

- Actuators – The domain of actuators ranges from simple electronic switches over motors and wheels to various forms of manipulators. Small sensor nodes can be used as actors by attaching relays for controlling electric devices, while large mobile robots are able to move around and to act on it by manipulating in the physical space. Active Radio Frequency ID (RFID) tags can also be seen as actuators that can, for example, be used as intelligent landmarks.

- Wireless communication – Besides low power radio chips, as depicted for sensor nodes, additional, possibly heterogeneous, communication techniques and devices can be used for SANETs. Depending on the application scenario and, especially, on the available space and energy, any wireless communication technology can be attached to a mobile-robot system. An often used example is Wireless LAN (WLAN).

As outlined above, mobile robots are often used in the literature to demonstrate the capabilities of SANETs. Unfortunately, there is no 'standard' robot system available. Therefore, we use the Robertino system[1] as a typical example, used in a number of research labs to outline the composition of such a robot system – which actually represents a typical SANET node. Figure 13.3 depicts such a Robertino system. An embedded standard PC is used to provide processing and storage components. Available sensors include infrared distance sensors and a video camera. The robot is able to move around and to use attached actuation

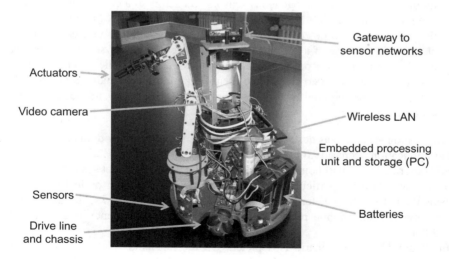

Figure 13.3 System architecture of a Robertino system; the main components are labeled to show the wide range of individual devices that a robot consists of

---

[1]http://www.openrobertino.org/.

devices such as manipulators. A block of batteries powers the whole system. Additionally, a sensor node is attached on top of the robot for seamless integration into a sensor network.

13.1.4 Application examples

In this section, we list a number of typical application scenarios for coordinated sensors and actors. This list is not intended to be complete and comprehensive. Instead, it should show the general application domains for SANETs:

- Temperature control – Perhaps one of the simplest application scenarios is autonomous temperature control. Only two kinds of systems are needed in this example: sensors measuring the temperature in a given environment and actors that are able to control the temperature, e.g. actuators attached to air conditioners and heaters. According to the particular application requirements, the sensors will deliver their readings to neighboring actors. In this way, a direct control loop can be created.

- Fire detection – A more complex scenario is a fire-detection application. Besides sensors measuring the temperature and smoke detectors, water sprinklers are typical actuators in this scenario. The higher complexity arises through the need to enable the SANET not only to detect fire and to turn on the sprinklers, but to connect to other networks for automated emergency calls. Therefore, heterogeneous networks need to be operated. Additionally, the SANET may be requested to report the exact location of the fire and further information to the fire department.

- Intruder detection – The most complex applications of SANETs include the use of mobile systems and heterogeneous sensors and actors in a collaborative scenario. An example is intruder detection, which will first rely on a WSN that passively observes the environment to detect movements, open doors or other unusual environmental changes. In the case of a positive detection, more complex systems will get alarmed, e.g. in form of more powerful sensors on wheels – mobile robot systems. Such systems might use video cameras to identify the intruder. Finally, a number of alarm devices will be required in this scenario to send emergency messages and to scare away the intruder.

Application scenarios as described above can be found in many domains. This includes all the application scenarios for WSNs, as discussed in Section 7.2.7 – for example, we already described the smart home scenario that includes mobile-robot systems as well. In the following, we outline a particular application for SANETs, which perfectly demonstrates the benefits and needs of heterogeneous Sensor and Actor Networks: the use of coupled-sensor networks and mobile robots as first responders in emergency operations (Kumar et al. 2004).

In Figure 13.4, the scenario is depicted. On the left-hand side, an early state of the operation is shown. Two mobile robot systems entered the building and began to deploy sensor nodes. All SANET nodes spontaneously connect themselves to form a network infrastructure for communication and cooperation. On the right-hand side, a later state is shown. The robots started to explore the unknown environment in order to locate sources of the emergency, i.e. a fire, and – most importantly – possible victims who are still in the building. All of this information will be reported to human fire fighters. Thus,

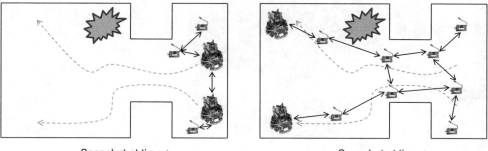

Snapshot at time t_1 Snapshot at time t_2

Figure 13.4 Using mobile robots and sensor nodes as first responders in emergency situations; the robots iteratively explore the unknown surroundings and deploy sensor nodes as
intelligent landmarks, providing a network infrastructure and continuous sensor measures

Table 13.1   Tasks to be achieved in the first-responder scenario by mobile robots and
sensor nodes, respectively

Task	Sensor nodes	Mobile robots
Network infrastructure	Establishment of an ad hoc network to provide reliable and fault-tolerant communication between arbitrary nodes	Deployment of sensor nodes to establish the network infrastructure; participation in the ad hoc network as simple network nodes
Monitoring of the environment	Periodical gathering of sensor readings and transmission to the robots and to an external base station	Employment of high-quality and more resource-expensive sensors such as video cameras
Support for victims and fire fighters	Network infrastructure for arbitrary communication	Explicit search for victims; actions on behalf of fire fighters (externally controlled)

additional sensor nodes have to be deployed in the whole area in order to obtain more
accurate information about the current environmental situations and to provide a fault-
tolerant network infrastructure.

The tasks to be achieved by the mobile robot systems and the sensor nodes can be
summarized as described in Table 13.1. Obviously, all tasks require efficient communication
methods and robots and sensor nodes have to collaborate on the common objectives.

A picture of an experimental setup used to verify several algorithms developed in the
context of SANETs, focusing on the first responders scenario, is shown in Figure 13.5.
Additionally, a temperature map, as obtained from collected distributed sensor readings, is
shown on the right-hand side. Based on this map, fire fighters may plan how to enter the
building in a safe way.

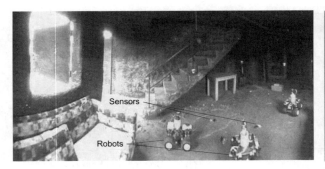 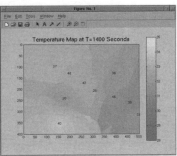

Figure 13.5 Mobile robots are used to deploy sensor nodes in a hazardous area (left) in order to provide a network infrastructure for robots and human fire fighters and to obtain current environmental information such as a temperature map (right) [Reproduced by permission of Sanjiv Singh (Kumar *et al.* 2004); © 2004 IEEE]

13.2   Challenges and research objectives

Studying research challenges and objectives in SANETs will include all aspects from WSNs and robotics. While many of these issues, as depicted in Table 13.2 are interesting and, of course, challenging, we will concentrate in the following on challenges unique to SANETs. These issues, as listed below, provide the basis for further studies in SANETs as discussed in the following chapters.

In addition, a number of further requirements can be identified, either unique in SANETs or at least more challenging compared with other domains. In the literature, a number of objectives are listed. The two most important requirements for efficient operation and coordination in SANETs are reliable communication mechanisms, supporting real-time operation. In this context, solutions have been proposed for event-to-sink reliable transport in WSNs (Akan and Akyildiz 2005; Sankarasubramaniam *et al.* 2003) and for real-time coordination and routing in SANETs (Gungor *et al.* 2007; Shah *et al.* 2006).

Regarding the degree of autonomous behavior, automated and semi-automated architectures need to be distinguished (Akyildiz and Kasimoglu 2004). This also refers to the idea to move all the processing and control activities into the network, i.e. to enforce in-network processing and control (Dressler 2006a). Communication and coordination have

Table 13.2   Selected research issues in the domains of sensor networks and mobile robots

Sensor network	Mobile robots
Ad hoc routing	Localization
Addressing issues	Navigation
Energy efficiency	Tracking
Coverage	Security
Security	Energy efficiency

been identified by Melodia *et al.* (2006) as the main criteria for efficient control of collaborative nodes in SANETs. Finally, coverage needs to be named as a characteristic of SANETs, which has the same meaning as already discussed for WSNs. Nevertheless, the use of mobile robots enables a generation of algorithms to optimize coverage while keeping all other characteristics of sensor nodes (Batalin and Sukhatme 2003a).

In the following, we will concentrate on two domains – communication and coordination, and collaboration and task allocation. Both are briefly outlined in the following and discussed in detail in the next two chapters. In order to prevent confusion, it needs to be mentioned that Akyildiz and Kasimoglu (2004) also denoted these two domains as sensor–actor coordination and actor–actor coordination, respectively.

13.2.1 Communication and coordination

Communication and coordination aspects in SANETs basically refer to all aspects relating to sensor–actor coordination. Besides the communication aspects as discussed for WSNs, time constraints in particular need to be regarded for coordination issues. Real-time communication is a strong requirement for sensor–actor coordination, especially motivated by the findings that in many scenarios, communication is delay-sensitive, e.g. for use in emergency situations. Time synchronization is hard to achieve in massively distributed systems such as WSNs and SANETs. Therefore, logical ordering of observed events is often used as a last resort.

Additionally, the communication relationship among systems in the entire network needs to be controlled and restricted to the necessary parts in order to achieve scalability and energy efficiency. Thus, the communication protocols need to focus on necessary communication relationships, e.g. by finding minimal sets of involved sensors and actors for a particular application. This task can be summarized as redundancy elimination (Akyildiz and Kasimoglu 2004). This also includes the necessity of moving operation and control into the network because external control is limited due to communication restrictions towards a central base station. These limitations are due to delayed communications, bandwidth restrictions and the need for disruption-tolerant coordination.

Based on the mentioned observations, the following four research issues can be identified: (We will discuss particular solutions in detail in Chapter 14.)

- Synchronization and coordination – Because hard synchronization is often infeasible in communication networks (Awerbuch 1985) and particularly in SANETs, coordination mechanisms are typically used for collaborative tasks (Melodia *et al.* 2005). Usually, a precise time estimation of an event is not really necessary. Instead, the ordered delivery of information collected by sensor nodes needs to be assured. This can be achieved, for example, by ordering events according to a logical time.

- Time synchronization – In special cases, rough coordination is not exact enough. In such cases, precise time synchronization among a set of nodes is desirable. Because standard time-synchronization techniques as used in the Internet cannot be directly applied to ad hoc and sensor networks, a number of approaches have been proposed in these domains (Elson and Römer 2003; Su and Akyildiz 2005).

- Distributed coordination – For efficient coordination, adequate communication mechanisms are needed. An example is the distributed sensing of physical phenomena.

The optimization goal is the redundancy elimination, i.e. the prevention of any unnecessary communications. The coordination with associated actors leads to further challenges in this domain. The communication aspect is, for example, addressed by directed diffusion according to the hypothesis that sensor-network-coordination applications may need to be performed by localized algorithms (Estrin *et al.* 1999b). Other approaches are, for example, using an event-driven clustering paradigm in which cluster formation is triggered by specific events (Melodia *et al.* 2005).

- In-network operation and control – External controllers, such as a base station that is coordinating all actions in the entire SANET, are often not feasible due to resource restrictions in the network or due to disrupted network operations. Therefore, all operation and control mechanisms need to be moved into the network. This is especially important for coordinated sensing and actuation tasks (Dressler 2006a).

13.2.2   Collaboration and task allocation

With respect to coordination aspects in SANETs, collaboration among multiple systems also needs to be investigated. Akyildiz and Kasimoglu (2004) termed this aspect of SANET operation actor–actor coordination. Without loss of generality, collaboration can be controlled by means of task and resource allocation. Usually, two groups of problems are distinguished in the literature, described as Multi-Robot Task Allocation (MRTA) (Gerkey 2003; Mataric *et al.* 2003):

- Single-task multi-robots – Sometimes, a particular task cannot be performed by a single robot system. Therefore, multi-robots need to be employed in this case, e.g. two robots collaboratively moving an object. The allocation of a single task to a group of robots requires perfect synchronization among the system, either in a direct way, e.g. by means of time synchronization, or indirectly, e.g. through stimulation by work (stigmergy).

- Multi-tasks multi-robots – The most commonly investigated problem is the allocation of multiple tasks to a large number of available robot systems. In multi-tasks multi-robots systems, typically tasks are considered that can be solved by a single robot. Nevertheless, heterogeneity issues will lead to differences in the quality and time of task execution. Therefore, either (optimal) scheduling algorithms are needed or a high degree of redundancy.

The primary optimization goals for MRTA in general are network lifetime and scalability. These objectives apply for application in SANETs. Among others, communication-related issues, energy limitations and time constraints in particular need to be considered for the development of task-allocation algorithms and the application in Sensor and Actor Networks.

In general, actor–actor coordination or MRTA can be reduced to an optimization problem with inherent time restrictions (Akyildiz and Kasimoglu 2004). In most cases, algorithms developed in the context of task allocation can be applied to the problem of generic resource allocation as well. Resources to be shared include, for example, memory and storage (where to store which data in a SANET) and computational resources (which node should perform which data-processing part).

According to Gerkey (2003), proposed solutions for task and resource allocation (in multi-robot systems) can be distinguished into the following two categories of MRTA:

- Intentional cooperation – If collaboration of multiple systems towards a common goal is performed according to a plan, which all the robots agreed on, we speak about intentional cooperation. Thus, the systems explicitly cooperate with purpose. Usually, task-related communication is used for task allocation. The best known example of intentional cooperation is auction-based allocation. In this context, negotiation and mediation algorithms are used to distribute the work among a group of available systems, each of which is able to perform specific actions in a given quality and time.

- Emergent cooperation – In contrast, emergent cooperation refers to non-intentional task allocation, i.e. the robot systems do not explicitly work together on a common goal. Instead, the cooperative behavior emerges from interactions among the systems and with the local environment. Emergent-task-allocation schemes are usually based on simple local rules and non-linear coupling of multiple systems, e.g. through stimulation by work (ant-like cooperation) or stimulation by state (positive and negative feedback).

13.3   Limitations

All the algorithms for coordination in SANETs are subject to a number of limitations, which apply to communication and coordination as well as to collaboration and task allocation. First of all, the *scalability* of distributed algorithms greatly depends on the necessary communication and control overhead. In sensor–actor coordination, the necessary synchronization or weak coordination in particular relies on message-exchange techniques for state maintenance. Depending on the algorithm, this state may refer to globally valid state information (e.g. topology and time) or to clusters of nodes in which less communication overhead is needed for coordination but additional effort is needed for group membership maintenance.

Actor–actor coordination is also subject to scalability issues according to the necessary communication overhead for task and resource allocation. In most cases, this overhead primarily contributes to the time needed for task-allocation processes. Thus, as actor–actor coordination strongly relies on *real-time operation*–each application inherently defines limitation for allocation algorithms according to the number of nodes and the real-time constraints.

Energy-related issues need to be considered in all kinds of SANETs. Even if energy-harvesting techniques are available, the amount of energy available for a particular task will always be limited. In general, such energy constraints can be mapped to the *network lifetime* (see Section 17.3), which is perhaps the most important metric from an application's point of view. Classically, only energy-related issues are incorporated into the network lifetime but communication aspects, e.g. connectivity and Quality of Service (QoS), and coverage parameters, e.g. communication and sensor coverage, also apply.

Besides the mentioned constraints, which actually represent limitations in terms of applicability of coordination techniques in SANETs, two additional limitations need to be mentioned, both of which are basically the result of developing distributed and self-organizing techniques:

- Selfish nodes – In a self-organizing environment, employed coordination and task-allocation algorithms usually require (and depend on) collaborative nodes. In other words, nodes are considered to cooperate according to the global objective instead of contributing but following self-serving interests. Hubaux *et al.* (2001) discussed the problems introduced by selfish nodes. The authors also pointed out how to counteract such selfishness, for example, through stimulation approaches.

- Cross-layer optimization – Another problem relates to cross-layer and cross-system optimization, respectively. In many cases, cross-layer optimizations are needed for improved energy management or QoS-related issues. Nevertheless, there might be implications of such optimizations affecting the negotiation scalability or changing the non-linear behavior of groups of nodes. Such side effects need to be detected and analyzed in order to improve the overall system behavior.

14

Communication and Coordination

The main requirements and research challenges specific to SANETs can be seen as sensor–actor coordination and actor–actor coordination. In this chapter, we focus on the first challenge–sensor–actor coordination. One of the first frameworks for distributed coordination in Sensor and Actor Networks has been described by Melodia *et al.* (2005). The authors especially outline the main requirements in SANETs, i.e. energy efficiency and real-time operation with respect to communication and coordination.

The interaction between sensor networks and integrated actors in particular demands features in the communication domain unique to SANETs. Therefore, we focus on communication aspects first. Besides the typical message exchange among neighboring nodes or nodes sharing a common network infrastructure, which has been already discussed in the context of WSNs, real-time operation strongly requires synchronization and coordination techniques.

In this chapter, we discuss the need for synchronization among a number of nodes and, more importantly, its limitations. Coordination (or weak synchronization) is considered to be the final answer for sensor–actor control. Such coordination can be established in multiple ways. Starting with the concept of logical time as introduced by Lamport (1978) to more sophisticated distributed coordination techniques, a number of case studies are provided.

Discussing communication and coordination in SANETs, we also need to introduce the concepts of in-network operation and control. The basic idea is a consequent follow-up to data-centric networking, as presented in Chapter 11. Besides data-centric communication, all – or at least an essential number of – operation and control issues are distributed to the networked nodes. Thus, the need for a centralized control is completely eliminated.

The structure of this chapter is primarily organized according to the needs of coordination in SANETs:

- Synchronization vs coordination – Starting with a general discussion on the needs of time synchronization, we introduce the principles of coordination (weak synchronization) as the basis for operating huge numbers of nodes in a common environment.

- Time synchronization – A number of approaches have been proposed in the context of WSNs. A brief discussion outlines the primary ideas and techniques that can be used in SANETs as well.

- Distributed coordination – The primary goal of this chapter is to describe coordination algorithms. Based on well known algorithms for scalable coordination and coordination frameworks, we present sensor–actor coordination mechanisms. At this time, we will also discuss limitations resulting from heterogeneous scenarios such as resource starvation by selfish nodes – and possible solutions.

- In-network operation and control – Coordination and control need to be distributed in WSNs and SANETs. In this part, we present and discuss the extreme situation in which all operation and control tasks are moved into the network, thus preventing any centralized entities and decision processes.

The efficiency of the communication and coordination techniques can be measured in terms of energy efficiency as the typical prime measure in massively distributed systems empowered by batteries or specific energy-harvesting techniques. Additionally, the coordination time, i.e. the real-time behavior of the entire system, needs to be considered. In the context of real-time systems, usually soft and hard real-time is distinguished. Similarly, applications for SANETs may request hard deadlines for specific coordination tasks.

14.1   Synchronization vs coordination

In this section, we discuss the need for synchronization techniques to achieve coordination among multiple autonomous systems. Additionally, the discussion will show the differences between hard synchronization and weak coordination. Therefore, we will especially discuss the need for a logical time describing the ordering of events in distributed and self-organizing systems such as SANETs for taking adequate decisions. We start with a brief description of the basic problems and the requirement of clock synchronization.

14.1.1   Problem statement

The primary problems of clock synchronization are depicted in Figure 14.1. The local clocks of two autonomous systems will always show the depicted behavior. There will be no common baseline and no common scale. In our example, System A will discover Event 1 at time $t = 10$ and Event 2 at time $t = 15$. Fortunately, System B discovers Event 1 at the same time but, due to the different time scale, it will see Event 2 at time $t = 14$. Thus, both systems are not synchronized.

Our example discovered the two – usually disconnected – problems for clock synchronization: a common absolute time (the same time basis) and a common relative time (the same time scale).

Absolute time

If a common time basis is required, a synchronization of the absolute time is needed. Simple time synchronization is usually not available in WSNs and SANETs. We discuss such time synchronization algorithms in detail in Section 14.2.

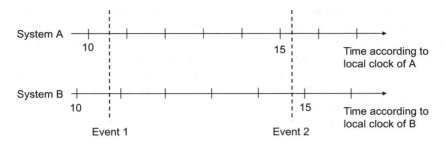

Figure 14.1 The need for clock synchronization – problem description; the local clocks of systems A and B have a different basis and different scales

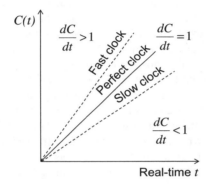

Figure 14.2 Clock drift in comparison to the perfect clock

Relative time

The relative time can usually be computed locally. Therefore, no time synchronization is required. Nevertheless, the scaling problem must be considered. Today's computing devices are equipped with a hardware oscillator-assisted computer clock, which implements an approximation $C(t)$ for real-time t. $C(t) = k \int_{t_0}^{t} \omega(\tau) d\tau + C(t_0)$ is a real valued function over real-time t, which depends on the angular frequency $\omega(t)$ of the hardware oscillator. k is a proportional coefficient (Römer 2001). Usually, different time scales are the result of drifts of the local clock, i.e. runtime differences of the used oscillator.

The basic problem is depicted in Figure 14.2. Shown is the real-time t on the x-axis and the local time available at the system $C(t)$ on the y-axis. The drift is usually denoted as ρ. It describes the slope of the curve in the depicted diagram.

For a perfect hardware clock, $\dfrac{dC}{dt}$ would be equal to 1; thus, the clock drift ρ equals 0. For practical implementations, it must be considered that all hardware clocks are imperfect, i.e. $\rho \neq 0$ and $1 - \rho \leq \dfrac{dC}{dt} \leq 1 + \rho$. The exact clock drift is hard to predict because it depends on environmental influences (e.g. temperature, pressure, power voltage). One can usually only assume that the clock drift of a computer clock does not exceed a maximum value ρ. A typical value for ρ achievable with today's hardware is 10^{-6}, i.e. the computer clock drifts away from real time by no more than 1 s in 10 days, which is still a significant value. Note that different computer clocks have different maximum clock drift values ρ_i. Typically, the maximum clock drift of a given clock ρ is known in advance and specified by the clock manufacturer. A more detailed description of clock drift can be found, for example, in Mills (1991).

14.1.2 Logical time

Obviously, some effort is needed to synchronize clocks and to counteract clock drifts. On the other hand, in most cases, only the internal consistency of the clocks matters. In a classic paper, Lamport (1978) showed that although clock synchronization might be possible, it need not be absolute. If two processes do not interact, it is not necessary that their clocks are synchronized. Furthermore, he pointed out that what usually matters is not that all processes agree on exactly what time it is, but rather that they agree on the order in which events occur. This idea was also investigated by Allen (1983).

The so called Lamport timestamps are based on these observations. Figure 14.3 (left) depicts an example. Shown are three systems (A, B and C) that exchange messages. Each system maintains a local clock depicted by the timestamps in vertical order. Obviously, the clocks are not synchronized, i.e. they drift away from each other. According to Lamport, events need to be ordered in the global context; thus, message exchanges are used to update the local clocks. In the shown example, message C is transmitted at $t = 54$ and received at $t = 42$. Similarly, message D is sent 'before' it was received by system A. Figure 14.3 (right) shows an updated version of the message exchange using the Lamport timestamps.

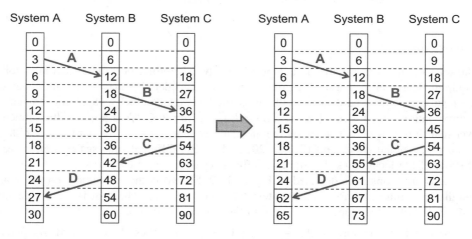

Figure 14.3 Update of local clocks according to the Lamport timestamps. The left figure depicts the unmodified local clocks of systems A, B and C. The right figure shows the updated version using Lamport timestamps

If the timestamp carried by the message is greater than the local clock, the local clock is updated. Thus, the logical ordering of events is ensured.

Lamport introduced the relation happens-before, written as $a \rightarrow b$ (a happens before b). This means that all processes agree that the first event a occurs and that, afterwards, event b occurs. Two observations can be concluded:

1. If a and b are events in the same process, and a occurs before b, then $a \rightarrow b$ is true.

2. If a is the event of a message being sent by one process, and b is the event of the message being received by another process, then $a \rightarrow b$ is also true. A message cannot be received before it is sent, or even at the same time it is sent, since it takes a finite, non-zero amount of time to arrive.

More formally, Lamport timestamps can be described as follows. For all events a assign time value $C(a)$ to event a. Thus, the following conditions are ensured:

- If $a \rightarrow b$ in the same process, then $C(a) < C(b)$.

- If a and b represent the sending and receiving of a message, respectively, then $C(a) < C(b)$.

- For all distinctive events a and b, $C(a) \neq C(b)$.

Logical clocks have been widely investigated in the context of distributed systems. The generation of global state information is a particular concern for Lamport timestamps. A global state is usually defined as the local state of each process in combination with the messages currently in transit (not yet delivered). Such global state information can, for example, be established using a distributed snapshot, as described by Chandy and Lamport (1985). The authors propose a consistent cut for such snapshots describing exactly the mentioned global state. The best known applications are in distributed systems, e.g. for termination detection. We will see that coordination in WSNs and SANETs can be optimized using such logical time as well.

14.1.3 Coordination

Coordination can be considered as weak synchronization. The basis for most coordination actions are variants of logical clocks and distributed snapshots. Usually, only the order of events becomes necessary. A major difference can be found in coordination issues in real-time systems. In such systems, a deadline (either soft or hard) also needs to be considered. Such real-time coordination represents a future research challenge.

In order to prevent the maintenance of global state information, coordination needs to be restricted to the directly involved processes or systems. Sometimes, a coordinator may be used, e.g. if clustering algorithms are used, as described in Chapter 12.

To conclude the discussion of synchronization vs coordination, the main properties, advantages and limitations are summarized in Table 14.1.

14.2 Time synchronization in WSNs and SANETs

In the previous section, we discussed the need for time synchronization as well as coordination among multiple systems according to a logical time. Based on this introduction,

Table 14.1 Synchronization vs coordination

	Synchronization	Coordination
Properties	Accurate synchronization to a given clock source or agreement on a common (average) time	Based on logical clocks and events as distributed among nodes
Advantages	Synchronized clocks are easy to use; provide capabilities for many distributed applications	Usually low communication overhead; applicable in large-scale networks (scalability)
Limitations	Message overhead (we tried to reduce the (global) state information in SANETs); imprecise synchronization in large-scale networks and in low-bandwidth networks (inadequate for sensor networks)	Distributed snapshots (global state) are hard to acquire; contradiction to energy-aware operation or quality-of-service requirements

we introduce the requirements for more precise time synchronization and a number of techniques in the following. Starting with a few well known general approaches for time synchronization in the Internet, we will focus on solutions specific to WSNs and SANETs.

14.2.1 Requirements and objectives

Independently of the specific algorithm used for time synchronization and its particular objectives, all techniques can be measured according to a number of characteristics:

- Precision – The precision refers either to the time dispersion among a group of nodes, or to the maximum error with respect to an external standard.

- Lifetime – In the context of time synchronization, the lifetime refers to the persistence of the synchronization effect. This can range from persistent synchronization that lasts as long as the network operates, to nearly instantaneous (useful, for example, if nodes want to compare the detection time of a single event).

- Scope and availability – Algorithms can be distinguished according to the geographic span of nodes that are synchronized, and the completeness of coverage within that region.

- Efficiency – Time and energy expenditures needed to achieve synchronization reflect the efficiency of the algorithm.

- Cost and form factor – Besides methodical measures, the cost and form factor can become particularly important in sensor networks that involve thousands of tiny, disposable sensor nodes.

Specifically required are energy efficiency and scalability in WSNs and SANETs. Obviously, this concern relies to time synchronization algorithms as well (Elson and Römer 2003).

The application range for efficient and precise time synchronization ranges from velocity estimation and the measurement of the Time of Flight (ToF) for localization to distributed beam-forming arrays and the suppression of redundant measurements of the same event. More specific applications of highly precise synchronization algorithms are Time Division Multiple Access (TDMA) schedules in Medium Access Control (MAC) protocols. New application domains are cryptographic schemes relying on synchronized clocks. In the context of coordination, the ordering of logged system events is needed.

14.2.2   Conventional approaches

A number of well studied solutions for time synchronization are already available for use in networks such as the Internet and even in mobile scenarios. Often, non-network-based solutions such as Global Positioning System (GPS) are preferred, as the synchronization quality does not depend on specific network configurations. Unfortunately, GPS cannot be used in all scenarios (indoor, environments with limited sight to the GPS satellites). Therefore, a number of network-based solutions have been investigated.

Complexity of network synchronization

In an early piece of work, Awerbuch (1985) studied the complexity of network synchronization. This complexity refers to, first, the communication complexity C, i.e. the total number of messages sent during the algorithm, and, secondly, on the time complexity T, i.e. the number of pulses (time slices) passed from its starting time until its termination.

For more convenient analysis, the time complexity T of an asynchronous algorithm is usually considered the worst-case number of time units from the start to the completion of the algorithm, assuming that the propagation delay and the inter-message delay of each link are, at most, one time unit.

Obviously, the total complexity of an algorithm depends on the overhead introduced by the synchronizer v, i.e. $C(v)$ and $T(v)$ for communication and time complexity, respectively. If we denote the initialization complexity as C_{init} and T_{init}, the complexity of an algorithm S with synchronizers v can be calculated according to Equation 14.1:

$$C_A = C_S + T_S \times C(v) + C_{init}(v)$$

$$T_A = T_S + T_S \times T(v) + T_{init}(v). \tag{14.1}$$

In this equation, C_A, T_A and C_S, T_S are the communication and time complexities of algorithms A and S, respectively. The synchronizer v is efficient if all the parameters $C(v)$, $T(v)$, C_{init} and T_{init} are small enough – which is strongly application dependent.

Cristian's Algorithm

Cristan's algorithm (Arvind 1989; Cristian 1989) is one of the simplest clock-synchronization algorithms used to synchronize the time on a machine with a remote time server. This is a very straightforward algorithm, and is quite easy to understand. The first approximation is to require all clients to sets their clocks to C_{UTC}. The procedure, which is also depicted in Figure 14.4, works as follows.

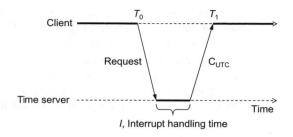

Figure 14.4 The principles of time synchronization using Cristian's algorithm

The client generates a request at time T_0, sends it to the time server and receives the time value $t = C_{UTC}$ in a message at time T_1. Time t is inserted in the message at the last possible point before transmission from the server. The error introduced by the message-transmission time can therefore be calculated using the round trip and the interrupt handling time: $(1/2T_{round}) - I$.

The major problem for time synchronization is always that the time must never run backwards. Thus, a gradual slowing down or advancing of the clock is needed, e.g. 1 ms per 10 ms. A minor problem is the non-zero transmission latency. This latency can be approximated as $1/2(T_1 - T_0 - I)$, which requires symmetric routes in terms of transmission latency.

Advanced clock synchronization using NTP

Perhaps the most popular protocol for time synchronization is Network Time Protocol (NTP) (Mills 1991). The current version is three, as specified in Mills (1992).

Basically, NTP works similarly to Cristian's algorithms. Let us consider the following scenario (as depicted in Figure 14.5). A client sends a message at T_1 to the server. The message arrives at T_2. After some processing time, the server will submit an answer at T_3. Finally, the response is received by the client at T_4. Based on these timestamps, which have been transmitted in the NTP messages, the client can estimate the round trip delay δ

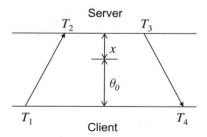

Figure 14.5 Message exchange needed for NTP based time synchronization

and the clock offset θ as shown in Equation 14.2:

$$\theta = 1/2[(T_2 - T_1) + (T_3 - T_4)]$$
$$\delta = (T_4 - T_1) - (T_3 - T_2). \tag{14.2}$$

This calculation is repeated periodically and a δ_0 is estimated as the minimum of the last eight delay measurements and the tuple (θ_0, δ_0) is used to update the local clock (Mills 1991). The true offset between the local clocks θ can be expressed by the following inequality:

$$\theta_i - \frac{\delta_i}{2} \le \theta \le \theta_i + \frac{\delta_i}{2}. \tag{14.3}$$

Considering the robustness of NTP, two major problems have to be considered. First, system failures and unreliable data communication must be taken into account when designing time-synchronization protocols. Secondly, misbehavior may also happen – and lead to time warps, i.e. unwanted jumps in time. As high robustness against both problems is required, several filters are employed by NTP to counteract both problems. The actual modification of the local clock is done by so-called Phase-Locked Loops (PLLs), which continuously correct the local oscillator phase and frequency variation relative to the received update. The complete process is depicted in Figure 14.6. Multiple servers are queried in parallel and filtering and selection algorithms are used to select the best measures for adjusting the local clock. A variable frequency oscillator (VFO) is used for a smooth clock update. The accuracy of NTP is, for example, discussed by Mills (1990).

Sources of errors

Typically, a number of sources of error can be expected, as described by Elson and Estrin (2001). It is the main task of time-synchronization algorithms to counteract these errors in order to improve the precision and efficiency:

- Skew in the receivers' local clocks – One way of reducing this error is to use NTP to discipline the frequency of each node's oscillator. Although running NTP all the time may lead to significant network utilization, it can still be useful for frequency discipline at very low-duty cycles.

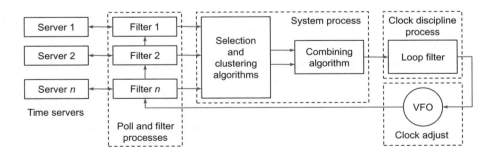

Figure 14.6 The process architecture as used by NTP

- Propagation delay of the synchronization pulse – Some methods assume that the synchronization pulse is an absolute time reference at the instant of its arrival – that is, that it arrives at every node at exactly the same time.

- Variable delays on the receivers – Even if the synchronization signal arrives at the same instant at all receivers, there is no guarantee that each receiver will detect the signal at the same instant. Nondeterminism in the detection hardware and operating system issues such as variable interrupt latency can contribute to unpredictable delays that are inconsistent across receivers.

14.2.3   Algorithms for WSNs

As shown in the last section, time synchronization is often hard to achieve, even if permanent connectivity is provided and all systems posses enough computational and energy resources. In this section, we discuss the possibilities of time synchronization in WSNs. Starting with an overview to general principles and requirements, we present and discuss a number of approaches for time synchronization in WSNs.

Principles and models

According to Elson and Römer (2003), the main design principles for time synchronization in WSNs can be summarized as shown in Table 14.2.

Further requirements for time synchronization in WSNs are discussed by Ganeriwal *et al.* (2005). All time-synchronization techniques used in today's networks are based on NTP or GPS. Both approaches are not suitable for time synchronization in sensor networks. Solutions in this domain need to be multimodal – the scheme should not only be able to adapt itself to the runtime system dynamics, but should also be able to optimize on different axes, e.g. scope, availability, precision and persistence of the time scale. Thus, another required capability is tunability.

Solutions for application in WSNs can be categorized according to the used model for time synchronization:

- Virtual clocks – Virtual clocks represent the simplest type of synchronization algorithms. Based on the concept of logical clocks, they concentrate on maintaining the

Table 14.2   Design principles for time synchronization in WSNs

Design principle	Description
Energy efficiency	The amount of work needed for time synchronization should be as small as possible
Scalability	Large populations of nodes must be supported in unstructured topologies
Robustness	The service must continuously adapt to conditions inside the network, despite dynamics that lead to network partitions
Ad hoc deployment	Algorithms for time synchronization must work without a priori configuration settings

relative notion of time between nodes based on the temporal order of events without reference to the absolute time. For example, Römer (2001) achieves a precision of 1 ms in establishing a correct chronology of events in WSNs.

- Internal synchronization – If only internal correctness is required, a class of synchronization algorithms can be used that actively maintain a common time in a single system or a group of nodes. Depending on the definition of 'internal', this may include the notion of virtual clocks in WSNs. Internal synchronization cannot be extended to maintain clocks for distributed coordination actions.

- External synchronization – External synchronization represents perhaps the most complex model. Every node maintains a local clock that is perfectly synchronized to a global and unique time scale.

- Hybrid synchronization – Internal and external synchronization can be used in a combined, hybrid way. This is usually less expensive and more simple compared with external synchronization. Basically, logical correctness is maintained over longer periods of time, accompanied by periodical synchronization to a global clock.

Post-facto synchronization

The post-facto synchronization algorithm proposed by Elson and Estrin (2001) assumes normally unsynchronized clocks. Whenever a stimulus arrives, the node records the time of the stimulus with respect to the local clock. Immediately afterwards, a third-party node – acting as a beacon – broadcasts a synchronization pulse to all nodes in its radio broadcast range. All nodes that are receiving this broadcast use it as an instantaneous time reference and can normalize their stimulus timestamp with respect to that reference. The principles of post-facto synchronization are depicted in Figure 14.7. A similar approach for temporal ordering has been investigated by Römer (2001).

Obviously, this approach is not applicable in all situations. It works within a local radio range only. Elson and Estrin (2001) reported a precision of about 1 μs for event rates higher than $10\,s^{-1}$. For lower event rates, the precision quickly degrades. A solution might be to use NTP-trained clocks, i.e. to use NTP in an initial phase to measure the accuracy of the local oscillator.

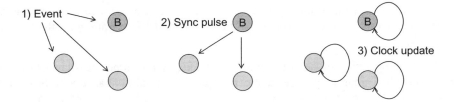

Figure 14.7 Principle of post-factor synchronization. First, an external event is received by all the nodes. This stimulates a dedicated beacon node to broadcast a synchronization pulse. Finally, all nodes update their local clocks

Timing-sync Protocol for Sensor Networks

The Timing-sync Protocol for Sensor Networks (TPSN) follows the sender–receiver model (Ganeriwal *et al.* 2003). Basically, a two-way message exchange is used, together with time stamping in the MAC layer of the radio stack. Measurements with an implementation for TinyOS[1] demonstrate an accuracy of 5 μs.

In order to establish a global time scale, TPSN creates a hierarchical structure in the network and performs pairwise synchronization along the edges of the root of this structure. In particular, two phases need to be distinguished in the protocol execution.

- Level discovery phase: The root node is assigned level 0 and initiates this phase by broadcasting a level discovery packet. Each node receiving this packet is assigned to level 1. Then, these nodes re-broadcast the discovery packet and so on and so forth.

- Synchronization phase: Pairwise synchronization is performed along the edges in the established tree based on sender–receiver synchronization according to the techniques described for NTP. In a two-way message exchange, which is periodically repeated – also described as the synchronization pulse–nodes can estimate the propagation delay δ and the clock drift θ. Based on the estimated clock drift, the receiver can adapt its local clock accordingly.

Reference Broadcast Synchronization

In contrast to TPSN, Reference Broadcast Synchronization (RBS) uses the receiver–receiver model (Elson *et al.* 2002). The fundamental property of this design is that it synchronizes a set of receivers with one another, as opposed to traditional protocols in which senders synchronize with receivers. Thus, it leverages the broadcast property inherent in wireless communication.

There are a number of prerequisites for proper operation of RBS. Essentially, all of them are inherently provided in typical WSNs and SANETs:

- Communications is performed as local broadcasts rather than unicasts between arbitrary nodes.

- Radio ranges are short compared with the product of the speed of light times the synchronization precision, i.e. each broadcast is seen essentially simultaneously by the receivers within range (1 μs roughly translates to 300 m).

- Delays between timestamping and sending a packet are significantly more variable than the delays between receiving and timestamping (due to waiting for the free radio medium).

In its basic operation mode, a transmitter periodically issues physical layer broadcasts without an explicit timestamp. Instead, receivers use its arrival time as a point of reference for comparing their clocks. Thus, RBS removes the sender's nondeterminism from the critical path and, in this way, produces high-precision clock agreement (about 2 μs, i.e. in the same order as TPSN). The critical path for sender–receiver and receiver–receiver synchronization is depicted in Figure 14.8.

---

[1]http://www.tinyos.net/.

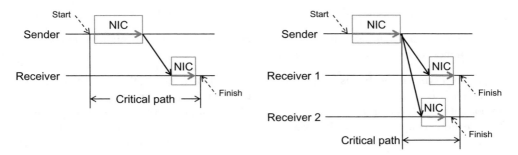

Figure 14.8   The critical paths for sender–receiver synchronization (left) and receiver–receiver synchronization (right); obviously, the critical path is much shorter in the latter case, as the nondeterministic latency at the sender to acquire channel access is removed

RBS can be used without external references, forming a precise relative times cale, or can maintain microsecond-level synchronization to an external time scale such as UTC. Such improvements to RBS have been, for example, studied by Karp *et al.* (2003).

In practice, some applications require all the sensor nodes to maintain a similar time within a certain tolerance throughout the lifetime of the network. Even though RBS can be extended for multi-hop time synchronization (using time translations), Su and Akyildiz (2005) pointed out that the impact of translation errors and delays of the synchronization still needs to be studied. Additionally, the energy dissipation end effects of node mobility in large-scale sensor networks, e.g. 300–2000 nodes, need to be addressed.

14.3   Distributed coordination

Coordination among a given number of nodes can be achieved in a local context based on a few message exchanges. In a global context, such coordination actions become complex and resource-expensive (communication and processing). A number of solutions have been proposed to address this problem. Basically, distributed coordination can be achieved by adequately grouping nodes into smaller clusters and coordinating only the nodes within such a group.

In this section, we outline the requirements for distributed and scalable coordination in massively distributed systems. Based on some selected examples, we show the current state of the art in sensor–actor coordination. Additionally, we discuss the problem of selfish nodes that try to exploit available resources while not contributing to the entire system.

14.3.1   Scalable coordination

The primary requirements in the context of scalable coordination can be summarized as follows (Bulushu *et al.* 2001):

- The algorithms need to be designed to support ad hoc deployment of SANET nodes, which continuously adapt to the environmental conditions.

- Untethered operation should be supported based on wireless radio communication.

- The coordination needs to be able to operate unattended as it might be infeasible to support continuous or periodic maintenance.

Estrin *et al.* (1999a,b) discuss scalable coordination based on their directed-diffusion approach (see Section 11.4). They outline the increased coordination complexity as the most important issue for distributed coordination. In particular, the number of involved systems, their heterogeneity in terms of hardware and software, and the node mobility (single-node vs large-scale movements; predictable vs unpredictable) need to be considered. It turned out that there is a substantial need for localized coordination algorithms supporting distributed coordination. Based on these observations, the following two resulting design choices can be identified:

- Data-centric communication – Sensor nodes may not have unique address identifiers. Therefore, pairs of attributes and values should be used to identify and to process received messages.

- Application-specific operation – Traditional communication networks are supposed to support a wide variety of applications. In contrast, WSNs and SANETs are often designed for specific purposes or configured for a particular purpose.

The coordination among collective agents has also been discussed by Mataric (1995b). In the design process of collective agents, communication is a major requirement. Such communication can be either direct, aimed at a particular, well specified destination, or indirect (stigmergic), based on observed behavior of agents. Then, cooperation is considered in this context as a special form of communication, again either explicit or implicit.

14.3.2   Selected algorithms

According to the observation that in high-density networks consisting of sensors and actors – a large number of such systems will be deployed due to further reduced cost and redundancy – manual configuration will become impossible and environmental dynamics will preclude design-time pre-configuration. Thus, self-configuration of a topology that provides communication under stringent energy constraints is demanded.

In the following, we investigate two approaches–Span and ASCENT, which provide distributed coordination to establish a basic network topology. The basic idea is that with increasing densities, only a subset of the nodes are necessary to establish a packet-forwarding backbone.

Span

The first approach is Span–a power-saving technique for generic ad hoc networks (Chen *et al.* 2002). It provides energy-efficient coordination for topology maintenance. Actually, Span is similar to Low-Energy Adaptive Clustering Hierarchy (LEACH), which we discussed in Section 12.2.1, but relies on localized coordination instead of random election schemes. The algorithm of Span adaptively elects coordinators from all nodes in the network responsible for forwarding. This coordinator is assumed to stay alive all the time.

In consequence, Span aims to achieve the following four goals in order to provide energy-efficient coordination.

- Span ensures that enough coordinators are elected to make sure that each node has a coordinator in its radio range.

- The coordinators are rotated to distribute workload.

- The algorithm aims at minimization of the number of coordinators in order to increase network lifetime.

- Span provides decentralized coordination relying on local state information only.

The protocol is based on proactive neighborship management using 'HELLO' messages. The basic coordinator-election algorithm works according to periodic broadcasts and to the coordinator eligibility rule. In essence, each non-coordinator node should become a coordinator if it discovers that two of its neighbors cannot reach each other, either directly or via one or more coordinators. This procedure ensures connectivity but does not minimize the number of coordinators or yield good capacity.

The announcement contention occurs when multiple nodes discover the lack of a coordinator at the same time – and all decide to become a coordinator. Span resolves this conflict by delaying coordinator announcements with a randomized back-off delay. Chen *et al.* (2002) propose a delay calculation specifically adapted to the environmental conditions.

Let N_i be the number of neighbors for node i and let C_i be the number of additional pairs of nodes among these neighbors that would be connected if node i would become a coordinator. Then, $C_i / \binom{N_i}{2}$ depicts the utility of node i (where $0 \le C_i \le \binom{N_i}{2}$). Obviously, the complete network benefits if nodes with high C_i become coordinators, as fewer coordinators will be needed to achieve connectivity.

Based on these observations, the back-off delay should be chosen randomly distributed and weighted by C_i. Additionally, it needs to be proportional to $N_i \times T$, with T being the round-trip delay between neighboring nodes. Thus, when all nodes have roughly the same energy resources, the back-off delay can be calculated as shown in Equation 14.4. The randomization is achieved by using the uniformly distributed parameter R in the interval $(0, 1)$.

$$delay = \left(\left(1 - \frac{C_i}{\binom{N_i}{2}} \right) + R \right) \times N_i \times T. \tag{14.4}$$

Nevertheless, the algorithm does not yet consider non-uniform energy resources. In order to incorporate the remaining energy into the decision process, the quotient E_r / E_m needs to be considered. E_r represents the remaining energy and E_m the maximum energy of a node. According to this observation, the back-off delay should be computed according to Equation 14.5:

$$delay = \left(\underbrace{\left(1 - \frac{E_r}{E_m} \right)}_{\text{remaining energy}} + \underbrace{\left(1 - \frac{C_i}{\binom{N_i}{2}} \right)}_{\text{utility of node } i} + \underbrace{R}_{\text{random value}} \right) \times \underbrace{N_i}_{\text{number of neighbors}} \times \underbrace{T}_{\text{round-trip delay}}.$$

$$\tag{14.5}$$

Fairness is achieved by Span because the likelihood of becoming a coordinator is reduced as the node uses up its energy. Please note, again, that the probability of becoming a coordinator is encoded in the back-off delay. The final decision process is quite simple. The first node that wishes to be become a coordinator, i.e. the one with the smallest back-off delay, wins.

ASCENT

Cerpa and Estrin (2002, 2004) developed an approach similar to Span, which was named Adaptive Self-Configuring Sensor Network Topologies (ASCENT). ASCENT nodes assess their connectivity and adapt their participation in the multi-hop network topology based on the measured operating region. In contrast to Span, ASCENT also incorporates the current network quality in terms of measured loss characteristics.

The following three operations are executed periodically by all ASCENT nodes in order to adapt to the current network conditions:

1. A node signals when it detects high packet loss, requesting additional nodes in the region to join the network in order to relay messages.

2. The node reduces its duty cycle if it detects losses due to collisions.

3. Additionally, the node probes the local communication environment and does not join the multi-hop routing infrastructure until it is 'helpful' to do so.

In order to achieve these objectives in a decentralized way, ASCENT provides adaptive techniques that permit applications to configure the underlying network topology based on their needs, e.g. to save energy and to increase the network lifetime. Additionally, the network conditions, e.g. the loss rate, are incorporated into the decision process. ASCENT relies on self-configuring techniques that react to operating conditions measured locally. In other words, these techniques ensure the proper self-election of nodes to become active, i.e. to provide network functions in a given part of the network.

The working behavior of ASCENT is depicted in Figure 14.9. All nodes are considered to be in one of the following four states: sleep, passive, test and active. Initially, each node is in test state. It observes its local environment, i.e. the number of active neighboring nodes and the current loss ratio. Additionally, it listens for help requests from local neighbors. After a pre-defined timeout T_t, the node automatically switches to active state.

The condition for switching between test and passive state is the current network behavior. If enough neighbors are active (threshold NT) or the loss ratio is increasing, i.e. the currently observed loss ratio is higher than the last measure, the node will switch to passive. On the other hand, if the number of neighboring nodes is decreasing below the threshold NT and the loss ratio is higher than LT, the node state will change to test for subsequent activation. Similarly, the node will switch to test if a help request has been received.

If in passive, ASCENT enforces a periodic sleep with duty-cycle $\dfrac{T_p}{T_p + T_s}$, i.e. the node state will change to sleep after T_p and back to passive after T_s. The complete process is repeated periodically.

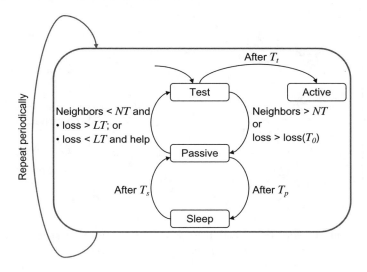

Figure 14.9 Working behavior of a single ASCENT node

Based on the available threshold parameters and timers, ASCENT allows tuning of the algorithm according to the needs of the application scenario. Additionally, the variability of the loss rate on particular connections is much higher for link loss rates of between 20 and 80 % (Cerpa *et al.* 2003). Thus, the efficiency of ASCENT can be improved if such knowledge is exploited.

14.3.3 Integrated sensor–actor and actor–actor coordination

Distributed coordination has also been investigated by Melodia *et al.* (2005, 2007). The authors propose an event-driven partitioning algorithm that associates sensors in a given area to distributed actors – from the sensor network point of view, these actors represent distributed sink nodes. Two algorithms were proposed to solve sensor–actor and actor–actor coordination, respectively.

The scenario is depicted in Figure 14.10. The objective is, first, to associate sensors to actors (sink nodes) and, secondly, to distribute-application specified tasks among the actors.

The sensor–actor coordination is based on the event-driven partitioning paradigm. The optimal strategy for this partitioning problem can be found by Integer Linear Programming (ILP). In particular, the Distributed Event-driven Partitioning and Routing (DEPR) protocol has been specified that constructs data-aggregation trees between the sources that reside in the event area and the appropriate actors in such a way as to provide the required reliability with minimum energy expenditure (Melodia *et al.* 2007). DEPR relies on local information and on greedy routing decisions, thus providing a good compromise between the energy efficiency of data-aggregation trees and the amount of topology information needed by each sensor node to take a routing decision.

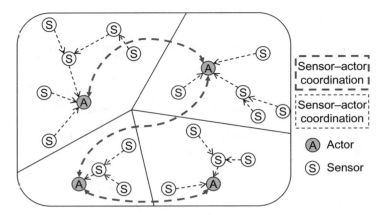

Figure 14.10 Sensor–actor coordination to associate sensors to a specific actor and actor–actor coordination for optimal task allocation in the SANET

For the sensor–actor coordination, Melodia *et al.* (2007) propose to rely on the reliability as the main measure. In order to support real-time requirements, a latency bound B is included in the used notion of reliability. If a message does not meet the latency bound B, the received event is expired and, thus, unreliable. In general, complying with pre-determined delay bounds requires some form of end-to-end feedback. In order to reduce the amount of necessary feedback information, collective feedback from the receiving actors is used.

DEPR relies on a few assumptions, which can be assured by accompanying protocols. First, each node needs to be aware of its own position. Additionally, it needs to know exactly the position of its neighboring actors. Finally, the clocks of all sensor nodes are assumed to be synchronized.

The event reliability r is the ratio of reliable data packets over all packets generated in a decision interval. The application can now describe its requirements in the form of a minimum reliability, i.e. the reliability threshold r_{th}. The main idea of DEPR is to continuously observe the reliability r and to compare it with the threshold. If $r > r_{th}$, it is possible to save energy. Thus, the node may enter a so-called aggregation state in which aggregation is exploited to reduce the number of messages in the network (based on the computed aggregation tree). If $r < r_{th}$, a speed-up procedure tries to increase the reliability by a greedy routing scheme to minimize the number of hops between source and actor. All decisions of DEPR additionally incorporate probabilistic state transitions in order to prevent synchronization effects and oscillations.

The main algorithm works as follows. For each decision interval, the actors compute the event reliability r as the ratio of unexpired packets over all generated packets and periodically broadcasts its value. The associated sensor nodes update their local behavior according to this reliability measure:

- $r < r_{th}^-$: If the advertised value r is below the so-called low event reliability threshold $r_{th}^- = r_{th} - \epsilon^-$, then it is necessary to speed up the data-delivery process by reducing the end-to-end delay. The speed-up state is chosen with a given probability $p_{speed-up}$.

- $r > r_{th}^+$: If the advertised value r is above the so-called high event reliability threshold $r_{th}^+ = r_{th} + \epsilon^+$, then there is reliability in excess that can be traded off for energy savings. Then, the aggregation state is chosen with probability $p_{aggregate}$.

The parameters ϵ^+ and ϵ^- basically define a tolerance zone around the required reliability threshold to reduce oscillations.

The objective of the speed-up state is to minimize the number of hops between sources and actors. This is achieved by applying a greedy routing scheme. Each node sends its data packets to the node closest to the destination within the transmission range. Obviously, this rule minimizes the number of hops in the path, the distance traveled by the packet and the number of transmissions of the same data packet. In the aggregation state, the node tries to reduce the overall energy consumption. This can be done by exploiting the aggregation trees for data transmission.

In addition, the actor–actor coordination problem is addressed, i.e. how to split the area among different actors. Two objectives are supposed to be supported by the coordination algorithm simultaneously: heterogeneous actors and distributed coordination. Both goals have been addressed using the Mixed Non-Linear Programming (MINLP) technique for the task-assignment problem. The resulting localized distributed solution for the coordination problem is basically comparable to auction as presented in the context of cooperation in SANETs. We will discuss auction-based task allocation in more detail in Chapter 15.

14.3.4 Problems with selfish nodes

In most scenarios, WSN and SANET nodes are considered equally 'interested' in participating in a particular application. Nevertheless, there might be networks consisting of devices under multiple administrative regulations or devices trying to increase their own lifetime by denying cooperative assistance (Hubaux et al. 2001). Such devices are called 'selfish', as they only focus on their own objectives, but not on the goals for the entire network. One solution is to enforce cooperation among the nodes by stimulation. Therefore, mechanisms to 'stimulate' nodes to participate on a given task are needed.

Buttyán and Hubaux (2003) developed a framework to stimulate nodes for packet forwarding. In particular, they addressed the problem of distinguishing between network lifetime and node lifetime. Selfish nodes will try to alleviate their own load by enforcing neighboring nodes to care for network-related operations, thus saving their own energy. This behavior may lead to inefficient operation in terms of optimized network lifetime (please refer to Section 17.3 for more information on lifetime estimation).

The solution as proposed by Buttyán and Hubaux (2003) is stimulation according to the provided work for the entire 'community'. Each device gets a tamper-resistant core–the nuglet–maintaining the credit of the node. Now, two simple rules can be used to enforce cooperation:

1. A node may send if it has enough credit, i.e. if its nuglet count is large enough. For an estimated n-hop transmission, the node requires n credits from the nuglet (the nuglet must not become negative).

2. Whenever a node forwards a packet, its credit is increased by one.

This principle is depicted in Figure 14.11 (left).

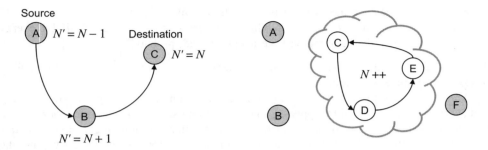

Figure 14.11 Nuglet management used for stimulated cooperation. Normal behavior (left) requires credits for sending at node A, node B earns credits for forwarding the packet. Groups of nodes may still act selfish by transmitting a packet in a loop (right)

The simplicity of the approach leads to a number of open issues. From a practical point of view, the needed tamper-resistant device may represent a major problem. While such devices can be developed, they must be installed on each and every node, and the nodes must be developed such that no bypassing is possible. A second problem is that border nodes will never be able to send after exhausting their credits, as they will never be allowed to forward packets. Thus, the algorithm needs to be adapted according to the relative position of the node with respect to the sink. This, again, provides problems with the tamper-proof device. If all other problems can be solved, selfish single nodes can be prevented. Unfortunately, groups of nodes may still act selfish. They may acknowledge each other credits or they may just send packets in a loop among themselves in order to receive credits – while not contributing to the network operation. This effect is shown in Figure 14.11 (right).

14.4   In-network operation and control

The concept of in-network processing in SANETs has been often discussed in the context of energy-aware operation. We already discussed the high costs of wireless radio communication in comparison with computational efforts. Studying algorithms for operation and control in massively distributed systems, two additional objectives have been identified that motivate the in-network operation in SANETs: scalability and timeliness. If messages are processed 'near' to the sources and if control loops are inherently integrated into the network nodes, then the degree of self-organization in such distributed systems can be increased.

In this section, we outline the basic principles of in-network operation and control. In addition, we depict one particular approach, namely RSN, to program SANETs according to the mentioned design objectives.

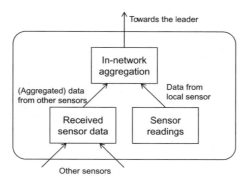

Figure 14.12 The cougar architecture; each node performs data aggregation in order to reduce the communication overhead and to validate the sensor data

Cougar

One of the proposals that specifically pointed out the need for in-network operation was the cougar approach (Yao and Gehrke 2002). The idea is to enforce in-network query processing in WSNs, for two reasons. First, computation is much cheaper compared with communication (energy); therefore, reduced communications lead to essential energy savings. Secondly, sensor readings might be failure-prone; therefore, adequate validation is needed.

The cougar architecture is depicted in Figure 14.12. Each node is able to perform local measurements (sensor readings) and to receive sensor data from other nodes. Based on this information, it can perform aggregation and validation operations.

We already discussed data aggregation in sensor networks in Section 11.5. Therefore, we concentrate on the validation function. Consider a given sensor node able to measure a physical value such as the temperature. Assuming that an error will always be introduced in this measurement and a Gaussian distribution of the error with a known standard deviation, then the real temperature T will lie in the interval $[T_1, T_2]$, with a given probability p. As long as multiple sensor nodes measure the same physical phenomenon, their readings can be aggregated to construct a 'super-node' whose temperature readings have a much lower variance, i.e. the aggregated temperature T_a lies in an interval $[T_{a1}, T_{a2}]$ with $T_{a1} \geq T_1$ and $T_{a2} \leq T_2$. Outliers can be identified if the particular measures do not match the aggregated confidence interval $[T_{a1}, T_{a2}]$.

Rule-based data processing

Dressler (2006a, 2007) proposed an architecture named Rule-based Sensor Network (RSN) for SANET programming and operation that consequently follows the data-centric communication approach and enforces a complete network-centric operation.

The typical operation in SANETs is depicted in Figure 14.13 (left). A number of sensor nodes are measuring physical values according to a scheme specified by the particular

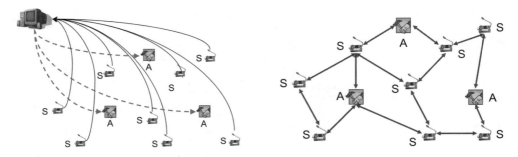

Figure 14.13 Centralized operation in SANETs (left) relies on a central base station to which all sensor information is transmitted. The actor control is communicated back into the network. In contrast, network-centric operation and control (right) rely on distributed sensing and acting without central helpers

application. This information is transported to a sink node for further processing and evaluation. Finally, the actors are controlled by dedicated sink–actor commands. Obviously, such an operation has a number of disadvantages, such as the high communication demands and high sensor–actor delays. In the last few sections, we have already discussed particular solutions for optimized sensor–actor communication such as that proposed by Melodia *et al.* (2007).

The overall goal is, therefore, to keep the control in the network. The intended operation principle is depicted in Figure 14.13 (right). Sensors and actors build an integrated network, generating and acting on sensor measures.

Principles of RSN   The key objectives motivating the development of RSN can be summarized as follows. Self-organized operation without central control will lead to reduced network utilization and accelerated response, i.e. actuation. Thus, the scalability and real-time capabilities are increased. Additionally, centralized 'helpers' may be allowed, e.g. for self-learning capabilities.

RSN is based on the following three design objectives that enable the mentioned objectives:

- Data-centric operation – Each message carries all necessary information to allow data-specific handling and processing without further knowledge, e.g. on the network topology.

- Specific reaction on received data – A rule-based programming scheme is used to describe specific actions to be taken after the reception of particular information fragments.

- Simple local behavior control – We do not intend to control the overall system, but focus on the operation of the individual node instead. Simple state machines have been designed, which control each node (being either sensor or actor).

RSN follows the approach used in data-centric communication schemes. Thus, instead of carrying address information, each message is encoded using a (type, content) pair.

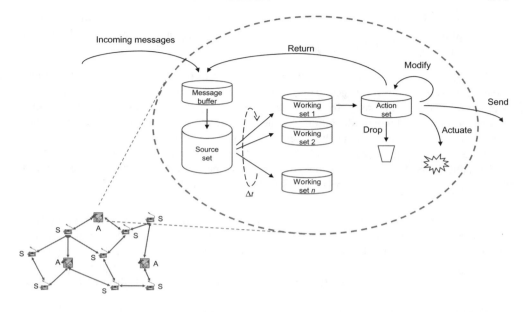

Figure 14.14 The working behavior of a single RSN node; received messages are stored in a buffer, selected to a working set according to specific criteria, and finally processed, i.e. forwarded, dropped, etc.

The type describes the message and the attached content. The data themselves will usually include a value and application-specific meta information, such as a geographical position or priority information.

Figure 14.14 depicts the working behavior of a single RSN node. After receiving a message, it is stored in a message buffer. Periodically (or after receiving a new message), the rule interpreter is started. The specific reaction on received data is achieved by means of predicates. RSN is able to select all messages of a given type or messages with specific content attributes. All selected messages are stored in so called working sets. From this set, RSN performs one of the following actions:

- modify – A message or a set of messages can be modified, e.g. to fuse the carried information with locally available meta information.

- return – Messages may be returned to the message buffer for later processing, e.g. for duplicate detection or improved aggregation.

- send – Obviously, a node needs to be able to send messages. This can be a simple forwarding of messages that have been received or the creation of completely new messages needed to coordinate with neighboring nodes.

- actuate – Local actuators can be controlled by received messages, e.g. to enable sensor–actor feedback loops.

- drop – Finally, the node needs to be able to drop messages which are no longer required, e.g. because they represent duplicates or because an aggregated message has already been created and forwarded.

The 'send' and 'return' actions may send or return the original message, but also create completely new messages. This action is used, for example, for data aggregation. The aggregated message may, for example, be used to carry the mean value of all source messages and the standard deviation. In the context of data aggregation, the execution period Δt of the rule interpreter also becomes important. If executed too often, typically, no more than a single message will be available, which makes aggregation techniques dispensable. If performed with a low frequency, the per-hop delay will be increased and, therefore, the real-time operation of the SANET will be limited.

Application examples   In order to demonstrate the capabilities, two simple application examples are depicted in the following. First, the probabilistic data-forwarding technique gossiping is reproduced in RSN (the algorithm is described in Section 11.2.2). The algorithm according to Haas *et al.* (2002a) forwards packets with a given gossiping probability p. In order to cope with special cases (problems) such as linear networks, the flooding is used for the first n hops.

Each message is assumed to be encoded in the following way:

```
M := { type, hopCount, content }
```

Then, the gossiping algorithm can be formulated as follows:

```
# infinite loop prevention
if $hopCount >= networkDiameter then {
    !drop;
}
# flooding for the first n hops
if $hopCount < n then {
    !sendAll;
    !drop;
}
# gossiping
if :random < p then {
    !sendAll;
    !drop;
}
# clean up
!drop;
```

In the first block, all messages are selected that have a hopCount greater or equal to networkDiameter. These messages are silently dropped (!drop). This command is included to prevent infinite loops. The second block selects all messages with hopCount smaller than n and forwards these messages (!sendAll). After processing the messages, they are discarded. The third block selects all remaining messages in the working set if an on-demand calculated random value (:random) is smaller than the gossiping probability p. These messages are forwarded and all remaining messages are dropped.

From this simple example, two mechanisms become obvious. First, each command operates on sets of messages instead of single messages. Secondly, messages remain in the working set until they are dropped. Thus, multiple commands may be applied to particular messages.

A second example should demonstrate more sophisticated applications. In this example, the sensors are used to measure the temperature. Data aggregation is performed to reduce the number of messages in the system. Additionally, critical temperature values are observed and alarm messages are created if a threshold has been exceeded.

The message encoding is similar to the previous example:

```
M := { type, position, content, priority }
type := ( temperate || alarm )
```

The complete algorithm can now be written as follows:

```
# test for exceeded threshold and generate an alarm message
if $type = temperature && $content > threshold then {
    !actuate(buzzerOn);
    !send($type := alarm, $priority = 1);
}
# perform data aggregation
if $type = temperature && :count > 1 then {
    !send($content := @media of $content,
          $priority := 1 - @product of $priority);
    !drop;
}
# message forwarding, e.g. according to the WPDD algorithm (simplified)
if :random < $priority then {
    !sendAll;
    !drop;
}
!drop;
```

In this example, the three command blocks actually perform different operations. The first block tests the temperature value and, if the threshold is exceeded, a local actuation is enforced (a buzzer is turned on – !actuate) and a new alarm message is generated with message priority set to one (!send). In the second block, all temperature messages are aggregated (if more than one has been received – :count). The content is set to the median of all temperature values and the message priority is increased (please refer to our discussion on aggregation in Section 11.5). Finally, the last block is in charge of message forwarding. Presented is a simplified version of Weighted Probabilistic Data Dissemination (WPDD) (see Section 11.2.3).

From these two examples, it can be seen that RSN provides a powerful set of commands to enable in-network operation and control for SANETs. Nevertheless, a number of open issues still exist:

- Handling of unknown messages – Which action should be performed if unknown messages, i.e. messages of unknown type, have been received? Basically, two decisions are possible–drop vs seamless forwarding–while not being appropriate in all application scenarios.

- Period of RSN execution Δt – The duration of messages stored in the local node introduces an artificial per-hop delay. The optimal value for Δt affects the aggregation quality vs. real-time message processing.

- Rule generation and distribution – So far, we have considered homogeneously programmed nodes. This is not necessarily the optimal case. Also, new rules may be
required during the lifetime of the network. The rule deployment needs further research in terms of diffuse or random distribution vs global optimization.

14.5 Conclusion

In this chapter, we discussed communication aspects for coordination in SANETs. The
most important issues for communication in this domain are energy efficiency and real-
time operation. Accordingly, mechanisms are needed for distributed coordination among
the networked nodes to establish a working network infrastructure that supports low-energy
operation and that also provides QoS features such as delay bounds for arbitrary communication.

Necessary features to develop such techniques are synchronization and coordination.
We outlined the typical synchronization problems in distributed systems – which also
apply to massively distributed systems such as WSNs and SANETs. Often, coordination
(or weak synchronization) is considered to be the final answer for sensor–actor control.
Such coordination can, for example, be achieved based on a distributed logical time basis.

In particular, we discussed the following issues and concepts that are strongly demanded
for communication and coordination in SANET, i.e. for sensor–actor coordination:

- Synchronization vs coordination – In our discussion on synchronization principles
among networked nodes, we figured out that clock synchronization is hard to
achieve – even in more reliable and deterministic networks compared with wireless
radio communication. The unpredictable clock drift of all local clocks of the synchronizing nodes requires permanent control and manipulation. Therefore, Lamport
(1978) proposed the notion of logical time to control at least the order of observable
events. The basic concept relies on dynamic manipulation of the logical time with
each message exchange. Coordination (weak synchronization) based on logical time
is the basis for operating huge numbers of nodes in a common environment.

- Time synchronization – In the context of the time-synchronization, we started with
a brief overview of Internet-based synchronization techniques. These mechanisms
provided a good starting point for further discussions of the time synchronization algorithm developed specifically for WSNs. In particular, we have seen the advantages
and drawbacks of three techniques: the maintenance of a logical time (post-facto
synchronization), the sender–receiver model, in which a single time server tries to
synchronize one or more clients to a given global time, and the receiver–receiver
model, in which the uncertainty of the sender (when to acquire access to the radio
channel) has been removed.

- Distributed coordination – The primary goal of this chapter was to describe coordination algorithms. We discussed a number of well known algorithms for distributed
coordination among sensor nodes and for sensor–actor coordination. Most of these
algorithms are based on clustering techniques, as presented in Chapter 12. Nevertheless, in the presented algorithms, the distributed coordination came to the fore. Local

coordination is provided in terms of energy efficiency and to control available network resources, e.g. to limit the maximum network delay. In addition, we discussed the problem of selfishness in distributed coordination algorithms and presented an example to explicitly stimulate coordination.

- In-network operation and control – In this part, we discussed the need to distribute all operation and control functions into the network. Actually, most of the distributed coordination techniques are based on this concept. As an example, we presented RSN–a rule-based programming paradigm for SANETs. This scheme allows the processing of messages and the initiation of actuation by consequently following the data-centric communication approach.

15

Collaboration and Task Allocation

Collaboration is one of the main objectives in SANETs. Usually, collaboration refers to the joint work of multiple systems on a common goal. In SANETs, such joint work can be required in many facets. Sensors can assist mobile robot systems for efficient navigation and communication, robots can assist the sensor network in terms of maintenance and repair and, finally, multiple sensors or robots might be required to solve a single task. In conclusion, it can be said that there are a great number of problems requiring collaboration on common goals or tasks.

The algorithms for solving such collaboration issues are usually known as task and resource allocation. Primarily, task and resource allocation has been studied in the context of groups of robots, i.e. multi-robot systems that are considered to perform multiple tasks in parallel in order to either speed up the problem-solving process compared with a single robot's doing the task, or to enable the solution of complex problems that cannot be solved (or executed) by a single system.

The allocation of tasks to single systems out of a number of available systems and the allocation of multiple systems to a single task are also known as Multi-Robot Task Allocation (MRTA). Similar applications can be found in WSNs and other types of SANETs, in which a number of tasks need to be assigned to multiple systems. The algorithms for MRTA proposed in the literature typically focus on task allocation. Nevertheless, they can usually be applied to general-purpose resource allocation as well. Examples of such resources related to SANETs are storage (where to keep which data) and communication facilities (which nodes are required for building an optimal network topology).

Similarly to other methods used in the context of SANETs, the main objectives for MRTA are network lifetime and scalability. Independently of particular application-defined strategies, MRTA needs to consider a number of system-inherent limitations. Additionally, domain-specific constraints need to be considered to cope with lifetime and scalability (please refer to Chapter 17 for more details):

Self-Organization in Sensor and Actor Networks   Falko Dressler
© 2007 John Wiley & Sons, Ltd

- Communication – Most task-allocation techniques rely on intensive communication among the available systems. Therefore, communication-related constraints and limitations need to be considered for the development and the evaluation of MRTA.

- Energy – Discussing MRTA, we usually rely on real robot systems as a typical application domain. Because robots are considered to be battery-powered, energy restrictions are quite important for any algorithm developed in this context.

- Time – Differently from many network applications, the execution time of a task needs to be considered for MRTA. Real-time collaboration requires synchronization and coordination techniques but also optimal scheduling strategies for dynamic task allocation.

Accordingly, MRTA can be reduced to an optimization problem with inherent time restrictions. Focusing on task-allocation problems typical of SANETs, the proposed solutions can basically be assigned to the following two categories of MRTA approaches (Gerkey 2003):

- Intentional cooperation – The model of intentional cooperation enforces robots to explicitly cooperate with purpose, often through task-related communication. This model is well suited to real-world tasks to be associated to robot systems. Furthermore, intentional cooperation has the potential to better exploit the capabilities of heterogeneous robots.

- Emergent cooperation – In contrast to intentional cooperation, robots do not explicitly work together for emergent cooperation. Instead, group-level cooperative behavior emerges from the interactions among the systems and with the local environment. An emergent system is usually able to perform a specific task efficiently while not providing general-purpose functionality.

In this chapter, we will first introduce task and resource allocation and its primary objectives. The typical problem spaces and possible strategies will be categorized in the form of a taxonomy. Afterwards, we will focus on the two primary-task allocation models–intentional cooperation and emergent cooperation. Based on a number of case studies, we discuss the advantages and limitations of both approaches for MRTA.

15.1   Introduction to MRTA

In our discussion of task-allocation mechanisms, we will specifically focus on MRTA. The discussed approaches and solutions are nonetheless, in most cases, applicable to general task and resource-allocation problems as well.

Therefore, we need to briefly introduce the research field of multi-robot systems. Basically, all problems can be considered a target for multi-robot systems if either multiple robots are needed to solve a single task or if a benefit, e.g. time savings, can be expected from using multiple systems. Arai *et al.* (2002) provided an overview to research issues in multi-robot systems. They identified a number of relevant primary research topics. In particular, the authors listed the following topics to be relevant in the area of multi-robot systems:

- biological inspirations;

- communication;

- architectures, task allocation and control;

- localization, mapping and exploration;

- object transportation and manipulation;

- motion coordination;

- reconfigurable robots.

Obviously, at least three domains, i.e. communication, task allocation and coordination, rely on the support of efficient communication and coordination techniques which we are discussing for SANETs. Task allocation is explicitly investigated in the domain of multi-robot systems.

The problem of MRTA is depicted in Figure 15.1. The tasks T_i have to be associated to a number of robots R_j. Depending on the requirements of the particular task, one or more robots need to be selected for performing the desired actions. Tasks may be annotated with task-specific constraints that influence the selection process. Examples are given deadlines, required resources at the executing robots or dependencies between multiple tasks. In the depicted example, the execution area is the only execution constraint.

Obviously, task and resource allocation represents a specific domain of the Optimal Assignment Problem (OAP) (Gale 1960), which was originally studied in game theory and then in operations research. A general introduction to multi-robot task allocation has, for example, been provided by Gerkey (2003). In the following, we will outline the primary objectives of task-allocation. Afterwards, we outline possible approaches for classifying task-allocation approaches according to the problem space and to the employed coordination model.

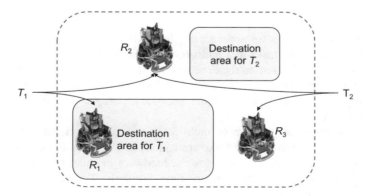

Figure 15.1 Simple example of the task-allocation problem; the tasks T_1 and T_2 need to be assigned to one out of the three robots R_1, R_2 and R_3 under the constraint to be executed in a specific area

15.1.1 Primary objectives

The primary objective of task-allocation algorithms is to identify an appropriate system (or a set of multiple systems) in a group of autonomous systems that has the required resources, these resources are available and the system is available to perform the requested task. Thus, a primary prerequisite for allocating a task to a particular system is the availability of resources. Such resources can be characterized by functional and non-functional properties.

Functional properties are, for example, the availability of a particular hardware or software module. On the other hand, non-functional properties describe system dynamics and the consumption of resources. Examples are energy consumption, QoS-related characteristics such as the execution time and the current position of a system within the environment.

Types of resources are special hardware and software components, the CPU capacity, memory and storage options, the available energy and the position of the system. Additional constraints are given through possible time dependencies. Tasks may require different execution times at different systems. This demands optimal scheduling strategies if multiple concurrent tasks are allowed. Additionally, tasks may be created at any time, even during the execution of previously allocated tasks. If preemptive task allocation is supported, tasks may be either interrupted or reallocated to another system. Thus, task-allocation techniques supporting preemption will provide better support for application domains with high dynamics.

In summary, the following conditions must be met in order to allocate a task to a system:

1. The destination system must own the necessary resources and be able to perform the task with the given conditions described by non-functional properties.

2. The required resources must be available at the time of task execution and all preconditions must be met to perform the task, e.g. the needed position of the system must be ensured.

3. The system must be available to execute the task – for non-preemptive task-allocation strategies, this might be the most challenging condition.

According to the concepts for approaching the OAP problem, Gerkey (2003) described two measures for allocating tasks in a multi-robot scenario. Basically, the fitness for every task needs to be evaluated based on the expected quality of task execution, e.g. the accuracy or the performance, and the expected resource cost given the spatio-temporal requirements of the task, e.g. the required power or the necessary storage. This fitness is often also referred to as the task utility.

Figure 15.2 depicts the problem of multi-robot task allocation. Shown are two tasks T_1 and T_2 ready for execution. Both tasks are annotated with specific requirements on the execution system. In particular, T_1 requests the hardware module HW-A and the software module SW-2 to be available at the execution time. Similarly, T_2 requests specific hardware (HW-A, HW-C) and software (SW-2) modules.

The required features are listed in form of profiles. In the context of SANETs, such profiles together with respective profile-matching techniques are often used to identify available features and configurations (Truchat et al. 2006), and for efficient system specific software management, respectively (Dressler et al. 2007b; Yao et al. 2006). The following

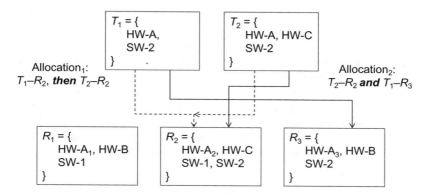

Figure 15.2 Task allocation in multi-robot systems. Depicted are two tasks T_1 and T_2, ready for execution. The arrows show two possible allocations. Either two systems (R_2 and R_3) can be employed for parallel execution or both tasks may be allocated to R_2, to be performed sequentially

example outlines the capabilities of the profiles. The profile can either describe the available resources at a system or the demanded resources by a particular task.

```
# hardware capabilities
processor {PowerPC, 8MHz}    // processor of type PowerPC with 8MHz
memory {128MB}               // memory size 128MB
chassis {indoor, 1m/s}       // indoor movement with a speed of 1m/s
camera {color, 1Mpixel}      // color camera with 1Mpixel resolution
# software capabilities
mapping software             // algorithms for dynamic map generation
JPEG encoder                 // JPEG picture encoder
face recognition             // face recognition software
object tracking              // CPU and memory expensive tracking
```

In our simple example, shown in Figure 15.2, three robot systems R_1, R_2 and R_3 are present in the system. All of them are available for task execution while providing different capabilities in terms of available hardware and software modules. Two possible allocation schemes are depicted in the figure:

- Parallel execution – The robots R_2 and R_3 can be used to perform tasks T_2 and T_1, respectively, in parallel.

- Sequential execution – Robot R_2 is able to execute both tasks. Thus, it may perform these tasks sequentially.

Depending on the application requirements on the non-functional properties – which are not depicted in our example – both solutions might be appropriate. Consider, for example, robot R_2 being able to perform both tasks twice as fast compared with R_3. The overall execution time for the two tasks will be the same for both allocations, while the power consumption might be different. A utility function is usually employed to describe the fitting of a task to a particular system (if a robot is unable to perform a specific task, the utility function provides a rating of zero for that job).

MRTA has been studied primarily for application in robots (multi-robot systems). Nevertheless, the developed algorithms are directly applicable to any multi-system architectures that need to distribute workload among a number of available systems. Distributed and self-organizing systems such as SANETs inherently rely on the availability of such task-and resource-allocation strategies.

One particular example to be mentioned here are multi-agent systems. This domain has been initiated in the field of Artificial Intelligence (AI) and many task-allocation algorithms have been developed in the context of multi-agent systems (Gage *et al.* 2004; Gerkey 2003). Basically, virtual auctions are created that enable systems to bid for a particular task. In order to improve the allocation quality and performance, the following schemes can be used in addition:

- Motivation-based – The exploitation of the needs of single systems to motivate them to participate on a given task.

- Mutual inhibition – The inhibition of specific actions according to the quality or task execution or as a strategic action.

- Team consensus – The exploitation of decisions in a group of autonomous systems for team-level allocation improvements.

15.1.2   Classification and taxonomy

As previously described, the allocation process is actually the identification of available nodes that show the required properties. As with most optimization problems, the allocation process will usually involve the following three steps:

1. (Self-)election of systems – Available systems are identified that are able to perform the given task. This step can be organized using distributed election algorithms or by means of self-election.

2. Allocation proposal – A proposal for the final allocation is provided through negotiation or distributed self-allocation, e.g. according to a probabilistic scheme.

3. Optimization – The allocation can be further optimized to either improve the allocation proposal or to include new tasks and systems in the decision process (allocation improvement).

Following the discussions by Akyildiz and Kasimoglu (2004) and by Gerkey (2003), the allocation process can be classified in two dimensions. First, the environmental conditions (number of tasks, number of robots) require different approaches and, secondly, the allocation procedure itself (instantaneous or continuous process, centralized or distributed operation) allows multiple solutions.

Environmental conditions

According to the number of tasks to be executed by a single robot and the availability of multiple tasks and robot systems, different allocation strategies as depicted in Figure 15.3, can be assigned. Here, ST and MT refer to a single task and multiple tasks, respectively. Accordingly, SR and MR denote single robot and multiple robots, respectively.

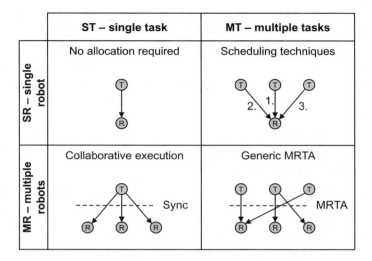

	ST – single task	MT – multiple tasks
SR – single robot	No allocation required	Scheduling techniques
MR – multiple robots	Collaborative execution	Generic MRTA

Figure 15.3 Taxonomy of MRTA according to the number of available robots and tasks

As denoted in Figure 15.3, ST–SR does not depend on any allocation algorithms and MT–SR can be addressed by typical scheduling algorithms (nevertheless, MT–SR is covered by MRTA as well). Conversely, ST–MR represents a class of tasks in which multiple robots need to cooperate on a single task. Usually, synchronization techniques are needed to coordinate either sequential or parallel processing of multiple systems on this single task. Finally, MT–MR is the generic class of MRTA problems in which multiple systems are available to perform multiple tasks concurrently. Depending on the objectives of each task, either a single robot might perform this task, which is also known as single-actor task (SAT), or multiple robots may interact on the task, which then refers to a multi-actor task (MAT).

Allocation procedure

The allocation procedure itself can be described according to a number of methodological properties. First of all, centralized decision procedures need to be distinguished from distributed decision algorithms. Actually, the most challenging scenario is MRTA with distributed decision taking, which requires self-organization among the available systems. Typically, nodes negotiate with each other and coordinate locally to select the 'best' fitting system (Akyildiz and Kasimoglu 2004).

A second classification parameter is the allocation time. For instantaneous assignments, all the information concerning the robots, the tasks and the environment must be available at allocation time. Instantaneous assignments allow no planning for future allocations. Typically, such allocations are non-preemptive, i.e. re-allocations will not be possible. In contrast, time-extended assignments support either the approximation of future arrivals or tasks to incorporate this information in the allocation scheme. Additionally, preemptive assignments may be supported to move tasks between robot systems during execution time.

The last, and most important, characterization used to classify allocation procedures is the differentiation of intentional and emergent cooperation. In the following sections, we will follow this classification for our discussion of selected case studies:

- Intentional cooperation – Allocation algorithms are, for example, negotiation and mediation as investigated in the context of multi-agent systems. This includes auction-based negotiations, center-based algorithms and mediation procedures for allocation improvement.

- Emergent cooperation – Examples are probabilistic methods with inherent calibration and self-optimization features. According to the working principles, such solutions are often called swarm-type cooperation (Parker 1998).

In addition to pure MRTA, more sophisticated approaches have been described in the literature. An example focusing on the application in SANETs is Distributed In-network Task Allocation (DINTA) (Batalin and Sukhatme 2003b,c). A static sensor network is used for allocation assistance in a distributed SANET environment consisting of heterogeneous nodes. Tasks, upon arrival, are allocated implicitly to robots by a pre-deployed static-sensor network. This provides the possibility of preventing expensive (real-time) communication between the robots. Additionally, it allows optimized solutions for robots to perform spatially and temporally distributed tasks.

The process of designing and understanding adaptive group behavior has been proposed by Mataric (1995a,b). She investigated a number of simple basic behaviors, as summarized in Table 15.1, which can be combined to perform complex actions.

Multi-Robot Task Allocation

In conclusion, MRTA refers to the allocation of n tasks to m robots according to a utility function $\mathcal{U}(T_i^*, R_j^*)$, which outlines the 'quality' of this mapping. If a robot is unable to perform a particular task, the utility function will return a rating of zero. This mapping can be described as show in Equation 15.1:

$$T_n \xrightarrow{\mathcal{U}(T_i^*, R_j^*)} R_m. \tag{15.1}$$

Table 15.1 A basis behavior set intended to cover a variety of spatial interactions and tasks for a group of mobile robots

Behavior	Description
Safe-Wandering	The ability of a group of agents to move about while avoiding collisions with obstacles and each other
Following	The ability of an agent to move behind another, retracing its path and maintaining a line or queue
Dispersion	The ability of a group of agents to spread out in order to establish and maintain some minimum inter-agent distance
Aggregation	The ability of a group of agents to gather in order to establish and maintain some maximum inter-agent distance
Homing	The ability to find a particular region or location

Based on the discussed issues, i.e. the environmental conditions, the allocation procedure and the utility function, a complete definition of MRTA can be given.

Definition 15.1.1 Multi-Robot Task Allocation (MRTA) – The task-allocation problem is a mapping of m robots to n tasks, each requiring at least one robot. According to a given utility function $\mathcal{U}$, which represents a non-negative efficiency rating estimating the performance of a (set of) robots performing a particular task, the goal is to assign the robots to the tasks so as to maximize the overall performance. This assignment incorporates the priorities or weightings of the tasks and the utility ratings.

It should be noted that all allocation problems are $\mathcal{NP}$-hard. This definition also applies for general purpose task and resource allocation in distributed and self-organizing systems. Then, the notion of robots and tasks needs to be replaced according to the domain-specific terminology.

15.2 Intentional cooperation – auction-based task allocation

In this section, we provide a number of case studies for intentional cooperation. Typically, auction-based task-allocation mechanisms are used in this context, focusing on negotiation among a number of distributed systems. Simple auctions have been discussed, for example, to provide a scheme for task allocation that guarantees the quality of the allocation (Lagoudakis *et al.* 2004). In the following, we will discuss a number of auction-based negotiation algorithms providing task allocation for multi-agent systems. Following the notion typically used in this domain, we refer to agents as systems able to perform a particular task.

15.2.1 Open Agent Architecture

The Open Agent Architecture (OAA) is a framework for constructing multi-agent systems (Cheyer and Martin 2001; Martin *et al.* 1999). The approach concentrates on distributed computing and the primary goal of OAA is to provide a means for integrating heterogeneous applications in a distributed infrastructure. Among the key objectives of OAA, the interoperation and cooperation of agents have been identified to be of high importance for the efficiency of the entire system.

According to Martin *et al.* (1999), interoperation refers to the ability of agents to communicate. On the other hand, cooperation stands for mechanisms that allow a community of agents to work together productively on some task.

OAA follows the centralized coordination approach, as depicted in Figure 15.4. The central entity that is responsible for task allocation is named the facilitator agent. A system is not necessarily limited to a single facilitator. Nevertheless, the assignment of agents, named clients in the context of OAA, to a facilitator needs to remain fixed for the complete task-allocation process.

The working behavior of OAA is based on two operations. First, agents are providing services. Basically, they describe their capabilities to solve tasks in the form of profiles to the facilitator. This state information can be periodically refreshed in order to provide fault

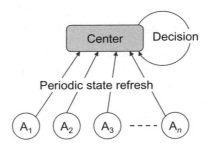

Figure 15.4 Centralized task allocation as used by OAA

tolerance. Secondly, agents can request services, i.e. the execution of one or more tasks. All task-allocation decisions are taken by the facilitator, based on the available information about the capabilities of the associated clients and the current state according to currently allocated tasks and other system information.

After the arrival of a new task, the facilitator performs the following two steps in order to distribute and to execute this task:

1. Facilitation – The facilitator decides how to distribute the new task to the available agents. This step also includes optimization techniques to improve the system behavior as well, e.g. according to approximated or learned task-execution times.

2. Delegation – Then, the facilitator delegates the task to one or more agents for execution. For reliable task allocation, the successful execution is monitored (or acknowledged) and, in the case of a failure, re-allocated to another agent.

Obviously, OAA can optimize the task allocation to a certain extent. This optimization is based on the periodic state updates that the facilitator received from all the available agents. Nevertheless, two problems need to be mentioned as well. First, the optimization process can only incorporate known tasks and available resources in the allocation process. Consider a stable system with previously allocated tasks. If a new task arrives or if new resources are made available to the system, the previous allocation might not be optimal for the changed conditions. Either the new task cannot be allocated because not all necessary resources are available or a different overall allocation scheme might be possible. On the other hand, the new resources would allow a much better allocation compared with the current scheme. Re-allocation could be a feasible solution, which is nevertheless not directly supported by OAA.

A second problem is the scalability of OAA. If the system grows in terms of available agents, the necessary communication for periodic state updates will grow as well. The same holds for the necessary memory at the facilitator to hold the complete state information and for the optimization algorithm that needs to process these data. Therefore, an upper bound for the system size can be predicted, given the available resources in terms of communication and computation facilities.

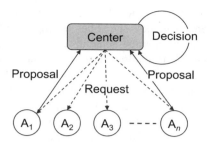

Figure 15.5 Center-based task assignment used by MURDOCH and dynamic negotiation

15.2.2 MURDOCH

An auction-based task-allocation system proposed in the context of MRTA is MURDOCH (Gerkey 2003). Similarly to OAA, MURDOCH supports the cooperation of heterogeneous systems that possess different skills. In contrast, communication is no longer considered reliable and the centralized coordinator is replaced by an auctioneer.

The architecture of MURDOCH is shown in Figure 15.5. The auctioneer requests the execution of a particular task by broadcasting a request message to all available agents. Accordingly, the agents will evaluate their willingness to participate and then propose to take over the task in their bid. Based on the received bids, the auctioneer can decide on the task allocation by means of available optimization strategies.

The MURDOCH auction protocol consists of the following five operations (Gerkey 2003):

1. Task announcement – The auctioneer, i.e. an agent working on behalf the task, publishes an announcement for this task. This announcement contains detailed information about the task. This message is either broadcasted to all available agents or submitted only to the agents being able to perform the task.

2. Metric evaluation – A metric-based evaluation is performed on the agent, providing a sound basis for achieving effective task allocation. The goal is to allocate each task to the best fitting agent.

3. Bid submission – After evaluating the appropriate metric(s), each candidate agent publishes its resulting task-specific fitness in the form of a bid message.

4. Close of auction – The auction is closed after sufficient time has passed. Then, the auctioneer processes the bids and determines the best candidate for executing the task. The winner is awarded a time-limited contract to execute the task and the losers return to listening for new tasks.

5. Progress monitoring/contract renewal – The auctioneer continuously monitors the task progress. Assuming sufficient progress is being made, the auctioneer periodically sends a renewal message to the winner until the task is completed.

Based on the achieved soft-state by continuously monitoring the task progress, MURDOCH provides system-inherent mechanisms for fault tolerance. On the other hand, this approach is limited by its own task scheduler. Since each task will be claimed each time by the most capable system, a greedy scheduler is implemented – showing the typical limitations of greedy algorithms. In some situations, resources may be exploited in a non-optimal manner such that not all the available tasks can be achieved. Centralized approaches usually avoid such situations.

An optimization has been proposed by Gage *et al.* (2004). This approach was named affective task allocation. The basic operation is similar to MURDOCH. Nevertheless, it introduces emotions and affective computing to reduce the communication costs of auction-based task allocation. Such emotions also lead to the introduction of a jitter to the allocation process, which helps to counteract the problems of greedy task schedulers.

A similar approach, relying on emotion-controlled task allocation, is ALLIANCE, as proposed by Parker (1994, 1998). Two types of internal motivations are modeled in ALLIANCE–robot impatience and robot acquiescence. The impatience motivation enables a robot to handle situations when other robots fail in performing a given task. The acquiescence motivation enables a robot to handle situations in which it, itself, fails to properly perform its task.

ALLIANCE utilizes a simple form of broadcast communication to allow robots to inform other team members of their current activities. According to the received messages, the impatience and the acquiescence levels are maintained. This minimum set of communication among the robots can help to improve the MRTA scheme used in negotiation algorithms.

15.2.3   Dynamic negotiation algorithm

Task allocation based on auctions has also been investigated by Ortiz *et al.* (2003) and Rauenbusch (2004). In particular, they studied and developed negotiation protocols in the context of multi-agent systems. Accordingly, agents must negotiate the assignment of resources to tasks in dynamically changing environments. The term 'negotiation' is used to refer to any distributed process through which agents can agree on an efficient apportionment of tasks among themselves. In particular, we discuss in this section center-based algorithms and a variant that has been named mediation (Ortiz *et al.* 2003).

In many application examples, the decision to allocate a task to a particular agent cannot be made independently from task allocations to another agent. Consider the following example, which is known as the sensor challenge problem. The scenario is depicted in Figure 15.6 (top). Two sensors are available to measure an effect. In order to save energy, idle sensors are deactivated. If a deactivated emitter is activated, the beam is unstable and will not give reliable measurements for 2 s. If one task is immediately followed by another in the same sector, the beam will not require the 2 s warm-up. The two tasks are said to interact positively.

Another example is depicted in Figure 15.6 (bottom). Consider that a scan with the detector takes at least 0.6 s. Then, two sequential tasks that are less than 0.6 s apart will need to be allocated to multiple sensors and, thus, interact negatively.

In many examples proposed in the literature, the context in which a negotiation is made is irrelevant to the negotiation. Thus, task costs can be assumed to be additive. In the following, we allow for arbitrary task interaction, particularly for the possibility of positive and negative task interactions, as depicted in the presented examples.

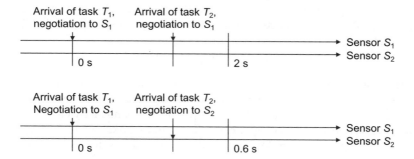

Figure 15.6 The sensor challenge problem. Positive interaction (top): considering a warm-up time of 2 s for sensor S_2, it might be better to allocate both tasks T_1 and T_2 to sensor S_1. Negative interaction (bottom): with an execution time greater than 0.6 s, the execution of two consecutive scans less than 0.6 s apart requires the use of both sensors

Center-based task assignment

The basic working principle of center-based task assignment is depicted in Figure 15.5. A center agent first submits a request describing a particular task to be executed. Then, it collects bids on proposed allocations. Each bid is meant to compactly encapsulate important local information. The final decision is taken by the center agent. Please note the difference from centralized coordination, as shown in Figure 15.4. The centralized coordinator maintains up-to-date information regarding the local states of the agents – which are used to compute optimal allocations. Depending on the size of the system, the amount of information (local state of the agents) may be very large. Thus, this operation can become infeasible in cases of communication limits or system failures.

According to Ortiz *et al.* (2003), the center-based task-assignment problem can formally be defined as follows. The task allocation system M is represented as a five-tuple $M = \langle A, T, u, \mathbf{P} \rangle$, where

$A = \{a_1, \ldots, a_n\}$ is a set of n agents with some agent designated as the mediator,
$T = \{t_1, \ldots, t_m\}$ is a set of m tasks,
$u : A \times 2^T \to \mathbb{R} \cup \{\infty\}$ is a value function that returns the value which an agent associates with a particular subset of tasks, and
$\mathbf{P}$ is an assignment of size n on the sets of tasks T such that $P = \langle P_1, \ldots, P_n \rangle$, where P_j contains the set of items assigned to agent a_j.

Each element of $\mathbf{P}$ is referred to as a proposal. For example $P_5 = \langle a_1, a_5, a_3 \rangle$ corresponds to the allocation in which task t_1 is assigned to agent a_1, task t_2 to agent a_5 and task t_3 to agent a_3.

The objective function $f(p, A)$ determines the desirability of an assignment based on the values that each agent ascribes to the items to which it is assigned:

$$f(p, A) = \sum_{a \in A} u(a, p) \text{ where } p \in \mathbf{P}. \tag{15.2}$$

The negotiation problem, i.e. the task allocation, is that of choosing an element $p^* \in \mathbf{P}$ that maximizes the objective function. Then, the chosen proposal p^* is called the outcome of the negotiation:

$$p^* = \arg\max_{p \in \mathbf{P}} f(p, A). \tag{15.3}$$

Solutions for center-based assignment algorithms include sequential auctions in which the item allocation is serialized. Thus, simple bidding rules can be used for the agents, which, in turn, bid on their expected contribution to the announced task. Unfortunately, sequential auctions provide no context, e.g. a list of other tasks to which an agent will be assigned in later auctions. Therefore, assumptions must be made about the outcomes of other, related auctions.

In contrast, combinatorial auctions give more flexibility to the agents as they bid. For exploring allocations of items that interact, the agents have the freedom to choose particular bunches of items. Combinatorial auctions allow an agent to pick certain bundles of tasks, which might interact in a favorable way. For each agent, there is an exponentially large state space of subsets on which an agent may base its bid (for m tasks, the space has a size of 2^m). Thus, the flexibility (and complexity) of bid generation lies at the agents and the center only needs to perform a simple selection. For inadequate allocations, re-allocation might be used to optimize the global task assignment.

Finally, mediation as described in the following subsection is a typical solution for the center-based assignment algorithms. In mediation, a mediator determines the order in which assignments are announced. The flexibility arises from a powerful mediator compared with agents using only simple rules for bidding.

Mediation algorithm

For mediation, one agent is selected to act as the mediator (Rauenbusch 2002). The algorithm makes use of a communication channel that is costly in terms of time (and other resources) but assumed to be lossless. The mediation procedure is an anytime algorithm, i.e. it can be halted at any time and will return the best proposal found so far. Therefore, the mediation is applicable even if the agents do not know in advance how much time they will have for negotiation.

The mediation algorithm is depicted in Algorithm 15.1. It implements an iterative hill-climbing search in the proposal space. It stores the best proposal b found so far for future comparison with other proposals and as a return value for termination. The algorithm works on the negotiation set $\mathbf{P}$ and a group of agents $G \subseteq A$ out of the available agents A. An update procedure is required such as the allocation-improvement mediation described below. The algorithm works as follows:

1. The mediator initializes $b \leftarrow \emptyset$ (representing the best proposal found so far) along with an initial value.

2. An update procedure generates a new proposal c (current proposal).

3. This proposal is broadcasted to the group of agents G.

4. Each agent responds with a message msg_i based on the proposal c.

5. The received response messages are combined to form a value c_{val}.

Algorithm 15.1 Mediation algorithm

Require: P, G, `UpdateProcedure`.
Ensure: outcome b.

1:   $b \leftarrow \emptyset$
2:   $b_{val} \leftarrow$ VALUE($\emptyset$)
3: **repeat**
4:     $c \leftarrow$ next value generated by `UpdateProcedure`
5:     broadcast c to G
6:     **for all** $G_i \in G$ **do**
7:        receive msg_i from G_i
8:     **end for**
9:     $c_{val} \leftarrow$ VALUE($msg_1, msg_2, \ldots, msg_n$)
10:     **if** $c_{val} \succ b_{val}$ **then**
11:        $b \leftarrow c$
12:        $b_{val} \leftarrow c_{val}$
13:     **end if**
14: **until** stop signal

---

6. If the value is preferred to the current proposal b, i.e. if $c_{val} \succ b_{val}$, b is updated with the current proposal.

The update procedure of the mediation algorithm is used for the actual negotiation over a task allocation among a number of agents. Rauenbusch (2002) proposed the allocation-improvement algorithm as an update procedure for mediation that supports task-allocation domains. In particular, a proposal $p \in \mathbf{P}$ is chosen, which corresponds to an allocation of agents in G to tasks in T. The complete procedure is depicted in Algorithm 15.2. The first proposal p is chosen randomly from $\mathbf{P}$. This proposal provides a context, from which subsequent proposals are generated. This context is common to all agents and ensures that each task is assigned to an agent. Subsequent iterations try to improve the allocation. The procedure returns proposals that result from making substitutions in p for i-tuples of tasks where i goes from $1 \ldots |T|$. The proposal p is always maintained to correspond to the best proposal in mediation.

To conclude the discussion on task allocation through negotiation, the remaining limitations need to be mentioned. So far, only sets with static resources have been investigated. What about the possibility of letting tasks and resources dynamically appear and disappear? The first solution, which is usually found in the literature, is to interrupt the ongoing negotiation and to initiate a re-allocation. Depending on the dynamics of the system, this approach might be feasible. A more practicable (and more sophisticated) approach is dynamic mediation. Actually, dynamic mediation represents a mixture of central coordination and mediation. The bids are enriched to include all relevant local state information. Thus, the complete negotiation space (the set of resources and tasks) is available at the mediator. This negotiation space might change according to additional resources and tasks.

Algorithm 15.2 Allocation improvement algorithm

Require: **P**, G, T.
Ensure: proposal p.

  1: **if** first run **then**
  2:     $p \leftarrow$ a random element of $\mathbf{P} \backslash \emptyset$
  3:     **return** p
  4: **else**
  5:     $q \leftarrow p$
  6:     **for** $i = 1 \ldots |T|$ **do**
  7:         **for all** $t \leftarrow$ every set of tasks of size i **do**
  8:             **for all** $a \leftarrow$ every possible assignment of agents in G to tasks in t **do**
  9:                **if** $q_{val} \succ p_{val}$ in mediation **then**
10:                    $p \leftarrow q$ {always maintain the best proposal so far in p}
11:                **end if**
12:                $q \leftarrow$ substitute a in p
13:                **return** q
14:             **end for**
15:         **end for**
16:     **end for**
17: **end if**

15.3   Emergent cooperation

In contrast to intentional cooperation, task allocation following the emergent cooperation approach is not based on a optimization scheme or based on an agreement among nodes (neither centralized nor distributed). Instead, simple local rules are employed by the autonomous systems based on the local state influenced by environmental conditions. In combination with direct or indirect interactions, an emergent behavior of the overall system can be observed.

In many cases, emergent cooperation is motivated by biological analogies, especially focusing on Swarm Intelligence (SI) (we will discuss the ideas of SI later, in Section 18.2). Therefore, the term 'ant-like cooperation' has been created to describe such emergent cooperation techniques. Optimization algorithms inspired by the ant system have been primarily investigates by Dorigo *et al.* (1996).

According to Krieger *et al.* (2000), ant-like cooperation, which relies on well studied principles, such as foraging and stimulation based on the colony energy, is often based on the following principles (assuming autonomous systems' sharing a common basis such as the nest in nature):

- First, colony-level information about the colony energy can be accessed and updated by individual systems. In the case of MRTA, the robots can, for example, be informed of the colony energy by radio broadcast messages from a control station installed in the basis or by other robots staying in the nest. No further information on colony energy is provided to the robot system during their missions out of the nest. Upon return, each robot may renew its energy reserves (which corresponded to a decrease in colony energy) and unload the collected 'food items' (if the foraging trip was successful), thereby increasing colony energy.

- Secondly, local decision processes need to be implemented to avoid negative interactions. Examples are collision prevention and the regulation of systems in a particular space, because robots, like foraging ants, cannot occupy the same space simultaneously.

- Thirdly, a local motivation level for individual systems can be realized according to individual variation in the tendency to perform a particular task. This directly models the different individual stimulus thresholds for task allocation that are observed in ants. For example, robots will not leave the nest simultaneously, but only when the colony energy drops below the pre-set threshold of a particular robot.

- Finally, recruitment strategies can be exploited mimicking the recruitment behavior observed in many ant species. By informing other systems about particular objectives, e.g. the identification of a resource-rich area, these systems can be encouraged to participate on the task.

In general, emergent cooperation can be further categorized into two classes. Both rely on stimulation effects among the participating systems. Either systems get stimulated by work or by state. In the following, we outline the ideas of these to cooperation principles.

15.3.1 Stimulation by work

Task allocation among a number of robot systems with limited communication capabilities is addressed by state-based stimulation techniques. One of the best known approaches has been inspired by foraging or prey-retrieval methods observed by ants. These concepts can be used, for example, in the context of swarm robotics for self-organized task allocation performed by autonomously acting robots, which are supposed to collectively perform a common task (Labella *et al.* 2004).

An example of prey-retrieval techniques used for MRTA is the algorithm proposed by Labella and Dorigo (2004). It employs positive and negative feedback and specifically prevents all direct or symbolic communication among the robot systems.

The main parameter considered in this approach is the efficiency of prey retrieval. Two components must be taken into account. First, the robot (or ant) needs to spend a non-zero amount of energy for prey retrieval. Additionally, further energy is required for moving around, e.g. for searching the food, as well as for staying idle (even though this amount of energy might be negligible). This energy consumption represents the costs for the colony. Secondly, successful prey retrievals contribute to the available energy, thus representing the income. Income and costs depend on the number of foragers N. The efficiency of the group can now be defined as:

$$\eta = \frac{income}{costs}. \tag{15.4}$$

In nature, ants rely on recruitment strategies and stigmergic communication to improve the efficiency of a swarm. Both techniques are difficult to realize in multi-robot scenarios. Therefore, Labella and Dorigo (2004) proposed a similar algorithm that shows the following effects, which can also be observed in nature:

- Efficiency increase – If too many robots are searching for prey, the probability of being successful would obviously decrease. Therefore, the unsuccessful robots need to decrease their probability to leave the nest P_l. On the contrary, i.e. if far more prey are available than foraging robots, some robots would eventually leave the nest, be instantly successful, and increase their P_l in order to spend more time on foraging. In both cases, the efficiency η of the group would improve without external intervention.

- Task allocation – So far, successful robots would continuously increase their P_l and – if enough food is available – eventually spend all their time searching for food (and, furthermore, increase P_l). Similarly, unsuccessful robots will tend to stay in the nest with low P_l.

The algorithm can be implemented in the form of a simple state machine, as depicted in Figure 15.7. This figure represents a simplified version of the approach presented by Labella and Dorigo (2004). If in the nest, the robot will leave with probability P_l in order to search for prey. If the robot is not successful in the time period τ, it will return to the nest and update its P_l accordingly. Similarly, in the successful case, the robot will try to bring the prey to the nest and, if successful, increase its P_l.

The adaptation algorithm is depicted in Algorithm 15.3. Two counters are introduced to either amplify or to control the adaptation of P_l–the number of successful tries and the number of failures. Depending on these counters, P_l is increased or decremented by a factor of $successes \times \Delta$ or $failures \times \Delta$, respectively. Additionally, P_{min} and P_{max} have been introduced to enable fixed bounds for the nest-leaving probability P_l.

15.3.2   Stimulation by state

Low et al. (2005) proposed a task-allocation architecture based on autonomic behavior of the individual systems partially controlled by neural networks, which are employed for target tracking in a given environment. The proposed MRTA technique relies on three concepts of ant behavior to self-organize the autonomously acting robot systems. First, an encounter pattern can be derived from a series of waiting times or the intervals between successive encounters, Secondly, the self-organization of the social dominance in the colony is exploited for task allocation, and, finally, the dynamic task-allocation scheme is based on the notion of response thresholds.

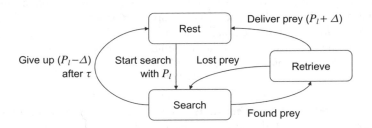

Figure 15.7   State machine for stimulation-based task allocation

---

Algorithm 15.3 Adaptation algorithm for prey retrieval

Require: P_{init} (initial nest leaving probability), P_{min} (lower bound) and P_{max} (upper bound).

Ensure: dynamically adapted P_l.

1:   $successes \leftarrow 0$
2:   $failures \leftarrow 0$
3:   $P_l \leftarrow P_{init}$
4:   **repeat**
5:     **if** prey successfully retrieved **then**
6:       $successes \leftarrow successes + 1$
7:       $failures \leftarrow 0$
8:       $P_l \leftarrow P_l + (successes \times \Delta)$
9:       **if** $P_l > P_{max}$ **then**
10:         $P_l \leftarrow P_{max}$
11:       **end if**
12:     **else if** unsuccessful search for prey (timeout τ) **then**
13:       $successes \leftarrow 0$
14:       $failures \leftarrow failures + 1$
15:       $P_l \leftarrow P_l - (failures \times \Delta)$
16:       **if** $P_l < P_{min}$ **then**
17:         $P_l \leftarrow P_{min}$
18:       **end if**
19:     **end if**
20:   **until** stop signal

Encounter pattern based on waiting time

The knowledge of the regional density of robots is crucial to regulating the division of labor. The proposed ant-based task-allocation scheme uses encounter patterns to predict the target density via local sensing instead of relying on global communication. Such encounter patterns can be derived from waiting times of the intervals between successive encounters. An encounter is defined as the reception of a message from another robot in the same region. For i robots in region r, the time interval between the $(k-1)$-th and the k-th encounters contributes to the waiting time for other robots $w_{ir}(k)$ and for targets $w'_{ir}(k)$. Such waiting times will be strongly influenced by stochastic variation. Thus, the average waiting time $W_{ir}(k)$ can be computed as depicted in Equation 15.5, where $n = \min(k, n_{max})$ and n_{max} is the maximum number of monitored encounters. Similarly, $W'_{ir}(k)$ is calculated.

$$W_{ir}(k) = \underbrace{\frac{1}{n} w_{ir}(k)}_{last\,waiting\,time} + \underbrace{\frac{n-1}{n} W_{ir}(k-1)}_{history}. \qquad (15.5)$$

In Equation 15.5, $\frac{1}{n} w_{ir}(k)$ represents the weighted waiting time for the n-th encounter and $\frac{n-1}{n} W_{ir}(k-1)$ depicts the historical evolution of the waiting time $W_{ir}(k)$.

Both waiting times are updated according to the environmental changes. Now, the average waiting time for encounters with other robots $W_{ir}(k)$ reflects the robot density in the given region r, and thus the degree of possible interference. The waiting time for encounters with targets $W'_{ir}(k)$ describes the target density. Therefore, the demand for the particular task $S_{ir}(k)$ can be determined locally by robot i, as shown in Equation 15.6:

$$S_{ir}(k) = \frac{W_{ir}(k)}{W'_{ir}(k)}. \qquad (15.6)$$

The task demand $S_{ir}(k)$ that is basically describing the ratio of robots to targets in a given region will further be used in the self-organization of social dominance and the dynamic task allocation.

Self-organization of social dominance

The social dominance of an individual robot system is used for self-organized task allocation according to the colony demands and the changing environment. In particular, the social dominance can be calculated according to the task performance of the particular robot system. The self-organization concept relies on periodic dominance contests at a regular interval τ among robots in the same coalition if they are in communication range. According to the outcome of the contest, the winner increases its tendency to leave the current region, i.e. it is continuing to perform the particular task.

Depending on the task demand $S_{ir}(k)$, the probability of robot i winning the contest after encountering robot j can be defined according to Equation 15.7:

$$P(\text{robot } i \text{ wins}) = \frac{n_i^2 S_{ir}^2}{n_i^2 S_{ir}^2 + n_j^2 S_{jr}^2}. \qquad (15.7)$$

In order to inject the influence of social dominance on the self-organization of robot coalitions, each time robot i wins a contest according to Equation 15.7, it increases its tendency to stay in the current region, which is represented by the response threshold $\theta_i(t)$ to be used for dynamic task allocation, shown in Equation 15.8, where δ is a small constant:

$$\theta_i(t) = \theta_i(t-1) + \delta. \qquad (15.8)$$

Similarly, each time the robot loses, it decreases its tendency to stay in the region (Equation 15.9). The response threshold $\theta_i(t)$ should stay in the range $[0, 1]$ to prevent robots from becoming extremely submissive or dominating:

$$\theta_i(t) = \theta_i(t-1) - \delta. \qquad (15.9)$$

Dynamic task allocation

The dynamic task-allocation scheme is primarily based on the described response thresholds. These thresholds can be used to create the emotion-like behavioral models. Robots with a low response threshold respond more readily to lower levels of task demand compared with robots with high response thresholds. On the other hand, if the task demand decreases, the latter will engage in performing the task.

The proposed dynamic task-allocation scheme for MRTA uses a stochastic problem-solving methodology. It is periodically repeated and calculates probabilities for specific robot actions according to memorized state information. In particular, each robot i maintains a memory of the task demand S_{ir} of each region r (initialized to zero) and the amount of time T_{ir} that it previously spent in region r.

The probability of a robot i to stay in its current region c can now be defined as:

$$P(\text{stay}) = \frac{S_{ic}^2}{S_{ic}^2 + (1 - \theta_i)^2 + T_{ic}^{-2}}. \tag{15.10}$$

On the other hand, the probability of a robot i leaving region c to go to region r is defined as (a robot that loses in the dominance contest in a coalition does not always leave the region – it may stay if it experiences a higher task demand in its region than in other regions):

$$P(\text{leave}) = \frac{S_{ir}^2}{S_{ir}^2 + \theta_i^2 + T_{ic}^{-2} + d_{cr}^2}, \tag{15.11}$$

where d_{cr} depicts the distance between region c and region r, i.e. the cost of task switching.

15.4 Conclusion

Collaboration and task allocation in SANETs represent a challenging research domain, especially motivated by the strong restrictions in SANETs such as the limited communication in large-scale networks consisting of a huge number of interacting nodes.

In this chapter, we discussed a particular class of algorithms addressing the so-called Multi-Robot Task Allocation (MRTA) problem. As we have shown in the presentation of the main characteristics of MRTA, such techniques can be used for general task and resource allocation in SANETs as well.

The taxonomy of MRTA solutions opens a multidimensional domain in which particular solutions can be categorized. We decided to follow the classification strategy in which multiple robots coordinate on multiple tasks (MR–MR). The approaches in this class can further be differentiated into intentional and emergent cooperation. Based on selected examples, we outlined the principles and methods used for MRTA according to the ideas of these two cooperation concepts:

- Intentional cooperation – Intentional cooperation refers to the concept of performing coordinated actions on purpose, i.e. based on a well organized decision process that is involving all available systems or at least all systems that might be capable of participating. Most intentional cooperation techniques rely on auction-based mechanisms, i.e. a central auctioneer requests and analyzes bids for a set of tasks and finally takes a centralized decision.

  In particular, we discussed OAA being one of the simplest solutions following the centralized task-allocation approach, i.e. a central system is maintaining state information for optimized task allocation among all available systems. In contrast, MURDOCH follows the auction-based approach, with a center agent as the decision taker. Non-optimal solutions can, for example, be prevented using emotion-controlled task allocation in combination with the center-based approach. Finally, we discussed the mediation algorithm, which performs a hill-climbing optimization to find the best allocation.

Advantages: Intentional cooperation supports heterogeneous robots as well as heterogeneous tasks. Basically, all solutions represent an optimization process. Reallocation of tasks is usually possible.

Disadvantages: The most challenging problem inherent to all intentional cooperation techniques is the communication overhead. For centralized task allocation, all systems need to periodically report their local state to a central system, and center-based task assignment relies on bids from systems able to perform a particular task.

- Emergent cooperation – Instead of explicit cooperation, emergent cooperation focuses on achieving a group-level behavior that emerges out of simple local algorithms and stimulation effects. Such stimulation can either be the work of other systems or the estimated (global) system state. Because algorithms are often inspired by the behavior of ants, emergent cooperation is sometimes also called ant-like cooperation.

The stimulation used for task allocation can be derived either from work in progress or from available state information, for both of which we discussed a simple example.

Advantages: The main advantage of emergent cooperation is its simplicity. Simple local algorithms lead to an observable global behavior. At the same time, none or limited communication requirements are necessary to complete the task allocation. Therefore, emergent cooperation provides support for really large-scale installations.

Disadvantages: Unfortunately, emergent-cooperation techniques provide good means only for distributed operation. A global optimization is not possible. Additionally, only limited support for heterogeneous environments (robots, tasks) can be provided.

In the evaluation of task-and resource-allocation techniques, which are also used for collaborative actions, three main measures need to be considered. First, the possible communication overhead might limit the scalability of the particular solution for task allocation in large-scale SANETs. Secondly, the energy consumption needs to be reflected, as most systems used in the context of SANETs will be battery-driven and have limited capabilities for energy harvesting. Finally, the time constraints of real-time tasks have to be evaluated. Dynamic task allocation involving quite optimal but long-lasting decision processes may not be feasible in real-time environments.

Appendix III

Coordination and Control – Further Reading

In the following, a selection of some interesting textbooks, papers and articles is provided that is recommended for further studies of the concepts of self-organization. Additionally, a list of major journals and conferences is provided either directly focused on self-organization and related aspects or broadly interested in publishing special issues or organizing dedicated workshops in this domain. Obviously, this list cannot be complete or comprehensive. It is intended to provide a starting point for further research.

Textbooks

- N. Bulusu and S. Jha, *Wireless Sensor Networks:A Systems Perspective*, Artech House Publishers, 2005.

- C. De Morais Condeiro and D. P. Agrawal, *Ad Hoc & Sensor Networks: Theory and Applications*, World Scientific Publishing Company, 2006.

- S. S. Iyengar and R. R. Brooks, *Distributed Sensor Networks*, Chapman & Hall/CRC, 2004.

- H. Karl and A. Willig, *Protocols and Architectures for Wireless Sensor Networks*, John Wiley & Sons Ltd, 2005.

- B. Krishnamachari, *Networking Wireless Sensors*, Cambridge University Press, 2006.

- V. Lesser, C. L. Ortiz and M. Tambe, *Distributed Sensor Networks: A Multiagent Perspective*, Kluwer Academic Pubishers, 2003.

- J. Liu and J. Wu, *Multiagent Robotic Systems*, CRC, 2001.

- C. S. R. Murthy and B. S. Manoj, *Ad Hoc Wireless Networks*, Prentice Hall PTR, 2004.

Self-Organization in Sensor and Actor Networks   Falko Dressler
© 2007 John Wiley & Sons, Ltd

- C. E. Perkins, *Ad Hoc Networking*, 1st edn, Addison-Wesley Professional, 2000.

- C. S. Raghavendra, K. M. Sivalingam and T. Znati, *Wireless Sensor Networks*, Kluwer Academic Publishers, 2004.

- M. Schumacher, *Objective Coordination in Multi-Agent System Engineering: Design and Implementation*, vol. LNCS 2039, Springer, 2001.

Papers and articles

- I. F. Akyildiz and I. H. Kasimoglu, 'Wireless Sensor and Actor Networks: Research Challenges,' *Elsevier Ad Hoc Networks*, vol. 2, pp. 351–367, October 2004.

- M. A. Batalin and G. S. Sukhatme, 'Sensor Network-based Multi-Robot Task Allocation,' Proceedings of IEEE/RSJ International Conference on Intelligent Robots and Systems (IROS2003), Las Vegas, Nevada, October 2003, pp. 1939–1944.

- L. Buttyán and J.-P. Hubaux, 'Stimulating Cooperation in Self-Organizing Mobile Ad Hoc Networks,' *Mobile Networks and Applications*, vol. 8(5), pp. 579–592, October 2003.

- B. Chen, K. Jamieson, H. Balakrishnan and R. Morris, 'Span: An Energy-Efficient Coordination Algorithm for Topology Maintenance in Ad Hoc Wireless Networks,' *ACM Wireless Networks Journal*, vol. 8(5), September 2002.

- F. Dressler, 'Network-centric Actuation Control in Sensor/Actuator Networks Based on Bio-inspired Technologies,' Proceedings of 3rd IEEE International Conference on Mobile Ad Hoc and Sensor Systems (IEEE MASS 2006): 2nd International Workshop on Localized Communication and Topology Protocols for Ad hoc Networks (LOCAN 2006), Vancouver, Canada, October 2006, pp. 680–684.

- F. Dressler, I. Dietrich and B. Krüger, 'Efficient Operation in Sensor and Actor Networks Inspired by Cellular Signaling Cascades,' Proceedings of First ICST/ACM International Conference on Autonomic Computing and Communication Systems (Autonomics 2007), Rome, Italy, October 2007.

- M. G. Lagoudakis, M. Berhault, S. Koenig, P. Keskinocak and A. J. Kleywegt, 'Simple Auctions with Performance Guarantees for Multi-Robot Task Allocation,' Proceedings of IEEE/RSJ International Conference on Intelligent Robots and Systems, Sendai, Japan, September/October 2004.

- K. H. Low, W. K. Leow and M. H. Ang, 'Autonomic Mobile Sensor Network with Self-Coordinated Task Allocation and Execution,' *IEEE Transactions on Systems, Man, and Cypernetics–Part C: Applications and Reviews*, vol. 36(3), pp. 315–327, March 2005.

- M. J. Mataric, G. S. Sukhatme and E. H. Ostergaard, 'Multi-Robot Task Allocation in Uncertain Environments,' *Autonomous Robots*, vol. 14, pp. 255–263, 2003.

- T. Melodia, D. Pompili, V. C. Gungor and I. F. Akyildiz, 'A Distributed Coordination Framework for Wireless Sensor and Actor Networks,' Proceedings of 6th ACM International Symposium on Mobile Ad Hoc Networking and Computing (ACM Mobihoc 2005), Urbana-Champaign, Il, USA, May 2005, pp. 99–110.

- T. Melodia, D. Pompili, V. C. Gungor and I. F. Akyildiz, 'Communication and Coordination in Wireless Sensor and Actor Networks,' *IEEE Transactions on Mobile Computing*, 2007.

- G. A. Shah, M. Bozyigit, Ö. B. Akan and B. Baykal, 'Real-Time Coordination and Routing in Wireless Sensor and Actor Networks,' Proceedings of 6th International Conference on Next Generation Teletraffic and Wired/Wireless Advanced Networking (NEW2AN 2006), St Petersburg, Russia, May 2006.

- W. Wang, V. Srinivasan and K.-C. Chua, 'Using Mobile Relays to Prolong the Lifetime of Wireless Sensor Networks,' Proceedings of 11th ACM International Conference on Mobile Computing and Networking (ACM MobiCom 2005), Cologne, Germany, August/September 2005, pp. 270–283.

- Y. Yao and J. Gehrke, 'The Cougar Approach to In-Network Query Processing in Sensor Networks,' *ACM SIGMOD Record*, vol. 31(3), pp. 9–18, 2002.

- M. Younis, K. Akkaya and A. Kunjithapatham, 'Optimization of Task Allocation in a Cluster-Based Sensor Network,' Proceedings of 8th IEEE International Symposium on Computers and Communications, Kemer-Antalya, Turkey, June 2003, pp. 329–340.

Journals

- *Ad Hoc Networks* (Elsevier)

- *Advanced Robotics* (Robotics Society of Japan)

- *Autonomous Agents and Multi-Agent Systems* (Springer)

- *Autonomous Robots* (Springer)

- *Communications Magazine* (IEEE)

- *Computer* (IEEE)

- *Computer Communication Review* (ACM)

- *Computer Networks* (Elsevier)

- *Journal of Selected Areas in Communications* (IEEE)

- *Robotics and Autonomous Systems* (Elsevier)

- *Transactions on Networking* (IEEE/ACM)

- *Transactions on Robotics and Automation* (IEEE)

- *Wireless Communications* (IEEE)

- *Wireless Communications and Mobile Computing* (Wiley Interscience)

- *Wireless Networks* (ACM/Springer)

Conferences

- AAMAS – International Conference on Autonomous Agents and Multiagent Systems

- EWSN – European Workshop on Wireless Sensor Networks

- ICRA – IEEE International Conference on Robotics and Automation

- INFOCOM – IEEE Conference on Computer Communications

- IROS – IEEE/RSJ International Conference on Intelligent Robots and Systems

- MASS – IEEE International Conference on Mobile Ad Hoc and Sensor Systems

- MobiCom – ACM International Conference on Mobile Computing and Networking

- Mobihoc – ACM International Symposium on Mobile Ad Hoc Networking and Computing

- Multi-Robot Systems – International Multi-Robot Systems Workshop

- Networking – IFIP International Conference on Networking

- SECON – IEEE International Conference on Sensor and Ad hoc Communications and Networks

Part IV

Self-Organization Methods in Sensor and Actor Networks

We have seen that self-organization is strongly demanded for building highly scalable and efficient systems. Networked embedded systems such as sensor and actor networks are only a first step towards a new scale of massively distributed systems. In this part, we conclude the discussions about self-organization and technical aspects of communication and control in sensor and actor networks. First, we revisit the investigated basic self-organization methods on the basis of networking and coordination aspects that are relevant for wireless sensor and actor networks. Secondly, we outline the most important evaluation criteria that need to be considered for comparing and evaluating novel algorithms and techniques in this domain.

Outline

- Self-organization methods – revisited

  The objective of this chapter is to summarize the discussions of the previous investigations. The self-organization methods outlined in Part I are revisited, focusing on the technical solutions studied in the context of sensor and actor networks. The networking aspects discussed in Part II, as well as the coordination and control concepts presented in Part III, are categorized according to the exploited self-organization techniques.

- Evaluation criteria

  The evaluation of algorithms and techniques proposed and developed in the context of self-organization in sensor and actor networks is extremely challenging due to possible interference between multiple solutions, e.g. protocols applied to different network layers might negatively interact. In this chapter, we summarize the most important measures in the context of communication and control in sensor and actor networks. First of all, scalability needs to be considered. Closely related is energy

Self-Organization in Sensor and Actor Networks   Falko Dressler
© 2007 John Wiley & Sons, Ltd

consumption. Recently, it turned out that most performance metrics (including scalability and energy) can be represented in the form of measurement of the network lifetime. This metric is strongly application-dependent but it allows estimatation of the performance of the entire network instead of evaluating single systems or single algorithms.

16

Self-Organization
Methods – Revisited

Self-organization mechanisms can be found in every stretch of our everyday life (von Foerster 1960). From an academic point of view, self-organization was first analyzed in biological systems (Camazine *et al.* 2003). This research was soon extended to technical systems and engineering in general (Eigen and Schuster 1979). Self-organization can be summarized as the interaction of multiple components on a common global objective (Ashby 1962). This collaborative work is done without central or de-central control. Instead, the interaction is done using a local context, e.g. the direct environment that can be changed and adapted by each individual and, therefore, affects the behavior of other individuals.

The primary objectives of self-organization are scalability, reliability and availability of systems composed of a huge number of subsystems. In computer networks, self-organization is especially important in the domain of ad hoc networks because of the spontaneous interaction of multiple heterogeneous components over wireless radio connections (Murthy and Garcia-Luna-Aceves 1996) without human interaction. This is especially the case in the areas of pervasive and ubiquitous computing, and, as primarily discussed in this book, in the context of Sensor and Actor Networks (SANETs). Eventually, self-organization is the only possible solution for many issues in massively distributed systems, but it definitely is not the universal remedy (Dressler 2006b).

In this chapter, we revisit the ideas and concepts of self-organization, with a special focus on issues and techniques discussed in the last two parts, i.e. in the context of ad hoc and sensor networks – networking aspects, and for SANET-specific problem spaces – coordination and control. In order to get started, we briefly summarize the ideas of self-organization as presented in Part I:

- **What is self-organization?** Self-organization can be regarded as a new control theory for massively distributed systems. In contrast to other control mechanisms, self-organization focuses on local decision processes performed in autonomously acting systems. Based on interactions among the systems and their local programmings, the overall system behavior emerges out of the individual actions. This emerging system behavior, as visible from a global perspective, can often be described in the form of a created pattern.

- **Why do we need self-organization?** The operation and control of autonomous systems require special methods and techniques. The emergent behavior – even though this is hard to engineer – enables further perspectives for application scenarios, especially in the context of Wireless Sensor Networks (WSNs) and SANETs.

- **Which methods are available?** Three basic methods have been identified that are used by self-organizing systems. In many cases, a combination of the three methods is preferred because this provides even better possibilities to achieve the demanded system behavior:

  - First, feedback loops are used to update the local system state and behavior according to observed changes in the environment. This includes the observable behavior of other systems. This feedback can be distinguished into positive feedback, providing amplification effects, and negative feedback for controlling the system state, i.e. to prevent over-reactions. For efficiently controlling the system behavior, it is typically required to employ both positive and negative feedback simultaneously.

  - Secondly, numerous interactions among the individual, autonomously acting systems enables coordinated behavior. Interactions can take place directly between neighboring systems or indirectly via artificial changes in the environment. The latter communication is also called stigmergic (stimulation by work). Finally, interactions with the environment help to adapt the system behavior according to the specific environmental conditions.

  - Thirdly, probabilistic techniques are employed, for two reasons. First, pure probabilistic control mechanisms often lead to astonishing results. In massively distributed systems, probabilistically controlled operations lead to similar execution times compared to centralized control. The main reason is the reduction in overhead through coordination and control to almost zero, as no state information needs to be maintained. Secondly, probabilistic techniques help to prevent synchronization effects, such as oscillations and being stuck in local optima.

- **Advantages of self-organizing systems** The main advantages of self-organizing systems are scalability, robustness and adaptivity. In this context, scalability refers to the supported number of (sub-)systems that participate in building the overall self-organizing system. Robustness stands for the ability of the system to work properly, even in cases of partial system failures or disrupted communication. Adaptivity finally means the capability to work in different or changing environments without further intervention and configuration by an administrator.

- **Limitations of self-organization** The most challenging limitation of self-organizing systems is the reduced determinism and controllability of the system behavior. In many cases, the algorithms will eventually lead to the desired solution or behavior but there is often only limited evidence of when this state will be reached.

We have seen a great number of examples of self-organizing systems in nature and technology. Focusing on the applicability of self-organization communication and coordination between networked embedded systems, we discussed particular solutions. In this chapter, we review these solutions in the context of used self-organization techniques. One of the most essential conclusions to derive from our investigations is that all the studied algorithms follow many different objectives, in many directions, but finally rely on similar solutions.

16.1 Self-organization methods in SANETs

In Parts II and III, we discussed a number of networking aspects and coordination and control techniques, respectively, both related to wireless sensor and actor networks. The primary objective was to introduce typical networking issues, as well as coordination strategies for massively distributed systems, to outline current research challenges and possible solutions. The discussion focused on methods and techniques which are providing improved scalability and efficiency in terms of energy and communication. So far, the analysis of the benefits of particular solutions has been discussed in a very technical way. More detailed information on typical measures for evaluating algorithms for communication and coordination in WSNs and SANETs will be provided in Chapter 17.

In this chapter, we study the basis methods for self-organization previously described in Part I by classifying all the presented technical networking and coordination solutions. Obviously, a great number of communication and coordination techniques are available in WSNs and SANETs that employ self-organization techniques, i.e. that aim at higher scalability and efficiency by reducing globally synchronized state information. Depending on the category within the solution space and the particular application scenario, the achieved speed-up can be different. The following objectives should be mentioned as examples:

- Manageability – On the one hand, the manageability of large-scale networks can be improved to a certain degree by the prevention of global state maintenance. Nodes try to identify their environment by just 'looking at it'. Therefore, little or even no central management is necessary to keep the network operational. On the other hand, it is even more challenging to enable an administrator to 'configure' the network or parts of the network from a central maintenance point.

- Scalability – Hopefully, only linear or less overhead is induced when increasing the network size. Self-organization helps to reduce the overhead required for state maintenance and operational overhead. Depending on the chosen methodology and the network behavior (e.g. traffic and network size), networks are allowed to increase to very large-scale dimensions.

- Overhead – As mentioned before, the overhead caused by maintenance and protocol overhead can be reduced by employing various self-organization mechanisms. It is important to keep analyzing the globally visible overhead. In the case of preventing the maintenance of global state information, it will always involve some amount of additional work to perform the desired operation such as routing in ad hoc networks. From a global point of view, there should be a visible reduction in necessary overhead if the right mechanisms for the particular applications have been chosen and they are properly configured.

- Reliability – The reliability of a network as visible to an application depends on many parameters. Here, we do not consider the reliable communication of a transport protocol, but the ability of the network to dynamically react to node or link failures and the employment of using multiple paths simultaneously. Basically, reliability is provided based on the inherent capability of self-organization mechanisms to adapt to changing environments.

The remainder of this chapter is organized as follows. Basically, we follow the categorization of self-organization methods as presented in Chapter 6. We distinguished the following three basis methods: positive and negative feedback, interactions among individuals and with the environment, and probabilistic techniques. For each of these methods, we outline the networking-related functions discussed in Part II and the coordination and control techniques presented in Part III. Thus, we build a global taxonomy of SANET approaches in the context of self-organization.

16.2   Positive and negative feedback

Feedback loops are used in many algorithms covering communication aspects in WSNs. This includes synchronization issues between neighboring nodes as well as topology control in multi-hop networks. Similarly, coordination and collaboration essentially demand such feedback loops. We follow the outline used in the technical parts on networking aspects and coordination and control to describe and discuss self-organization techniques relying on positive and negative feedback.

Table 16.1 summarizes our discussion on the use of feedback loops in networking algorithms and techniques for (mobile) ad hoc and sensor networks as well as in coordination and control approaches.

Medium Access Control

Medium Access Control (MAC) protocols for wireless ad hoc networks must ensure the coordinated and fair utilization of the wireless medium. In many cases, this includes the synchronized access in order to prevent collisions and starvation of single transmissions. Time Division Multiple Access (TDMA) techniques have been discussed in the context of WSNs, while the primary issue for management and control is the distributed autonomous establishment of synchronous transmission windows.

Sensor MAC (S-MAC) uses special SYNC messages between neighboring nodes to set up so-called synchronized islands, i.e. a number of nodes following the same schedule. This can be regarded as positive feedback. New nodes that are joining the network are 'motivated' to take over the synchronization. S-MAC also encourages existing islands to join a neighboring SYNC.

Similarly, Power-Control MAC (PCM) uses feedback information for controlling the used transmission energy for individual transmissions according to previously measured signal qualities. During the RTS/CTS handshake, PCM measures the received signal strength p_r. This value is used in comparison to a well known minimum threshold Rx_{thresh} for

Table 16.1   Summary of positive and negative feedback used SANET algorithms

Problem domain	Employed feedback loops
Networking aspects	
MAC	Positive and negative feedback for controlling the used transmission energy, e.g. in PCM; enforcement of synchronization between multiple nodes to a common schedule
Ad hoc routing	Positive feedback for route discovery in most table-driven routing protocols; negative feedback for suppression of further data messages over erroneous paths, both used, e.g. in AODV and DYMO
Data-centric networking	Positive feedback in the form of interest messages controlling the behavior of sensor nodes, e.g. in directed diffusion; energy levels and timeouts as negative feedback to suppress unnecessary communication, e.g. in rumor routing
Clustering	Feedback is provided, for example, in the form of energy levels controlling system-inherent parameters such as the probability to become a clusterhead in HEED
Coordination and control	
Communication and coordination	Feedback loops are inherently used by all time-synchronization techniques; positive and negative feedback enables adaptive coordination among nodes, e.g. for optimizing the utility by Adaptive Self-Configuring Sensor Network Topologies (ASCENT) or to ensure a latency bound by DEPR
Collaboration and task allocation	Positive feedback in the biddings in auction-based task allocation, e.g. in MURDOCH, and negative feedback through re-allocation; feedback-based probability adaptation in the case of emergent cooperation

further adaptations to the current environmental conditions, i.e. the distance between the two communicating nodes.

In the used equation $p_{desired} = \dfrac{p_{max}}{p_r} Rx_{thresh} \times c$, the factor c can be regarded as a regulating measure to enable PCM to smoothly adapt the used transmission power. If c is increased, the transmission power gets larger until the maximum is reached. Thus, an adaptation to higher mobility or rough environmental conditions can be enforced. Similarly, the reduction in c leads to reduced energy consumption, with less support for unpredictable errors in the wireless communication.

Ad hoc routing

In standard ad hoc routing techniques, pure feedback solutions will not be found. Instead, combinations of multiple self-organization techniques are used. Typically, ad hoc routing techniques exploit neighborhood information over a larger geographical region to set up and to maintain topology information. Nevertheless, the on-demand establishment of specific path information and the maintenance in terms of error management are clear examples of positive and negative feedback loops, respectively.

While proactive routing protocols only rely on the management of neighborhood information and the exchange of these data in the entire network, reactive routing protocols essentially focus on the enforced search for paths just in time if new data packets are available for transmission. We particularly discussed the on-demand routing protocols Dynamic Source Routing (DSR), Ad Hoc on Demand Distance Vector (AODV) and Dynamic MANET on Demand (DYMO). All these protocols have in common that during the path setup phase, a route request is flooded though the entire network (limited only by a pre-configured maximum network diameter). This kind of positive feedback encourages all targeted nodes to set up and to maintain destination-specific state information. Similarly, the (usually unicasted) route reply will participate in this process. On the other hand, the detection of route failures, e.g. due to transmission problems or due to node mobility, results in a constraining route error message. This kind of negative feedback is first suppressing any further data messages' being forwarded along the broken path and, secondly, stimulating a new path discovery.

A particularly good example of the successful exploitation of feedback techniques in ad hoc routing is the Route-Lifetime Assessment Based Routing (RABR) protocol. This protocol relies on the calculation of the link affinity a_{nm}. This measure basically describes the estimated stability of a particular link between two adjacent nodes n and m. If two nodes are moving further, i.e. the distance between them is increasing, the link affinity a_{nm} stimulates the local energy management to adjust the transmission power level according to the new estimate.

Data-centric networking

The presented methods for data-centric networking primarily concentrate on reducing or even preventing state information by relying on the transmitted data themselves for taking forwarding decisions. Feedback is used in almost all presented data-centric networking methods.

For example, most of the discussed gossiping optimizations employ feedback loops for adapting the gossiping probability p according to the behavior of other nodes and to changing environmental conditions. So, the node density can be used as positive feedback similarly to the average reception percentage τ_{arp}. Another approach exploited the distance estimation between two nodes similar to the PCM protocol for adaptive approximations of the gossiping probability p.

Feedback can also be found in agent-based techniques, as discussed for rumor routing. The information carried by an agent that is (randomly) traveling through the entire network represents positive feedback information. This feedback is instantly used if the agent arrives at a node that knows less suitable routes to particular destinations. Time represents negative feedback in this example, as old or expired information does not further promote a particular destination.

Finally, directed diffusion relies on feedback loops as well. Received interest messages act as promoters at sensor nodes that are able to produce specific sensor messages. Again, time represents negative feedback that is suppressing further actions after a given timeout. Several optimizations for directed diffusion have been proposed. We discussed, for example, mobility support based on two feedback techniques. First, aggressive diffusion focused on the timeout handling. Instead of relying on static timeouts, simple heuristics are used for this estimation. Secondly, anticipatory diffusion is used to represent relative interest. Instead of using uncontrolled amplifications (sensor data generation), locality information is used to coordinate the interest propagation, i.e. coordinated stimulation is achieved.

Clustering

We also discussed clustering techniques as support functions for various algorithms and protocols used for communication in ad hoc and sensor networks. In particular, we introduced two general clustering algorithms—k-means and hierarchical clustering—as typical examples. In general, clustering techniques are thought to be self-organizing. Actually, feedback can be provided in most algorithms, e.g. in k-means and hierarchical clustering, by adaptive distance measures.

In particular, we discussed Low-Energy Adaptive Clustering Hierarchy (LEACH) and Hybrid Energy-Efficient Distributed Clustering Approach (HEED)–two well known cluster-based routing protocols. These protocols exploit the remaining energy in comparison with the energy level of other nodes for decisions about the current clusterheads. So, HEED re-calculates its probability to become clusterhead CH_{prob} in every round, based on the remaining node energy $E_{residential}$ and the maximum node energy E_{max} for fully loaded batteries.

Communication and coordination

In the context of communication and coordination principles, we first investigated time-synchronization algorithms and solutions for weak coordination based on logical time. Furthermore, we discussed a number of coordination approaches. Most of these solutions rely on multiple feedback loops, concentrating on the main objective–to reduce the communication overhead for better scalability and improved energy efficiency.

Among the presented time-synchronization algorithms, the Timing-sync Protocol for Sensor Networks (TPSN)–a sender–receiver synchronization approach–and the Reference Broadcast Synchronization (RBS), which is employing the receiver–receiver model, are exploiting feedback information. In particular, TPSN uses an update mechanism similar to that developed for the Network Time Protocol (NTP). The variation between multiple time-synchronization messages is included in the calculation, to control the adaptation process of the local clock, thus, it represents negative feedback.

The approaches for scalable coordination are based on data-centric communication and application-specific operation. We discussed Span and ASCENT as typical examples that provide energy-efficient coordination for topology maintenance. Feedback is used by Span for optimizing the utility $C_i/\binom{N_i}{2}$. The sources of feedback are the current energy level E_r/E_m and the number of neighboring nodes N_i. Similarly, ASCENT uses the current packet loss ratio for adaptation. The Distributed Event-driven Partitioning and Routing (DEPR) algorithm developed for sensor–actor coordination uses the latency bound B as system-inherent feedback among the packet forwarding nodes.

Positive and negative feedback are also exploited to stimulate the cooperative behavior of selfish nodes. In particular, positive feedback is provided by increasing the credit of a node after forwarding a message and negative feedback results from costly data transmissions.

In-network operation and control, as supported by the Rule-based Sensor Network (RSN), provide all the necessary means for creating feedback loops. For example, RSN can be used to reproduce the behavior, as proposed by the mentioned coordination techniques.

Collaboration and task allocation

We discussed algorithms for collaboration based on task- and resource-allocation algorithms, specifically using examples from the Multi-Robot Task Allocation (MRTA) domain. Both intentional and emergent cooperation techniques rely on sophisticated feedback loops.

While centralized task allocation using approaches such as the Open Agent Architecture (OAA) perform a global optimization process based on available network-wide state information, auction-based solutions rely on distributed decision processes. For example, MURDOCH uses the description of tasks, i.e. the specific requirements in terms of available resources and time constraints, as feedback information. Additionally, the possible reallocation of tasks provides a means of negative feedback. On the other hand, the dynamic negotiation according to the mediation algorithm explicitly uses a hill-climbing optimization process based on the maximization of an objective function: $p^* = \arg\max_{p \in \mathbf{P}} f(p, A)$.

In emergent cooperation, feedback is provided either by neighboring systems to adapt the local probability to leave or to stay in the nest– P(leave) and P(stay), respectively–or by observable system state, such as the group efficiency η.

16.3   Interactions among individuals and with the environment

Interactions among neighboring nodes is one of the key characteristics in all types of ad hoc networks. Additionally, environmental observations can be exploited in the special case of WSNs and SANETs. In this section, we summarize the techniques and methods based on node interactions used for communication and coordination in ad hoc and sensor networks.

Table 16.2 summarizes our discussion on the use of interactions among individuals and with the environment in networking algorithms and techniques for (mobile) ad hoc and sensor networks as well as in coordination and control approaches.

Medium Access Control

The complete MAC layer inherently relies on intensive interactions among neighboring nodes and observations of the surrounding (radio) environment. In order to detect or to prevent collisions, various solutions such as overhearing or idle listening have been developed, based on the passive analysis of the radio channel. Simple protocols such as Multiple Access with Collision Avoidance (MACA) or MACAW already provide techniques addressing the hidden terminal and exposed terminal problems by introducing the RTS/CTS handshake, i.e. a brief synchronization step between directly connected neighbors prior to forthcoming data transmissions.

S-MAC also uses this RTS/CTS handshake for synchronizing data transmissions among nodes sharing a common radio range. Additionally, two separate mechanisms are provided by S-MAC. First, the synchronization of the communication windows for supporting busy–sleep cycles is based on information exchange between neighboring nodes. All nodes

Table 16.2   Summary of interactions among nodes and with the environment used in SANET algorithms

Problem domain	Employed interactions
Networking aspects	
MAC	Intensive protocol-inherent interactions between neighboring nodes to detect or to prevent collisions, e.g. in all MACA-based protocols; synchronization according to local message exchanges; indirect information exchange using signal strength measurements, e.g. in PCM
Ad hoc routing	State and topology maintenance for address-based routing; interactions among neighboring and remote nodes to search shortest path information, e.g. in AODV and DYMO; duplicate address detection based on node interactions, e.g. in Passive Duplicate Address Detection (PDAD) and Dynamic Address Allocation (DAA)
Data-centric networking	Optimized gossiping strategies exploiting the local topology information; agent based approaches relying on stigmergic information exchange and on local interactions between neighboring nodes, e.g. in rumor routing; adaptation of source–sink relationships according to remote interactions in directed diffusion
Clustering	Interaction provides the basis for all clustering techniques; transmission power estimation and cluster affiliation according to local interactions, e.g. in LEACH and HEED
Coordination and control	
Communication and coordination	Time synchronization based on exchanged data or specific time protocol messages (used by all the discussed algorithms); topology maintenance and clustering techniques based on local interactions, e.g. in Span, ASCENT and DEPR
Collaboration and task allocation	Intentional coordination based on directed communication to a central decision taker, either for periodic state maintenance, e.g. in OAA, or for auction systems, e.g. in MURDOCH and mediation; local interactions among neighboring agents and stigmergic communication in emergent cooperation

in a common neighborhood are synchronized by S-MAC to a common sleep schedule. Secondly, the adaptive listening mechanism allows the adaptive modification of the sleep cycle according to forthcoming data transmissions. Again, neighboring nodes exchange necessary information to update their local states accordingly.

For energy-efficient communication, PCM provides a means for dynamic adaptation of the necessary transmission energy. In this case, the RTS/CTS handshake is exploited to estimate the distance between two nodes. According to the measured received signal strength of the RTS and the CTS, respectively, the nodes adapt the transmission power to minimize the necessary transmission energy. Basically, this estimation represents an indirect information exchange between the corresponding nodes, which allows measurement of environmental conditions (possible radio interferences) as well.

Ad hoc routing

Address-based ad hoc routing requires the identification of paths according to proactively or reactively obtained topology information. Basically, this procedure includes the exchange of state information among at least the nodes involved in a particular route from a given source to a particular destination. In most cases, the path setup involved almost all nodes in the entire network, as flooding techniques are employed for on-demand route setup and periodic exchange of topology information between all nodes in the proactive case.

In particular, proactive protocols such as Destination Sequenced Distance Vector (DSDV) rely on periodical exchanges in the neighborhood state that incorporates global topology information. Thus, interactions between neighboring nodes finally allow the setup of globally valid topology information. In the reactive case, protocols such as DSR, AODV and DYMO search adequate paths on demand, i.e. if data messages are available for transmission. In this case, the most efficient method of path finding in terms of minimized route setup time is flooding. A route request is sent through the entire network (limited only by a maximum network diameter) to find the searched destination. Thus, remote interactions between distant nodes is achieved by helping (relaying) intermediate nodes.

In the context of address-based routing, we also described a number of dynamic address-allocation techniques that are needed if pre-configured unique address identifiers are not available for all nodes in the entire network. Most of the discussed techniques rely on Duplicate Address Detection (DAD) algorithms. These algorithms exploit the interactions among nodes in the network to detect possible address duplicates.

Data-centric networking

The most simple data-centric networking approaches, such as flooding, gossiping or Weighted Probabilistic Data Dissemination (WPDD), do not inherently rely on interactions among nodes – in this context, we do not consider the successful transmission of a data message as interaction, as it does not influence the data-dissemination method itself. Nevertheless, there are optimized versions of gossiping that indeed exploit such interactions. First of all, mobility and energy-efficient enhancements exploit available knowledge about the local topology and about previously observed gossips to reduce the gossiping overhead. This kind of meta information is directly extracted from local interactions among neighboring nodes.

In the rumor-routing technique, a number of autonomous agents 'crawl' the ad hoc network in order to find suitable, i.e. shorter and more efficient, paths towards particular destinations. During their travel around the entire network, the agents interact with visited nodes by exchanging routing information. Obviously, this is a kind of local interaction allowing remote information transfer. As agents are just specific data messages, we need to clarify which kind of interaction happens in this case. Actually, two different explanations are possible – while, finally, both hold simultaneously. From the nodes' point of view, the agents represent the environment, i.e. they are not part of the sensor network. Therefore, the modification of the environment (the agents) and the reaction to observations of the environmental changes (again, the information carried by the agents) represent interactions between nodes using the environment, i.e. a form of stigmergy. Secondly, the interaction can be reduced to two neighboring nodes exchanging information particles (the agents), i.e. direct interaction among neighboring nodes.

In directed diffusion, nodes interact for setting up and maintaining the gradients used for message delivery. Additionally, the sink node interacts with the (perhaps with multiple) source node over a multi-hop network. Both nodes exchange information on which type of messages are needed, in which time interval and precision, and so on.

Clustering

General-purpose clustering algorithms such as k-means or hierarchical clustering do not incorporate information exchange directly into the algorithm. Instead, they assume an existing information channel used to set up and to maintain state information regarding distance measures and cluster memberships. Therefore, communication, i.e. interaction among nodes, is a prerequisite for successful clustering. Depending on the particular solution, either direct communication is necessary or indirect communication using the environment, i.e. the use of stigmergic communication, may suffice.

The studied examples of clustering for efficient communication–LEACH and HEED– incorporate these requirements directly into their algorithms. Basically, both approaches rely on local communication among nodes near a self-elected clusterhead. This interaction is exploited for achieving and maintaining cluster affiliations according to an energy-efficient scheme.

Communication and coordination

All the discussed coordination techniques inherently demand a certain degree of communication among the coordinating nodes. The main differentiating factor is the necessary number of messages and the degree of interconnectedness.

For example, most of the time, the synchronization approaches discussed perform message-exchange operations in order to update state information needed to calculate a common time basis and to correct the local clock. Post-facto synchronization exploits timestamps carried by regular data messages and the TPSN approach uses the sender–receiver model based on bidirectional communication between the sender and the receiver. In contrast, RBS follows the time broadcast (receiver–receiver) model that is also used by the Global Positioning System (GPS). The sender is usually considered part of the overall system, thus, interaction among the systems can be observed.

Similarly, all the coordination approaches rely on direct interactions among the involved systems. So, Span, ASCENT and DEPR update their local state information according to exchanged information messages. Based on this information, the network topology can be updated accordingly.

In-network operation and control techniques such as Cougar allow the exploitation of interactions with other systems, i.e. the transmission of (perhaps aggregated) sensor readings, combined with interactions with the environment in order to validate sensor measures. RSN can again be used to formulate all the discussed operations in the form of simple programs for autonomously working but interacting systems.

Collaboration and task allocation

The interaction among the collaborating nodes and with the environment is the most important self-organization technique in the context of collaboration and task allocation. MRTA is
inherently built on communication strategies used between the robots or agents to contend
for the available tasks and resources.

Intentional cooperation based on centralized decision processes or on auction-based
systems relies on the message exchange at least among possible candidates for a particular task. For example, centralized coordination as proposed by OAA forces the agents to
periodically submit their local status to a centralized decision taker. Based on the received
information, the optimization process can allocate tasks and resources to the participants.
Similarly, auction-based solutions such as MURDOCH and the mediation algorithm negotiate – using inter-agent communication – on an optimal task assignment after the arrival of
a new task.

The stimulation techniques used by emergent cooperation rely on either stimulation by
work, i.e. message exchange between neighboring systems or stigmergic communication
via the environment, or on stimulation by state, i.e. the observation of the environment to
adapt the local decision process.

16.4   Probabilistic techniques

The third instrument to enable self-organization is randomness. According to its main
characteristic–unpredictability–it allows the design of methods and algorithms that can be
assumed not to starve in local optima. While probabilistic techniques alone will not lead to
controllable system behavior, randomness is often used in combination with feedback and
local interactions. In this section, we summarize the stochastic processes used in networking
and coordination protocols for SANETs.

Table 16.3 summarizes our discussion on the use of probabilistic functions in networking algorithms and techniques for (mobile) ad hoc and sensor networks as well as in
coordination and control approaches.

Medium Access Control

In the domain of wireless MAC protocols, randomness is heavily used to prevent collisions
on the wireless medium. In the presented examples, probabilistic schemes are predominantly
employed in the RTS/CTS handshake. Protocols such as MACA and its relatives Wireless
LAN (WLAN) and S-MAC use a fixed schedule for local communication. During the
RTS/CTS phase, all nodes that want to transmit a data message in the next time slot will
send a RTS message to announce this forthcoming transmission (and to gain access by
the following CTS). If all nodes start at the very beginning of the RTS/CTS window to
transmit their RTS messages, the probability of a collision among these packets would be
very high – even if the RTS messages are rather small compared with data messages.

Therefore, randomized initial delays are enforced before a node is allowed to broadcast its RTS. This probabilistic measure has two objectives. First, mutual exclusion can be
achieved, as (depending on the size of the RTS/CTS window and the number of competing

Table 16.3   Summary of probabilistic techniques used in SANET algorithms

Problem domain	Algorithms relying on probabilistic techniques
Networking aspects	
MAC	Reduced collision probability through randomized medium access; fairness and mutual exclusion are achieved by using random startup delays for the RTS/CTS handshake, e.g. in all MACA-based protocols
Ad hoc routing	Gossiping techniques to reduce the flooding overhead in reactive routing approaches, e.g. in optimized AODV; dynamic address allocation based on stateless random address selections in combination with DAD algorithms, e.g. in PDAD and DAA
Data-centric networking	Probabilistic data forwarding in gossiping approaches; agent-based techniques relying on random waypoint strategies, e.g. in rumor routing
Clustering	Randomized clusterhead selection to maximize the network lifetime and to provide fair distribution of the energy load, e.g. in LEACH and HEED
Coordination and control	
Communication and coordination	Randomization through variation of network latencies; randomly distributed back-off delay, e.g. in Span; probabilistic state transitions, e.g. in DEPR
Collaboration and task allocation	Probabilistic decision processes and task allocation according to estimations for winning the contest for a new task or the nest leaving probability

nodes) collisions will happen with a very low probability. Secondly, fairness can be guaranteed, as all nodes running the same randomized algorithm will have the same chance to start first with their RTS message.

Ad hoc routing

Address-based routing protocols usually rely on deterministic decision processes. Therefore, probabilistic techniques are rarely found in these algorithms. Therefore, the discussed table-driven protocols (either proactive or reactive) do not directly employ randomness. Nevertheless, some optimizations indirectly incorporate probabilistic techniques for efficient route discovery in on-demand approaches. Reactive routing solutions flood route requests through the entire network to find (shortest) paths to a given destination. Optimized solutions change this behavior to gossiping in order to reduce the overhead of route-request flooding.

In the context of address-based routing approaches, we also discussed dynamic address-allocation schemes. Basically, all of the presented algorithms, e.g. PDAD and DAA, use random address selections, locally initiated by all participating nodes in conjunction with subsequent detections of possible duplicates, i.e. DAD algorithms. This solution ensures high scalability, as no global state needs to be maintained, as well as fast convergence times due to low probabilities of address collisions in multiple rounds (we already discussed this fairness factor of randomized solutions).

Data-centric networking

In contrast to address-based routing solutions, most data-centric networking approaches inherently rely on stochastic processes and randomized decisions. The basic idea to improve the scalability of communication in ad hoc and sensor networks is to reduce the global state that needs to be maintained and synchronized and replace it by local decision processes. Based on the available information, these processes usually estimate or guess optimal configurations and improve the behavior according to continuous measurements and observations. In this process, randomized initial configuration settings provide an adequate starting point to prevent global synchronization effects and oscillations.

Gossiping represents a class of routing algorithms that is completely based on pure probabilistic techniques. All forwarding decisions are taken according to a random process. The gossiping probability p represents a threshold for controlling the statistical properties of gossiping. Some optimizations, e.g. gossiping with adaptive probabilities or WPDD, still completely rely on the random process while providing means for optimizing the gossiping probability p according to location information or message priorities.

Randomized decisions are also exploited in agent-based solutions such as rumor routing. In this particular example, agents crawl the network to update routing tables and to iteratively optimize path information. In order to enable a self-organized, i.e. uncoordinated, action of multiple agents in an unknown network topology, the agents follow a path through the network according to the random waypoint mobility model. This allows optimized coverage of the entire network, regardless of the number of active agents.

Clustering

Clustering is a deterministic process. Nevertheless, specific solutions also employ probabilistic techniques, e.g. the discussed cluster-based communication protocols LEACH and HEED. The main objective is the 'optimized' selection of clusterheads. In standard clustering algorithms such as k-means or hierarchical clustering, the clusterhead selection is the result of the basis clustering algorithm. Therefore, the actual distance measures will lead to a static clustering (in non-mobile environments).

If the cluster structure is explicitly used for communication aspects, the clusterheads will usually perform more tasks compared with other nodes (e.g. higher-level topology maintenance, routing and data forwarding). Therefore, these nodes will discharge their batteries much faster – and finally stop further operations.

In contrast, LEACH and HEED are examples of clustering algorithms that enforce a uniform distribution of the energy consumption. This goal is achieved by randomly distributing the role of clusterheads to all nodes in the network. Actually, both protocols perform a round-based scheme enforcing all nodes in the entire network to participate in this process. The periodic re-allocation of clusterheads enables, first, a fair distribution of the energy load and, secondly, it ensures the maximization of the lifetime of the entire network.

Communication and coordination

Similarly to other domains of SANET algorithms, probabilistic techniques are frequently used to prevent unbalanced utilizations and oscillation effects. In the context of communication and coordination, this is strongly required for equal distribution of communication and synchronization costs.

Time synchronization is not directly performing probabilistic calculations. Nevertheless, the stochastic variation through uncertainties in network latency always leads to randomized behavior for all the participating systems.

In contrast, the coordination aspect discussed in the context of scalable coordination based on data-centric communication and application-specific operation strongly relies on probabilistic techniques. Span, for example, uses a randomly distributed back-off delay (weighted by C_i) to prevent single nodes from being disproportionally utilized. ASCENT incorporates the network quality in terms of packet loss in the decision process. Thus, a stochastic factor is explicitly exploited. DEPR (sensor–actor coordination) uses probabilistic state transitions in order to prevent synchronization effects and oscillations.

In RSN, the :random operator was introduced to enable probabilistic decision processes for in-network operation and control.

Collaboration and task allocation

In the context of collaboration and task allocation, there is a strong demand to prevent single systems from taking all the load while other participants stay idle. This objective is approached differently by the discussed MRTA techniques.

Intentional cooperation does not necessarily use randomized algorithms. Nevertheless, sometimes, a probabilistic decision is used as a tie-breaker for allocations with similar properties.

On the other hand, probabilistic techniques are strongly exploited by emergent cooperation approaches. We presented two examples–one for stimulation by work and one for stimulation by state. Both rely on the calculation of a probability of the robot's performing a particular action.

The stimulation-by-work approach finally estimates P(robot i wins), the probability of a robot to win the contest on a new task. Instead of using this estimation as a final decision taker, it is used in a further probabilistic decision process. Similarly, the stimulation-by-state approach calculates and continuously adapts the local nest-leaving probability P_l according to the current environmental conditions.

17

Evaluation Criteria

The evaluation of self-organizing systems can be performed in multiple dimensions and according to a broad number of criteria. Besides classical performance measures, two parameters in particular are of interest: *scalability*, i.e. the supported number of interacting systems or the maximum size of a system, and *energy*, i.e. the energy constraints of an application scenario compared with the energy consumption of the system. We discussed these two properties in many facets in this book, especially during the investigation of technical solutions in the context of networking aspects in Part II and in the context of coordination and control in Part III.

It turned out that most solutions that are primarily addressing the scalability of a self-organizing system often lead to limitations in other performance aspects of the system, such as reduced determinism and predictability of the system behavior and less power consumption.

On the other hand, energy-related issues have been investigated in multiple domains. Energy-efficient algorithms are usually concerned with the optimization of a single process or mechanism. Conversely, energy-aware systems consider a multitude of interactions among parts of a single system and among systems in the entire network.

In general, scalability and energy constraints are indirectly proportional. Therefore, the evaluation of the system efficiency can become difficult. For example, the comparison of algorithms shows one with improved scalability and higher energy consumption and another one that is less scalable but leads to reduced energy consumption. Which should be considered the best one?

A promising solution is to take the application demands into account. The developed network architectures, protocols and methods are obviously meant to be used in a particular application scenario. Therefore, the overall system efficiency needs to be tested with respect to this scenario.

The primary application constraint is the *network lifetime*, i.e. the ability of the network to fulfill all application requirements for a given (limited) time or to maximize network lifetime in order to enable further operation independently of any external maintenance. In this chapter, we demonstrate that all typical performance metrics relevant to WSNs and SANETs can be represented by a single measure – the network lifetime.

Self-Organization in Sensor and Actor Networks   Falko Dressler
© 2007 John Wiley & Sons, Ltd

The structure of this chapter follows the outlined considerations. Thus, we will discuss the evaluation of self-organizing systems according to three criteria:

- Scalability – We briefly revisit scalability issues such as protocol overhead, restricting factors in wireless networks, and limitations in the form of reduced determinism in self-organizing systems.

- Energy considerations – Furthermore, we summarize energy constraints and energy-saving techniques in general ad hoc networks and in SANETs.

- Network lifetime – Finally, we introduce the network lifetime as the main application-driven metric for evaluating the performance of WSNs. We discuss properties and concepts for a concise definition of the network lifetime.

17.1   Scalability

We have already learned that the scalability of a system can be measured along at least three different dimensions. First, the size of a system with respect to the number of involved subsystems is one important measure. This includes the possibility of easily adding systems and resources to the system without degrading the performance. Secondly, scalability includes the geographical size, i.e. the distance between individual subsystems. For example, the message delay, which is directly proportional to the distance, must be dealt with. Thirdly, and most importantly in our discussion of system control, the manageability must be considered. This includes administrative boundaries as well as the number of interconnected systems and resources.

Regarding the scalability of techniques and mechanisms used for communication and cooperation in SANETs, a number of issues must be considered. Some of these issues are directly addressed by self-organization techniques, whereas others are basically introduced at the same time. Three of the most important aspects are outlined in the following.

- Protocol overhead – The main objective of self-organizing systems is to counteract the protocol overhead of typical communication protocols. Actually, protocol overhead can be measured according to various metrics. First, it may be associated with the amount and size of state information that must be stored and maintained at each node in the network. If, for example, a globally synchronized state is necessary, the communication requirements will become a problem with the increasing size of the system. Secondly, the direct-communication overhead must be considered. Usually, the goodput is used as a measure describing the number of successfully transmitted application data compared with all packet transmissions in the entire network.

- Capacity of wireless networks – A major limiting factor for any techniques and mechanisms developed in the context of SANETs is the maximum capacity of wireless networks. Gupta and Kumar (2000) have shown that the capacity of wireless networks with n randomly located nodes, each capable of transmitting at W bits per second and employing a common communication range, and each with randomly chosen and therefore likely far away destinations, is $\Theta\left(\dfrac{W}{\sqrt{n \log n}}\right)$. Even in the best case, i.e. assuming optimally located nodes and optimized transmission ranges, the throughput is limited to $\Theta\left(\dfrac{W}{\sqrt{n}}\right)$ bits per second. This principle of an upper bound for the network capacity also applies to arbitrary wireless networks.

- Reduced determinism – In Section 3.7, we discussed the problem of reduced determinism and predictability as a general property of most self-organizing systems. It turned out that the more scalable a system becomes using self-organization techniques, the less control in the collaboration of all participating entities is possible. The primary conclusion is that the predictability of the system behavior must be reduced for such a self-organizing system. Essentially, it is only possible to address the probability of achieving the designated goal, i.e. show the demanded behavior, but not to guarantee a particular system behavior.

17.2   Energy considerations

The energy capacity is one of the most cited constraints in WSNs and SANETs. Sensor nodes usually do not have unlimited energy reserves at their disposal – please remember the primary reason for ad hoc networking: untethered communication without a fixed network infrastructure. Additionally, there is an increasing gap between power consumption and power availability observable. In the following, we briefly summarize the requirements and (network-related) solutions to energy savings.

17.2.1   Energy management

The basic problem is provided due to the difficulties in replacing the batteries of nodes in a WSN. First of all, the envisioned application scenarios include hazardous environments as well as environments inaccessible for humans. Even if the nodes can be accessed, the trend to wards networks' consisting of thousands of nodes makes manual battery replacement infeasible.

Energy harvesting is often proposed as a potential solution. Actually, this is obviously the best approach. Unfortunately, in many scenarios, energy harvesting is either infeasible (form factor, environmental properties) or does not produce enough energy to operate the device based on gathered energy only. On the other hand, there are some constraints on the battery source that need to be regarded. For example, the battery size is directly proportional to its capacity. Thus, the operation of more powerful systems requires really large energy sources, which tend to be heavy and may lead to environmental problems.

Therefore, a number of approaches have been proposed to prolong the energy capacity. Among others, the following concepts are investigated in the sensor network community:

- Battery management – Battery management aims at exploiting the inherent property of batteries to recover their charge when kept idle. Thus, the efficiency of wireless communication can be enhanced through the use of improved battery-management techniques. The main criteria describing the quality of a battery are its energy density, the recharging cycle, environmental impact, safety, cost, available supply voltage, charge and discharge characteristics. Because, in idle conditions, the charge of the battery recovers, it is possible to use the theoretical capacity of a battery by increasing the idle time. This can be provided by battery scheduling techniques, which schedule between multiple available batteries in a mobile node.

- System power management – The idea of system power management is to exploit possible power-saving states of the system to minimize the average energy consumption. Typically, modern micro controllers support at least the three states active, idle, and sleep. Similarly, the radio transceiver can operate with the transmitter and/or receiver turned off. Switching between these modes is often complicated due to uncertainty about the length of the desired sleep time. Therefore, alternatives supporting ultra-low supply voltages and clock frequencies are investigated.

- Optimal transmission power – Energy consumption increases with an increase in the transmission power, which, in turn, is also a function of the distance between the communicating nodes. Unfortunately, the necessary energy relates at least quadratically to the distance ($E \sim d^2$ up to $E \sim d^4$). Thus, modifications of the transmission power have a great influence on the reachability of a node. The optimization of the transmission power therefore considers the trade-off between coverage or connectivity and power consumption.

  An optimally chosen transmission power also decreases the interference among nodes, which, in turn, increases the number of simultaneous transmissions.

- Channel utilization – As mentioned before, a reduction in the transmission power increases the number of simultaneous transmissions. Thus, it opens up the possibility of frequency re-use and finally leads to better channel utilization. Therefore, power control becomes especially important in CDMA-based systems.

17.2.2 Transmission power management

The component primarily contributing to the overall energy consumption is the communications module. Energy is not only needed for the transmission of data. Also, idle listening requires an essential amount of energy. Therefore, in the literature, receiving is often cited as being as expensive as sending according to the idea that a node transmits quite infrequently compared with its idle time.

In Part II, we discussed a number of solutions enabling individual nodes to completely turn off their transceiver to prevent idle listening. Concerning the operation at the MAC layer, we investigated, for example, the duty-cycle of S-MAC (Section 9.2). Other solutions are the clustering techniques LEACH and HEED (Sections 12.2.1 and 12.2.2, respectively).

In general, the necessary energy consumption for transmitting n bits can be calculated as depicted in Equation 17.1 (Karl and Willig 2005):

$$E_{tx} = T_{init} \times P_{init} + \frac{n}{R \times R_{coding}} \left(P_{txElec} + \alpha_{amp} + \beta_{amp} P_{tx} \right). \qquad (17.1)$$

Basically, this equation consists of the following three parts (for simplification, the modulation is not considered):

- The energy required to leave the sleep mode: $T_{init} \times P_{init}$, where T_{init} corresponds to the necessary time and P_{init} denotes the average initialization power.

- The time to transmit n bits: $\dfrac{n}{R \times R_{coding}}$, where R is the nominal data rate and R_{coding} is the coding rate.

- The amplifier power: $P_{amp} = \alpha_{amp} + \beta_{amp} P_{tx}$, where P_{tx} is the radiated power, and α_{amp} and β_{amp} are constants, depending on the transmitter model. The efficiency of the transmitter can be described as $\eta = P_{tx}/P_{amp}$. In addition, transmitter electronics need some power P_{txElec}.

We discussed a number of approaches to control the transmission power. At the MAC layer, for example, the PCM protocol (Section 9.3) needs to be mentioned. Similarly, the RABR protocol (Section 11.3) provides transmission power control at the network layer.

Finally, the trade-off comparing computation and communication energy costs needs to be considered. Applications for in-network processing and data aggregation exploit the discrepancies between these two costs. It is much more energy-expensive to sending one bit compared with computing one instruction. According to the literature, the transmission of 1 kilobyte is usually considered the same as computing up to three million instructions (Meguerdichian *et al.* 2001b; Raghunathan *et al.* 2002). This ratio needs to be considered for developing energy-efficient system architectures for SANETs.

17.3 Network lifetime

Network lifetime has become one of the key characteristics to be used for evaluating SANETs in an application-specific way (Akyildiz and Kasimoglu 2004; Blough and Santi 2002; Dressler and Dietrich 2006). A variety of parameters have been included in discussions on network lifetime. Examples are the availability of nodes, the sensor coverage and the connectivity. Recently, Quality of Service (QoS) measures have also been reduced to lifetime considerations.

In this section, we review the lifetime definition proposed by Dietrich and Dressler (2006). It basically represents a summary of the typical lifetime considerations used to concisely evaluate the performance of WSNs and SANETs.

17.3.1 Definition of 'network lifetime'

The lifetime definition by Dietrich and Dressler (2006) supports a wide range of application characteristics relevant to sensor networks. We only provide a brief review of the formal definition presented in this paper. The definition inherently supports heterogeneity among the nodes. The main purpose is to enable a comparative study of algorithms and techniques developed for sensor networks. This comparison is based on the demands of particular applications.

Prerequisites

In the next paragraphs, we use the following notations. The total number of available sensor nodes is n. R represents the region of deployment. Examples include rectangles ($R = [0, a_1] \times [0, a_2]$, $|R| = a_1 * a_2$), cuboids ($R = [0, d_n]^n$, $|R| = \prod d_n$) and circles ($|R| = \pi r^2$).

Each sensor node can be equipped with one or more different sensors in the set of possible sensors $Y = \{y_1, \ldots, y_k\}$. S^{y_i} is the set of all nodes with sensor y_i attached. Similarly, P^{y_i} is the set of targets to be observed by nodes with sensor y_i. The set of all existing sensor nodes is then called S^Y ($|S^Y| = n$).

We define the set of all nodes that are alive at a time t as $U(t)$. In Equation 17.2, $u_i^{Y_i}$ is a sensor node from the set of all sensor nodes, which is equipped with the sensors of type Y_i, and whose energy is not yet depleted. Similarly, $V(t)$ represents all nodes that are active at time t, i.e. those which are not in sleep mode.

$$U(t) = \left\{ u_1^{Y_1}, \ldots, u_m^{Y_m} \mid u_m^{Y_m} \text{ alive} \right\}, \, U(t) \subset S^Y, |U(t)| = u(t) \tag{17.2}$$

$$V(t) = \left\{ v_1^{Y_1}, \ldots, v_l^{Y_l} \mid v_i^{Y_i} \text{ active} \right\}, \, V(t) \subset U(t), |V(t)| = v(t). \tag{17.3}$$

In some network settings, sink nodes or base stations might be ordinary sensor nodes acting as base stations for other nodes. For this reason, the definition retains the possibility of sink nodes' failing or sleeping just like any other node. Therefore, the set of sink nodes $B(t) \subset S^Y$ may vary over time, and it is also possible that there are no sink nodes present in the network at some point in time.

$$B(t) = \{b_1, \ldots, b_k\} \subset S^Y. \tag{17.4}$$

The set of target points to be sensed by the network can be defined as $P^Y(t)$. Each target point can be sensed only by a certain collection of sensor types, denoted by the subsets $Y_i \subset Y$. Target points outside the region of deployment R are not allowed.

$$P^Y(t) = \left\{ p_1^{Y_1}, \ldots, p_m^{Y_m} \mid p_i^{Y_i} \in R \wedge Y_i \subset Y \wedge Y_i \neq \emptyset \right\}. \tag{17.5}$$

The area covered by all sensors of a certain type y is $A^y(t)$. $A_{v_i}^y$ denotes the area that is covered by the sensor of type y of node v_i. The shape of this area can be any shape representing the sensing range of a sensor. This could be, for example, a circle centered at v_i or a circle section originating at v_i.

$$A^y(t) = \bigcup_{\forall v_i \in V(t)} A_{v_i}^y(t) \cap R, \, y \in Y. \tag{17.6}$$

If $E(t)$ represents the set of all edges between active nodes, i.e. the set of connections, all possible communication graphs in the network can be denoted by the undirected graph $G(t) = (V(t), E(t))$. Note that only active nodes are included in the communication graph. $\kappa(t, m_1, m_n)$ is used to express the ability of two arbitrary nodes m_1 and m_n to communicate at a time t. Formally, the connectivity between all nodes on the path between m_1 and m_n needs to be checked. The number of hops needed for the communication is $n - 1$.

$$\kappa(t, m_1, m_n) \equiv \begin{cases} \forall i \in \{1, \ldots, n-1\} : m_i \in V(t) \wedge (m_i, m_{i+1}) \in E(t) & m_1 \neq m_n \\ 1 & m_1 = m_n \end{cases}. \tag{17.7}$$

We are now ready to define a series of criteria that may influence network lifetime at least in some network settings. Each criterion can be excluded from the final definition of lifetime by setting its modification factor correspondingly. In the following equations, these parameters are denoted by c_{**}.

Area coverage

The area coverage requirement is that the area covered by all sensors of type y must be greater than a certain portion of the deployment region. In other words, the fraction of the deployment region covered by type-y sensors $A^y(t)/|R|$ must be greater than the parameter c_{ac}^y:

$$\zeta_{ac}^y(t) \equiv A^y(t) \geq c_{ac}^y * |R|, y \in Y. \tag{17.8}$$

Target coverage

Similarly, the target coverage criterion requires that for each type of sensor y, a certain portion c_{tc}^y of all targets, which can be sensed by type-y sensors, must be within the area covered by those sensors. In this definition, it is not relevant if the targets are stationary or mobile. At each point in time, the current position of the targets is evaluated. Between the evaluations, the target positions may be updated.

$$\zeta_{tc}^y(t) \equiv \exists P_m^y \subset P^y \wedge |P_m^y| \geq c_{tc}^y * |P^y| \wedge P_m^y \in A^y(t), y \in Y. \tag{17.9}$$

k-coverage

The k-coverage criterion requires that each point in the region of interest has to be within the sensing range of at least k active sensors. The k-coverage parameter c_k^y indicates the magnitude of k. The function τ returns 1 if a certain point x is within the sensing radius of the type-y sensor of node v. The function σ indicates by how many active sensors a point x is covered.

$$\tau(x, v^y) = \begin{cases} 1 & x \in A_v^y \\ 0 & x \notin A_v^y \end{cases} \tag{17.10}$$

$$\sigma(t, x) = \sum_{i=0}^{v(t)} \tau(x, v_i^y). \tag{17.11}$$

The k-coverage criterion is fulfilled if σ is not smaller than the k-coverage parameter c_k^y for all points in the region of interest. There are two variants, depending on the kind of region of interest: one for an areas (Equation 17.12) and one for targets (Equation 17.13):

$$\zeta_k^y(t) \equiv \forall x \in R : \sigma(t, x) \geq c_k^y \tag{17.12}$$

$$\zeta_k^y(t) \equiv \forall p \in P : \sigma(t, p) \geq c_k^y. \tag{17.13}$$

Number of active nodes

At least a portion of c_{an} times the number of existing nodes must be active at any time (sleeping nodes are not considered active):

$$\zeta_{an}(t) \equiv v(t) \geq c_{an} * n. \tag{17.14}$$

Number of alive nodes

The portion of alive nodes, including sleeping nodes, must be greater than c_{ln} times the number of existing nodes at any time. This means that the parameter c_{ln} can never be truly switched off because the true greater relation ensures that the lifetime of the sensor network is constrained to be, at most, the time of the failure of the last alive node:

$$\zeta_{ln}(t) \equiv u(t) > c_{ln} * n. \tag{17.15}$$

Availability (service disruption tolerance)

A service disruption of at most c_{sd} seconds is tolerated. This parameter is included in the final lifetime definitions (Equation 17.30).

Latency

For each type of packet in the network, all packets of the type must arrive at a sink node within a period of c_{la} after the initial sending:

$$\zeta_{la} \equiv \forall \text{packets} : \text{packet latency} \leq c_{la} \tag{17.16}$$

Loss and error

At most, a portion of c_{lo} packets of all data packets sent in the network may be lost or unusable due to packet loss or error. This is equivalent to demanding that at least a portion of $1 - c_{lo}$ packets must be correctly received by a sink node, i.e. that the packet delivery ratio must be at least $1 - c_{lo}$.

$$\zeta_{lo} \equiv \frac{\text{lost packets} + \text{erroneous packets}}{\text{total packets}} < c_{lo}. \tag{17.17}$$

Connectivity

Usually, it is not important to ensure connectivity between all sensor nodes, but rather to ensure connectivity towards the sink nodes. In order to support arbitrary communication (instead of following the base station model), we use the notation of sink nodes for all destination nodes for ongoing communications at a given time t. The function $\chi(v_j, t)$ indicates if a node v_j has a connection to any active sink node in $B(t)$ at time t. If there is no active sink node, the indicator function returns false because a connection to a sink node does not exist.

$$\chi(v_j, t) \equiv \exists b_i \in B(t) \wedge (v_j, b_i) \in V(t) \wedge \kappa(t, v_j, b_i). \tag{17.18}$$

One criterion to evaluate connectivity in a sensor network is to require that at least a certain portion c_c of all active nodes have a connection to a sink node.

$$\zeta_c(t) \equiv \exists V_c \subset V(t) : |V_c| \geq c_c * |V(t)| \wedge \forall v_c \in V_c : \chi(v_c, t). \tag{17.19}$$

Connected coverage

In addition to independently investigating connectivity and coverage, it is often necessary to analyze both criteria simultaneously. This is required because the nodes covering the area could be different from those able to communicate (Xing *et al.* 2005). For the connected coverage criteria, it is useful to redefine the covered area $A^y(t)$ as $A^y_*(t)$. The difference between the two definitions is that $A^y(t)$ uses all active nodes, whereas $A^y_*(t)$ uses only those active nodes with a path to the sink.

$$A^y_*(t) = \bigcup_{\forall v_i \in V(t): \chi(v_i,t)} A^y_{v_i} \cap R. \qquad (17.20)$$

Based on $A^y_*(t)$, we can now define the criteria for connected area coverage $\zeta^y_{cac}(t)$ and connected target coverage $\zeta^y_{ctc}(t)$. For area coverage, the area covered by those active sensor nodes with a path to a sink must be greater than a specified portion c^y_{cac} of the whole area (Equation 17.21). Similarly, for target coverage, the portion of targets covered by active sensor nodes with a path to a sink node has to be at least a specified percentage c^y_{ctc} of all targets (Equation 17.22).

$$\zeta^y_{cac}(t) \equiv A^y_*(t) \geq c^y_{cac} * |R|, \, y \in Y \qquad (17.21)$$

$$\zeta^y_{ctc}(t) \equiv \exists P_m \subset P \wedge |P_m| \geq c^y_{ctc} * |P| \wedge P_m \in A^y_*(t), \, y \in Y. \qquad (17.22)$$

Global coverage criteria

The previously depicted coverage criteria considered the functionality according to a specific sensor type. In order to obtain more comprehensive global knowledge about the functionality of the network concerning all sensors, these single metrics can be aggregated. Obviously, a global coverage criterion is only considered functional if the conjunctive combination of all single node criteria is fulfilled:

Global area coverage: $$\zeta_{ac}(t) = \bigwedge_{\forall y \in Y} \zeta^y_{ac}(t) \qquad (17.23)$$

Global target coverage: $$\zeta_{tc}(t) = \bigwedge_{\forall y \in Y} \zeta^y_{tc}(t) \qquad (17.24)$$

Global k-coverage: $$\zeta_k(t) = \bigwedge_{\forall y \in Y} \zeta^y_k(t) \qquad (17.25)$$

Global connected area coverage: $$\zeta_{cac}(t) = \bigwedge_{\forall y \in Y} \zeta^y_{cac}(t) \qquad (17.26)$$

Global connected target coverage: $$\zeta_{ctc}(t) = \bigwedge_{\forall y \in Y} \zeta^y_{ctc}(t). \qquad (17.27)$$

Network lifetime

We can now begin to integrate the presented single criteria into a final definition of network lifetime. First, we define an aggregate criterion – the liveliness of the network $\zeta(t)$ as the conjunctive combination of all single criteria:

$$\zeta(t) \equiv \zeta_{ac}(t) \wedge \zeta_{tc}(t) \wedge \zeta_k(t) \wedge \zeta_{an}(t) \wedge \zeta_{ln}(t) \wedge \zeta_{la} \wedge \zeta_{lo} \wedge \zeta_c(t) \wedge \zeta_{cac}(t) \wedge \zeta_{ctc}(t).$$

(17.28)

We then define T to be the ordered sequence of all points in time at which the liveliness $\zeta(t)$ changes its value (from true to false or vice versa). This can be done by checking ζ at time t and at time $t - \epsilon$, i.e. just before time t:

$$T = \{t_i | (\zeta(t_i - \epsilon) \wedge \neg\zeta(t_i)) \vee (\neg\zeta(t_i - \epsilon) \wedge \zeta(t_i))\}, t_i < t_{i+1}, i \in \mathbb{N}_0.$$

(17.29)

To simplify the following lifetime definitions, we define e to be the minimal index in T after which a service disruption of more than c_{sd} seconds follows. If such an index does not exist (e.g. if the service disruption tolerance is infinite), e is taken to be the last index in T, i.e. $|T|$:

$$e = \begin{cases} \min(i \in [0, |T| - 1] : \neg\zeta(t_i) \wedge (t_{i+1} - t_i) > c_{sd}) & \text{if such } i \text{ exists} \\ |T| & \text{otherwise.} \end{cases}$$

(17.30)

For further simplification, we define the periods of time during which the network is lively as t_i^a:

$$\forall i \in [0, e] : t_i^a = \begin{cases} t_{i+1} - t_i & \text{if } \zeta(t_i) \\ 0 & \text{otherwise.} \end{cases}$$

(17.31)

The first metric depicts the *accumulated network lifetime* Z_a. It is the sum of all time periods in which $\zeta(t)$ is fulfilled, i.e. exactly the intervals t_i^a, stopping only when the criterion is not fulfilled for longer than c_{sd} seconds:

$$Z_a = \sum_{i=0}^{e} t_i^a.$$

(17.32)

The second metric is the *total network lifetime* Z_t. It represents the first point in time at which the liveliness criterion is lost for a time period longer than the service-disruption tolerance c_{sd}:

$$Z_t = t_e.$$

(17.33)

Both lifetime metrics are intended to depict the network lifetime in seconds. The metrics probably become more expressive when used together.

17.3.2 Scenario-based comparisons of network lifetime

For the evaluation of the various network lifetimes, Dietrich and Dressler (2006) used the networking setup described in Xing *et al.* (2005). A simulation with 160 sensor nodes placed in an area of 400×400 m^2 was performed to obtain sample data for the evaluation. In this example, the nodes only consumed energy for communication. Additionally, the nodes only had a very small, randomly distributed supply of energy at their disposal. The nodes followed random sleep cycles and communicated regularly with a base station in the middle of the simulation area during their awake periods. A number of sensor types was randomly selected from the three types available for each node.

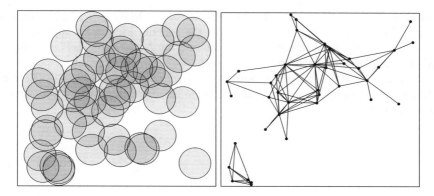

Figure 17.1 Coverage (left) and connectivity (right) for the sample setup at time $t = 300$; please note that the communication range is about twice as large as the sensing range

Figure 17.1 shows the distribution of the nodes in the sample setup at time $t = 300$. On the left-hand side, the sensor coverage is depicted for a common sensing radius of 40 m. On the right-hand side, the communication graph is shown for a communication range of 100 m.

During the simulation, the positions of the nodes, their failure times, their sleep periods and the types of sensors available on each node were recorded. We conducted only a single run of the simulation, as the purpose was not to obtain statistically significant simulation output, but to obtain sample values for the evaluation of the network lifetime definitions.

To illustrate the concepts of the presented lifetime definition, Figure 17.2 shows how service disruption tolerance and k-coverage influence the lifetime achievable if target coverage is the only criterion. The exact amount of target coverage required is varied in the range [0, 1] inside each of the box plots. If the targets are required to be covered by more than one sensor, the lifetime of the network decreases significantly, in some cases

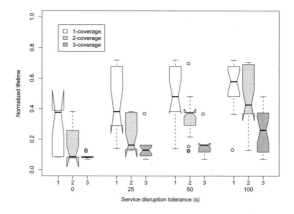

Figure 17.2 Target coverage depending on service disruption tolerance and k-coverage

by more than 20 % of the maximum achievable lifetime. On the other hand, allowing for some amount of service disruption can increase the network lifetime by approximately the same amount. As an example, compare the lifetime for 2-covered targets without service disruption tolerance with the lifetime for 3-covered targets with 25 s of service disruption tolerance. The medians of both lifetimes are nearly equal, demonstrating that a tolerance towards service disruptions can compensate for higher requirements in other parts of the system to some extent.

Part V

Bio-inspired Networking

In nature, we can find interesting solutions to communication and control in highly distributed systems. Therefore, the use of biological self-organization methods as a basis for technical self-organization seems to be appropriate. We investigate selected techniques in this part to provide a better understanding of bio-inspired systems and algorithms. Here, we emphasize the applicability for sensor and actor networks and discuss the concepts for completely autonomous decision taking to solve a complex global tasks. The design process of bio-inspired methods is outlined, starting with a biological model and the structural comparison with technical problems, and finishing with the adaptation of the model to the technical process. The conclusion of the investigation of bio-inspired systems can be that bio-inspired solutions are just one, but powerful, approach.

Outline

- Bio-inspired systems

  Primarily, the focus of this section is to demystify the concepts of bio-inspired networking. Based on selected approaches, the objectives and solution paths of biologically inspired methods are depicted in more detail. In this chapter, we provide an overview to the main concepts investigated in the various bio-inspired research domains. Afterwards, we discuss three of those domains in more detail: swarm intelligence as used for efficient ad hoc routing, the artificial immune system used for misbehavior detection, and cellular signaling pathways showing opportunities for efficient in-network operation and control.

- Bio-inspired networking – further reading

Self-Organization in Sensor and Actor Networks   Falko Dressler
© 2007 John Wiley & Sons, Ltd

Part V

Bio-inspired Networking

18

Bio-inspired Systems

The turn to nature for solutions to technological questions has brought us many unforeseen great concepts. This encouraging course seems to hold on for many aspects of technology. After the first approaches to classify bio-inspired solutions and to explicitly encourage further research in this domain by (Eigen and Schuster 1979), numerous algorithms and techniques have been proposed, which are based on bio-inspired research.

The most prominent domains as addressed in computer science have been genetic algorithms and neural networks. Both represent optimization algorithms used in domains in which a full search is either inappropriate due to time or resource limitations or not feasible at all. Later, a good number of other approaches joined this effort, based on varying biological mechanisms as inspiration.

In the late 1990s, a new domain emerged out of this bio-inspired systems research, which is usually known as *bio-inspired networking*. It turned out that a number of solutions studied in biology have similar counterparts in the networking domain. Especially when looking at massively distributed systems, as discussed in this book, focusing on Wireless Sensor Networks (WSNs) and Sensor and Actor Networks (SANETs), the efficiency and simplicity of biological systems are providing solutions that are as fascinating as they are efficient.

With the beginning of the 21st century, a completely new research community started to grow. Their focus is on the investigation and development of bio-inspired algorithms applicable for the operation and control of technical systems that are either hard to maintain due to the huge number of subsystems involved, or that work in environments with uncertain (and changing) conditions.

In Section 4.3, we differentiated between bio-inspired algorithms and the domain of self-organizing systems. Nevertheless, the capability of self-organization is inherent in many of the bio-inspired solutions. Thus, we will outline a number of these techniques in this part. The main focus is, of course, still self-organization. Therefore, only a few selected domains of bio-inspired research are presented in the following chapters.

For a comprehensive study of biologically inspired solutions, we use this chapter on bio-inspired systems to present a general overview of the techniques with biological backgrounds that are typically used in system-related problems.

Self-Organization in Sensor and Actor Networks   Falko Dressler
© 2007 John Wiley & Sons, Ltd

18.1   Introduction and overview

In order to become familiar with bio-inspired systems, we first discuss the ideas and concepts behind the scenes. This introduction allows understanding of the typical concepts used in bio-inspired research domains. Particular examples of bio-inspired solutions in these different research areas are provided, focusing on selected application examples related to SANETs. In the following sections, three domains are outlined in more detail, i.e. swarm intelligence, the artificial immune system and cellular signaling pathways.

18.1.1   Ideas and concepts

Many biological systems have capabilities and features that are strongly needed in technical systems. For example, the social behavior of ants or bees introduces collaborative work beyond the capabilities of current technological solutions. Additionally, scientists have been fascinated by the often highly elegant problem-solving techniques shown by biological entities. This and many other reasons led to the emergence of a new research domain of biologically inspired solutions – in short, bio-inspired systems.

The term 'bio-inspired' has been introduced to demonstrate the strong relationship between a particular system or algorithm, which has been proposed to solve a specific problem, and a biological system, which follows a similar procedure or has similar capabilities. Most of the bio-inspired solutions are problem-oriented, i.e. they are about problem solving. As an example, foraging ants should be mentioned. Ants are grandmasters in search and exploration. They collectively master to find the shortest paths in dynamic environments. We outline the principles of ant routing in detail in Section 18.2.

Usually, such bio-inspired algorithms are not based on theoretical modeling. Instead, the solutions are rather mimicking particular features and mechanisms known from biology. Typical examples are artificial neural networks and evolutionary algorithms. Since the very beginnings of bio-inspired systems research, numerous research domains have been created, each addressing different problem spaces and relying on different biological paradigms.

Basically, the following application domains of bio-inspired solutions to problems related to computing and communications can be distinguished:

- Bio-inspired computing represents a class of algorithms focusing on efficient computing, e.g. for optimization processes and pattern recognition.

- Bio-inspired systems rely on system architectures for massively distributed and collaborative systems, e.g. for distributed sensing and exploration.

- Bio-inspired networking is a class of strategies for efficient and scalable networking under uncertain conditions, e.g. for delay-tolerant networking.

The primary principles of investigating and exploiting biologically inspirations is depicted in Figure 18.1. The process consists of the following three steps. First, analogies between biological and technical systems, such as computing and networking systems, must be identified. It is especially necessary that all the biological principles are understood properly, which is often not yet the case in biology. Secondly, models must be created for the biological behavior. These models will later be used to develop the technical solution. The translation from biological models to the model describing bio-inspired technical systems is

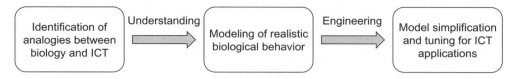

Figure 18.1 The process of investigating and exploiting biological analogies for developing efficient information and communication systems (ICT)

a purely engineering step. Finally, the model must be simplified and tuned for the technical application. It should be mentioned that biologists already started looking at 'bio-inspired systems' to learn more about the behavioral pattern in nature (Webb 2000). Thus, the loop closes from technical applications to biological systems.

Bio-inspired algorithms can be effectively used for optimization problems, exploration and mapping, and pattern recognition. Based on a number of selected examples, we will see that bio-inspired approaches have some outstanding capabilities that motivate their application in a great number of problem spaces.

In the context of SANETs, scalability issues or limitations in terms of available resources, such as processing and storage capacities, can be addressed in particular. Additionally, autonomously acting systems with limited communication capabilities are enabled to collaboratively work on global objectives. According to the main issues in SANETs, i.e. limited scalability and network lifetime, bio-inspired solutions may enable further progress in building systems consisting of an even larger number of individuals and being able to work unsupervised for a longer time period.

Public interest in bio-inspired approaches has been raised, for example, by presentations of solutions for unmanned flying vehicles that will explore other planets such as Mars. Thakoor *et al.* (2004, 2002) described such a scenario. In particular, they proposed using a variety of nature-tested mechanisms for bio-inspired engineering of small autonomous fliers designed to explore unknown environments. In this example, biological sensory has been used as a source for developing flight stabilization and localization techniques.

One of the first more sophisticated systems in the context of communication networks was the bio-networking architecture developed by Wang and Suda (2001). It represents a middleware for the design and implementation of scalable, adaptive and survivable network applications. Basically, it tries to emulate a number of biological concepts. The bio-networking architecture is built of so-called cyber entities. These units can die and reproduce, they compete for resources such as energy, they communicate locally, and natural selection and evolution are emulated in the form of mutations in the reproduction. While being an interesting approach, bio-networking architecture was only a first step towards efficient bio-inspired solutions in the context of communication networks.

In recent years, a great number of approaches have been proposed that claim improved efficiency by using bio-inspired techniques. The solutions range from simple algorithms and mathematical models to comprehensive architectures (Dressler and Carreras 2007). It turns out that the research community meanwhile agreed on a number of basic solutions that can be applied in multiple problem domains.

In the following, we outline the primary concepts of some of the best known bio-inspired research fields. The given examples concentrate on application domains related to

SANETs. Nonetheless, a much broader applicability is envisioned by bio-inspired systems research.

18.1.2 Bio-inspired research fields

In Section 4.3, we briefly described the typical research fields in the context of bio-inspired techniques. In this section, we pick up this discussion. The main idea is to get familiar with typical bio-inspired research, the main concepts and models. While there is, of course, more research focusing on adopting techniques and mechanisms as observed in biological systems in order to improve the working behavior of technical systems, in our overview, we concentrate on domains that emphasize either directly on networking issues in SANETs or provide means for improved solutions in this context. Furthermore, additional information is provided in the following sections to study selected examples in three bio-inspired research domains that explicitly address general wireless ad hoc networks, WSNs and SANETs.

Evolutionary Algorithms

Most Evolutionary Algorithms (EAs) are rooted on the Darwinian theory of evolution. Darwin proposed that a population of individuals capable of reproducing and subjected to genetic variation followed by selection results in new populations of individuals increasingly fit more into their environment. Based on this theory, the most natural computing approach has been developed, i.e. EAs. According to de Castro (2006), EAs can be categorized into the following classes:

- Genetic Algorithms (GAs);
- evolution strategies;
- evolutionary programming;
- genetic programming;
- classifier systems.

Basically, EAs represent a set of search techniques used in computing to find the optimal or an approximate solution to optimization problems. Specific examples are hill-climbing and simulated annealing algorithms. Every search task can be described by three parameters. First, a search space is defined, including the initial state. Secondly, an objective function is provided that describes the purpose of the search, e.g. min $f(x)$. Finally, an evaluation function needs to be defined that returns a specific value describing the quality (fitness) of the selected candidate solution.

The EA works as follows. The evolution starts from a population of randomly generated individuals and happens in generations. In each generation, the fitness of every individual in the population is evaluated; multiple individuals are stochastically selected from the current population (based on their fitness), and modified (mutated or recombined) to form a new population. This generational process is repeated until a termination condition has been reached. The solution is then represented by the fittest individual.

An example for using EAs, or more precise GAs, in wireless networks is the location management and channel-assignment procedure described by Das *et al.* (2006). The objective is to optimize the paging cost, the update cost and the update frequency in cellular networks, i.e. a multi-dimensional optimization process. Using a GA, an approximation of the optimal configuration can be obtained in a short period of time and with minimal computational resources.

Artificial Neural Networks

Traditionally, a neural network is used to refer to a network of biological neurons. In modern usage, the term is often used to refer to an Artificial Neural Network (ANN). The primary objective of an ANN is to acquire knowledge from the environment. This process is known as self-learning. In most cases, an ANN is an adaptive system that changes its structure based on external or internal information that flows through the network (Bishop 1996). It can be used to model complex relationships between inputs and outputs or to find patterns in data.

A typical artificial neuron representing the basic processing element of ANNs is depicted in Figure 18.2. The inputs, i.e. the artificial synapses, are characterized by their weight values w_i. All inputs are connected to a summing junction, which sums up all inputs x_i weighted (multiplied) by the particular w_i. A neuron-dependent parameter, the neuron's bias b is also connected to the summing junction. Finally, an activation function $f(u)$ is used to limit the amplitude of the output y of the neuron. This activation function is also referred to as the squashing function. In many cases, $f(u)$ is nonlinear, which makes ANNs as complex as they are interesting.

Mathematically, the neuron k that connects n inputs can be described as:

$$y_k = f(u_k) = f\left(\sum_{j=1}^{n} w_{kj} x_j + b_k\right).$$
(18.1)

So-called feed-forward ANNs perform signal processing without possible feedback; thus, the signals flow from the inputs to the outputs of the network through several layers of neurons. The weights of the neurons are to be determined in a learning process. This learning is either supervised or unsupervised. Samples with inputs and outputs are needed in the supervised learning, while, in the unsupervised learning only the inputs are needed.

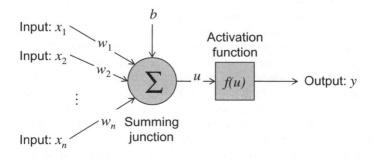

Figure 18.2 A single processing element of feed-forward ANNs

Swarm Intelligence

Swarm Intelligence (SI) is an Artificial Intelligence (AI) technique based on the observations of the collective behavior in decentralized and self-organized systems (Bonabeau *et al.* 1999). Such systems are typically made up of a population of simple agents interacting locally with one another and with their environment. According to de Castro (2006), five basic principles of SI systems can be distinguished:

- Proximity – Individuals are able to interact to form social links.

- Quality – Individuals can evaluate their interactions among themselves and with the environment.

- Diversity – Diversity improves the capability of the system to react to unknown and unexpected situations.

- Stability – Individuals should not alter their local behavior in response to environmental fluctuations.

- Adaptability – The entire system should be able to adapt to environmental changes.

Local interactions between autonomously acting agents often lead to the emergence of global behavior. Examples can be found in nature, including ant colonies, bird flocking, animal herding, bacteria molding and fish schooling.

In most cases, SI-based algorithms such as the Ant Colony Optimization (ACO) are inspired by the behavior of foraging ants (Bonabeau *et al.* 1999). Ants are able to solve complex tasks by simple local means. There is only indirect interaction between individuals through modifications in the environment, e.g. pheromone trails are used for efficient foraging. Ants are 'grand masters' in search and exploration.

The best known applications of swarm principles are routing in communication networks (Di Caro and Dorgio 1998) and the broad field of multi-robot systems (swarm robotics) (Dorigo *et al.* 2004). We will discuss the topic of SI in more detail in Section 18.2.

Artificial Immune System

An Artificial Immune System (AIS) is a type of optimization and pattern-recognition algorithm inspired by the principles and processes of the mammalian immune system (Hofmeyr and Forrest 2000). The algorithms typically exploit the immune system's characteristics of self-learning and memorization. The immune system is, in its simplest form, a cascade of detection and adaptation, culminating in a system that is remarkably effective.

Basically, the immune system can be seen as a huge number of guards representing a massively distributed compound that continuously checks every particle to see whether it proves to be innocuous. Additionally, all guards are continuously trained to be able to recognize further particles. If a guard detects hazardous material, it induces multiple processes. First, the material is immediately destroyed. Additionally, the new pattern is incorporated into the learning process of other guards.

Thus, the most interesting working behavior is the self-optimization and learning process of the immune system. In nature, two immune responses were identified. The primary one is to launch a response to invading pathogens, leading to an unspecific response (using leucocytes). In contrast, the secondary immune response remembers past encounters, i.e. it represents the immunologic memory. It allows a faster response the second time around, showing a very specific response (using B-cells and T-cells).

To aid the design phase of AISs, de Castro and Timmis (2002) proposed the immune-engineering framework. The immune engineering process that leads to a framework to design AISs is composed of the following basic elements:

- representations of the system components;

- mechanisms to evaluate the interaction among individuals and with the environment, i.e. stimulation pattern and fitness functions;

- adaptation procedures to incorporate the system dynamics.

Data analysis and anomaly detection represent typical application domains (de Castro and Timmis 2002). The complete scope of AISs is widespread. Applications have been developed for fault and anomaly detection, data mining (machine learning, pattern recognition), agent-based systems, control and robotics. One of the first AISs was developed by Kephart (1994).

Based on this work, misbehavior detection and attack- or intrusion-detection systems were developed based on the working principles of the natural immune system (Kim and Bentley 2001, 2002; Le Boudec and Sarafijanovic 2004). Besides network-security applications, the operation and control of multi-robot systems were addressed by AIS approaches. Selected application examples in the context of SANETs are presented in Section 18.3.

Cellular signaling pathways

The functionality of a eukaryotic cell relies on the complex network of biochemical processes (Alberts *et al.* 1994). Within these processes, single reactions take place in a coordinated fashion. They can take place simultaneously and successively. Thus, these processes must be highly regulated and controlled. This also means that these mechanisms are very specific for the given result.

The main goals of cellular processes are to regulate the intracellular metabolism and to communicate with their environment (Pawson 1995). We need to differentiate between intracellular and intercellular communication. Such communication, also known as signaling pathways, is an example of very efficient and specific communication. A first application in computer networking was described by Krüger and Dressler (2005).

- Intracellular signaling – The signal from the extracellular source is transferred through the cell membrane. Inside the target cell, complex signaling cascades are involved in the information transfer (signal transduction), which finally result in gene expression or an alteration in enzyme activity and, therefore, define the cellular response.

- Intercellular signaling – Cells can communicate via cell-surface molecules. In this process, a surface molecule of one cell or even a soluble molecule, which is released by one cell, directly binds to a specific receptor molecule on another cell. Soluble molecules such as hormones can also be transported via the blood to remote locations.

Basically, two communication paradigms can be differentiated. The regulation of the concentrations, e.g. of anions, allows diffuse message transmission. On the other hand, specific reactions can be triggered by exchanging particles, e.g. proteins, that can be detected by the receiving cell and that may activate powerful signaling cascades. This behavior can be directly applied to different aspects of computer communications (Dressler and Krüger 2004; Krüger and Dressler 2005). We will further outline the principles and application examples of cellular signaling pathways in Section 18.4.

Molecular computing

Molecular computing or Deoxyribonucleic Acid (DNA) computing is a form of computing which uses DNA and molecular biology instead of the traditional silicon-based computer technologies. The first experimental use of DNA as a computational system was demonstrated by Adleman (1994). He solved a seven-node instance of the Hamiltonian Graph problem—an $\mathcal{NP}$-complete problem similar to the traveling salesman problem. In general, molecular computing may not be considered a bio-inspired technique. Instead, biology is directly used as an appliance for more efficient computations similar to quantum computing. Nevertheless, this technology is included in this list, as it represents an example of investigating nature for developing efficient technical systems.

The basic idea is that it is possible to apply operations to a set of (bio)molecules, resulting in practical performances. So far, problem solving using molecular computing is executed in the form of brute-force search strategies in which the operations are simultaneously applied to all molecules. This shows also the main advantage of DNA computing: its high-speed information processing and storage.

DNA computing was originally proposed as a problem-solving technique. The massive parallelism and miniaturization of DNA suggests a number of problems that can be solved by DNA computing, particularly $\mathcal{NP}$-complete problems. So far, molecular computing has successfully been used for graph coloring, integer factorization and others.

18.2   Swarm Intelligence

The collaborative work of a multitude of individual autonomous systems is necessary in many areas of engineering. Swarms of small insects such as bees or ants address similar issues. For example, ants solve complex tasks by simple local means. There is only indirect interaction between individuals through modifications in the environment. The most prominent example is the efficient path exploration based on pheromone trails used by ants during foraging. Obviously, the productivity of all involved ants is better than the sum of their single activities and ants are 'grand masters' in search and exploration (Bonabeau *et al.* 1999).

18.2.1   Principles of ant foraging

The basic principles are simple. All individuals, i.e. the systems that collaborate on an overall task, follow simple rules that lead to an impressive global behavior, i.e. emerging

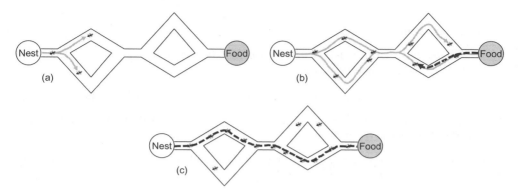

Figure 18.3 Collective foraging by ants. Starting from the nest, a random search for food is performed by foraging ants (a). Pheromone trails are used to identify the path for returning to the nest (b). The significant pheromone concentration produced by returning ants marks the shortest path (c)

behavior based on the simple rules and indirect interactions between the systems via the environment.

The principles of the foraging process are depicted in Figure 18.3. Ants perform a random search (random walk) for food. The way back to the nest is marked with a pheromone trail. If successful, the ants return to the nest (following their own trail). While returning, an extensive pheromone trail is produced, pointing towards the food source. Further ants are recruited that follow the trail along the shortest path towards the food. The ants therefore communicate based on environmental changes (pheromone trail), i.e. they use stigmergic communication techniques for communication and collaboration.

Dynamics in the environment are explicitly considered by the ant-foraging scheme. The pheromone slowly evaporates. Thus, if foraging ants are no longer successful, the pheromone trail will dissolve and the ants will start the search process again. Additionally, randomness is also a strong factor during successful foraging. As depicted in Figure 18.3(c), a number of ants will continue the random search for food. This adaptive behavior leads to an optimal search-and-exploration strategy.

The procedure of shortest-path finding in uncertain and dynamic environments according to the ant-foraging principle is also known as Ant Colony Optimization (ACO). Working on a connected graph $G = (V, E)$, the ACO algorithm is able to find the shortest path between any two nodes (Dorigo et al. 1999). Basically, the algorithm features the following capabilities. It employs a colony of ants to build a solution in the graph. A probabilistic transition rule is used for determining the next edge of the graph on which an ant will move. This moving probability is further influenced by a heuristic desirability. Finally, the 'routing table' is represented by a pheromone level of each edge, indicating the quality of the path.

The complete algorithm is described in Dorigo et al. (1999) and Bonabeau et al. (1999). The most important aspect of this algorithm is the transition probability p_{ij} for an ant k

to move from i to j, as given by Equation 18.2. Accordingly, this move depends on the following parameters:

- J_i^k is the tabu list of not-yet-visited nodes, i.e. by exploiting J_i^k, an ant k can avoid visiting a node i more than once;

- η_{ij} is the visibility of j when standing at i, i.e. the inverse of the distance;

- τ_{ij} is the pheromone level of edge (i, j), i.e. the learned desirability of choosing node j and currently at node i;

- α and β are adjustable parameters that control the relative weight of the trail intensity τ_{ij} and the visibility η_{ij}, respectively.

$$p_{ij}^k = \begin{cases} \dfrac{[\tau_{ij}(t)]^\alpha \times [\eta_{ij}]^\beta}{\displaystyle\sum_{l \in J_i^k} [\tau_{il}(t)]^\alpha \times [\eta_{il}]^\beta} & \text{if } j \in J_i^k \\ \\ 0 & \text{otherwise.} \end{cases} \qquad (18.2)$$

After completing a tour, each ant k lays a quantity of pheromone $\Delta\tau_{ij}^k(t)$ on each edge (i, j) according to the following rule (Equation 18.3), where $T^k(t)$ is the tour done by ant k at iteration t, $L^k(t)$ is its length and Q is a parameter (which only weakly influences the final result):

$$\Delta\tau_{ij}^k(t) = \begin{cases} Q/L^k(t) & \text{if } (i, j) \in T^k(t) \\ 0 & \text{otherwise.} \end{cases} \qquad (18.3)$$

This method cannot perform well without pheromone decay. This effect is provided by the pheromone update rule depicted in Equation 18.4, where $\Delta\tau_{ij}(t) = \sum_{k=1}^m \Delta\tau_{ij}^k(t)$. The decay is implemented in the form of a coefficient ρ with $0 \le \rho < 1$:

$$\tau_{ij}(t) \leftarrow (1 - \rho) \times \tau_{ij}(t) + \Delta\tau_{ij}(t). \qquad (18.4)$$

According to Dorigo *et al.* (1996), the total number of ants m is an important parameter of the algorithm. Too many ants would quickly reinforce suboptimal tracks and lead to early convergence to bad solutions, whereas too few ants would not produce enough decaying pheromone to achieve the desired cooperative behavior.

Algorithms and methods as studied in the domain of swarm intelligence are used in many domains. So, the ACO algorithm has been demonstrated to work successfully for routing and task allocation in ad hoc networks and SANETs. The most prominent examples are the AntNet approach (Di Caro and Dorgio 1998) and its counterpart, AntHocNet (Di Caro *et al.* 2005), which was developed for use in ad hoc networks. Both approaches are described in the following sections. Additionally, an approach for integrated routing and task allocation is presented (Labella and Dressler 2006, 2007), which is also based on the ACO algorithm.

Furthermore, many solutions are proposed in the literature that are based on ant routing and other SI principles. For example, combined routing and data fusion have been discussed by Muraleedharan and Osadciw (2003). Similar techniques have been used to address the problem of selecting components and redundancy levels to maximize the system reliability have been investigated by Liang and Haas (1999) and Liang and Smith (1999).

18.2.2   Ant-based routing

Routing in communication networks based on techniques inspired by the foraging behavior of ants has been addressed by the AntNet approach (Di Caro and Dorgio 1998). AntNet's agents concurrently explore the network and exchange collected information. The communication among the agents is indirect, following the stigmergy approach, and mediated by the network itself.

AntNet

In particular, mobile agents are periodically launched towards randomly selected destination nodes. The main objective is to find a minimum-cost path, i.e. a shortest path, between the source and the destination. The way through the network is determined by a greedy stochastic policy according to the ACO algorithm. So called forward ants randomly search for 'food'. After locating the destination, the agents travel backwards (now called backward ants) on the same path used for exploration. All traversed nodes are updated with the most current information about the destination node.

The most important parameters to be maintained by the agents are the routing tables of all networked nodes. The routing table $\mathcal{T}_k$ defines the probabilistic routing policy currently adopted for node k. For each destination d and for each neighbor n, $\mathcal{T}_k$ stores a probabilistic value P_{nd} expressing the quality (desirability) of choosing n as a next hop towards destination d. The outgoing probabilities are constrained according to Equation 18.5, where $d \in [1, N]$ (N is the total number of nodes in the network):

$$\sum_{n \in \mathcal{N}_k} P_{nd} = 1, \mathcal{N}_k = \{neighbors(k)\}. \tag{18.5}$$

Two specific capabilities of AntNet need to be mentioned explicitly. First, AntNet incorporates algorithm-inherent mechanisms to counteract network congestion. Instead of focusing on a fixed period for launching agents, AntNet maintains a probability p_d for creating an agent towards node d at node s according to Equation 18.6, where f_{sd} is a measure of the data flow from s to d. In this way, the ants adapt their exploration activity to the varying data-traffic characteristics:

$$p_d = \frac{f_{sd}}{\sum_{d'=1}^{N} f_{sd'}}. \tag{18.6}$$

Secondly, AntNet relies on a reinforcement strategy for maintaining the routing table $\mathcal{T}_k$. The returning agent updates the routing table entry for destination d', i.e. the probability $P_{fd'}$ according to Equation 18.7. The factor r is a dimensionless value, $r \in (0, 1)$ used by the current node for positive reinforcement:

$$P_{fd'} \leftarrow P_{fd'} + r(1 - P_{fd'}). \tag{18.7}$$

In order to satisfy the constraint in Equation 18.5, the probabilities $P_{nd'}$ for destination d' of the other neighboring nodes n implicitly receive a negative reinforcement by normalization:

$$P_{nd'} \leftarrow P_{nd'} - r P_{nd'}, n \in \mathcal{N}_k, n \neq f. \tag{18.8}$$

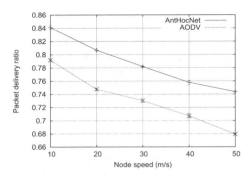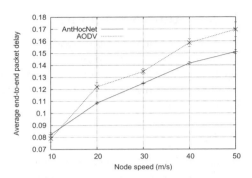

Figure 18.4 Performance comparison of AntHocNet and AODV; depicted are the delivery ratio (left) and the average packet delay (right) for various speed values for RWP mobility [Di Caro *et al.* (2005); 2005 © John Wiley & Sons Limited. Reproduced with permission]

AntHocNet

Similarly to AntNet, AntHocNet (Di Caro *et al.* 2005) is based on the ACO algorithm used in the context of ad hoc networks. AntHocNet sets up paths when they are needed at the start of a session. Thus, AntHocNet represents a reactive routing approach (see also Section 10.1.2).

Again, routing paths are represented as pheromone trails, indicating their respective quality. After setting up a path, data packets are stochastically routed as datagrams over the different paths using the pheromone information. While a data session is active, paths are proactively probed, maintained and improved. Thus, link failures can be quickly detected. The probabilities in the routing tables are calculated in a similar way to the maths presented for AntNet.

Di Caro *et al.* (2005) performed a number of performance studies in order to investigate the network behavior of AntHocNet. As an example, the performance comparison of AntHocNet and Ad Hoc on Demand Distance Vector (AODV) is depicted in Figure 18.4. Shown are the delivery ratio and the average packet delay for various speed values for Random Way Point (RWP) mobility in a network consisting of 100 nodes. As can be seen, AntHocNet outperforms AODV in this example.

18.2.3   Ant-based task allocation

Combined task allocation and ad hoc routing according to the ACO paradigm has been investigated by Labella and Dressler (2006, 2007). Such integrated techniques are of special interest in the domain of SANETs. The proposed architecture is completely based on probabilistic decisions. During the lifetime of the SANET, the nodes adapt the probability to execute one task out of a given set. Reinforcement strategies are exploited to optimize the overall system behavior. It needs to be mentioned that the integrated task-allocation and routing approach represents a cross-layer solution. Application layer and network layer are both responsible for operating the entire SANET.

Task selection

Task selection is performed by the nodes according to a probabilistic scheme. It is assumed that all the agents know a priori a list of possible tasks $T_{agent} = \{T_1, T_2, \ldots, T_n\}$ that they can perform. Heterogeneity is inherently supported. Therefore, the task lists of different agents will be different.

For task selection, each agent k associates a task $i \in T_{agent}$ to a real number τ_i^k, which represents the pheromone level. Then, according to ACO, the probability of choosing task i, $P(i)$, can be described as shown in Equation 18.9, with $\beta_{task} \geq 1$ used for improved exploitation of good paths:

$$P(i) = \frac{(\tau_i^k)^{\beta_{task}}}{\sum\limits_{j \in T_{agent}} (\tau_j)^{\beta}_{task}}. \tag{18.9}$$

All agents initialize their pheromone level $\tau_i^k = \tau_{init}$. Afterwards, this level is updated according to the achieved task:

$$\tau_i^k = \begin{cases} \min(\tau_{max}, \tau_i^k + \Delta\tau) & \text{if task } i \text{ was successful} \\ \max(\tau_{min}, \tau_i^k - \Delta\tau) & \text{otherwise.} \end{cases} \tag{18.10}$$

The routing is performed similarly to the techniques proposed in AntNet and AntHoc-Net, except for one major difference. In order to support the task-specific communication, the routing table is extended to cover different forwarding probabilities for the defined tasks, i.e. a class parameter c is added for each routing entry for destination d. Accordingly, the forwarding probability is denoted as $_c\mathbf{R}_{nd}$. This allows the exploitation of task-specific communication paths. Basically, this technique can be also used for supporting different message priorities in the routing process.

Evaluation

For evaluation, Labella and Dressler (2006) created a simulation environment for task allocation and routing in a SANET consisting of stationary sensor nodes and mobile robots. Four tasks are considered in this example, $T_{agent} = \{T_1, T_2, T_3, T_4\}$ whereas task T_4 can only be executed by the robot systems. Then, the distribution of the tasks was investigated according the proposed ACO based allocation and communication scheme.

A snapshot depicted in Figure 18.5 outlines the working behavior of the ant-based algorithm. As an example, the figure (left) shows the distribution of task T_3 and the used communication paths after some time performing the algorithm. Obviously, some nodes achieved higher probabilities $P(3)$ to execute the task due to positive reinforcement, while others tended to not to perform task T_3 (negative reinforcement). Most of the nodes seem to switch between multiple tasks according to their mean task-execution probability $P(3)$. The communication paths used for data related to task T_3 are shown in Figure 18.5 (right). The produced sensor data are transmitted to a destination in the upper-left corner. Please note that the paths highly correlate with the task execution probability.

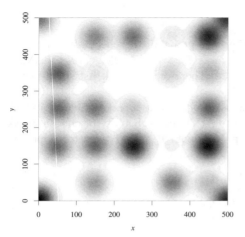 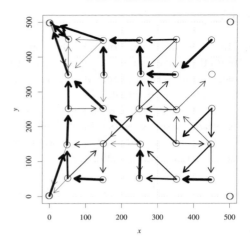

Figure 18.5 Distribution of task T_3 (left) and communication paths used to transfer the generated sensor data to a base station in the upper-left corner (right) [Reproduced from (Labella and Dressler 2006); © 2006 IEEE]

18.3   Artificial Immune System

The term 'Artificial Immune System' (AIS) relates to terminology that refers to adaptive systems inspired by theoretical and experimental immunology with the goal of problem solving (de Castro and Timmis 2002). The primary goal of an AIS is to efficiently detect changes in the environment or deviations from the normal system behavior in complex problem domains. In the following, we briefly outline the principles of the immune system and introduce some selected applications of AISs.

18.3.1   Principles of the immune system

The role of the mammalian immune system can be summarized as follows. It protects the body from infections by continuously scanning for invading pathogens, e.g. exogenous (non-self) proteins. The most interesting concept of the immune system is the self-optimization and self-learning process. Additionally, it features the following capabilities that are of eminent interest in various domains of computing and computer communication:

- Recognition – Ability to recognize patterns that are (slightly) different from previously known or trained samples, i.e. capability of anomaly detection.

- Robustness – Tolerance against interference and noise.

- Diversity – Applicability in various domains.

- Reinforcement learning – Inherent self-learning capability that is accelerated if needed through reinforcement techniques.

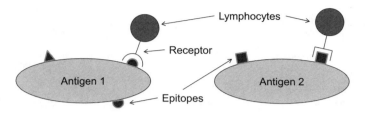

Figure 18.6 Pattern recognition in the immune system; specific antibodies (leucocytes) are able to detect antigens according to matching receptors and epitopes

- Memory – System-inherent memorization of trained pattern.

- Distributed – Autonomous and distributed processing.

The mammalian immune system

The main objective of the immune system is to recognize pathogens, i.e. exogenous non-self proteins. In general, objects to be tested are called antigens, of which pathogens represent the potentially dangerous subclass. The pattern-recognition capabilities of the immune system are based on the complementarity between the binding region of a receptor and a portion of an antigen called an epitope. This concept is depicted in Figure 18.6.

It is important to notice that antibodies present a single type of receptor, while antigens might present several epitopes. This means that different antibodies are able to recognize a single antigen.

The immune system needs to be able to differentiate between self and non-self antibodies. This ability of self/non-self recognition is a unique characteristic of the immune system. Basically, this is based on antigenic encounters that may result in cell death. Thus, they represent some kind of *positive selection* and some element of *negative selection*.

In the mammalian immune system, lymphocytes are responsible for the detection of antigens. Lymphocytes are called negative detectors because they are trained to bind to non-self antigens. They are created with randomly generated receptors. Thus, they can bind to either self or non-self antigens. One class of lymphocytes–T-cells–are tolerized in a single location–the thymus. Immature T-cells develop in the thymus and die through programmed cell death (apoptosis) if they are activated during development. Therefore, T-cells that survive the maturation and leave the thymus will be tolerant to self proteins. This process is called negative selection (Hofmeyr and Forrest 2000).

This developmental process is depicted in Figure 18.7, showing the lifecycle of a single detector, i.e. of a single T-cell. The activation and memorization part is described below.

Two immune responses were identified. The primary one is to launch a response to invading pathogens, leading to an unspecific response (using leucocytes). The primary response is slow because only a few lymphocytes bind to the new type of pathogen – the immune response will not be efficient. In order to increase the efficiency, activated lymphocytes clone themselves – increasing the population of lymphocytes that can detect the new pathogen. This adaptation process is also known as affinity maturation. The activation

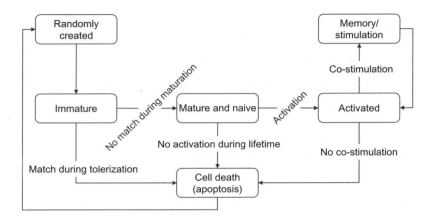

Figure 18.7   Lifecycle of a T-cell. After creation, a first training period leads to mature T-cells. After activation, i.e. after recognizing the detector-inherent pattern, further memorization and stimulation are initiated

requires co-stimulation from another signal. Usually, chemical signals from other cells are used for this procedure. The main reason is to prevent auto-activated T-cells that accidentally detect self proteins to initiate the affinity maturation.

In contrast, the secondary immune response remembers past encounters, i.e. it represents the immunologic memory. It allows a faster response to pathogens that are entering for the second time. Thus, the secondary immune response shows a very specific response. In particular, the secondary response works as follows. T-cells recognize the antigen (a huge number of such lymphocytes result from the affinity maturation in the first immune response) and get activated. Furthermore, they stimulate the production of further antibodies by B-cells, which are developed in the bone mark. These B-cells can clone further B-cells with different isotypes according to the detected antigen. This is also known as isotype switching and enables the immune system to choose between various effector functions.

In summary, the ability of the immune system to learn and to memorize self and non-self patterns is based on the following capabilities. Primary and secondary immune responses and the ability to remember encounters lead to the efficient reinforcement learning and the immune memory. Antigens are repeatedly checked by multiple antibodies. Finally, the memorization and the associative memory prevent the need for the immune system to start from scratch.

Developing an AIS

According to de Castro and Timmis (2002), the following basic elements are needed to develop an AIS–a representation of the system components and mechanisms to evaluate the interactions, i.e. an affinity measure.

Representation (shape-space)   Recognition in the immune system occurs mainly through shape complementarity between the receptor and the antigen. This shape recognition can be modeled using the concept of shape-space, which describes the general shape of a molecule, the interactions between them and the binding degree. Due to the simplified implementation, most AISs quantify the degree of similarity, whereas the mammalian immune system measures complementarity.

The most common data structures used to produce artificial shape-spaces are the following (de Castro 2006):

- Real-valued shape-space – the attribute strings are real-valued vectors.

- Integer shape-space – the attribute strings are composed of integer values.

- Hamming shape-space – composed of attribute strings built out of a finite alphabet of length k.

- Symbolic shape-space – usually composed of different types of attribute strings where at least one of them is symbolic, such as 'age', 'height', etc.

In general, an antibody can be represented by a string **Ab** and the antigen by a string **Ag**. Without loss of generality, antibodies and antigens are assumed to be of the same length L:

$$\mathbf{Ab} = \langle Ab_1, Ab_2, \ldots, Ab_L \rangle$$

$$\mathbf{Ag} = \langle Ag_1, Ag_2, \ldots, Ag_L \rangle$$

The interaction of antibodies with antigens can then be evaluated by a distance or similarity measure that is also known as the affinity measure.

Affinity measure   The affinity of two measures is directly related to their distance. A number of distance measures are available to describe the affinity of an antibody **Ab** and an antigen **Ag** (de Castro 2006). For real-value shape-spaces, for example the Euclidean and the Manhatten distances can be used as depicted in Equations 18.11 and 18.12, respectively.

$$D = \sqrt{\sum_{i=1}^{L}(Ab_i - Ag_i)^2} \tag{18.11}$$

$$D = \sum_{i=1}^{L}|Ab_i - Ag_i|. \tag{18.12}$$

Similarly, other measures such as the Hamming distance can be used. In particular, the Hamming distance allows estimatation of the similarity between any two strings represented by symbols over a finite alphabet. The distance measure is shown in Equation 18.13:

$$D = \sum_{i=1}^{L} \delta_i, \delta_i = \begin{cases} 1 & \text{if } Ab_i \neq Ag_i \\ 0 & \text{otherwise.} \end{cases} \tag{18.13}$$

According to the given shape-space and the affinity measure, an AIS can be used efficiently for general-purpose anomaly detection. The normal behavior of a system is often characterized by a series of observations over time. Then, the problem of detecting novelties or anomalies can be viewed as finding deviations in a characteristic property in the system. In the domain of communication networks, the identification of computational viruses and network intrusions is considered one of the most important anomaly-detection tasks.

18.3.2   Application examples

The scope of AISs is widespread. There are applications for fault and anomaly detection, data mining (machine learning, pattern recognition), agent based systems, control and robotics. One of the first AISs was developed by Kephart (1994) for adaptive virus detection. Based on this work, misbehavior detection and attack- or intrusion-detection systems were developed based on the working principles of the natural immune system (Kim and Bentley 2001, 2002). Besides network-security applications, the operation and control of multi-robot systems were addressed by AIS approaches. The collaborative behavior of robots collecting objects in an environment is difficult to optimize without central control. It was shown that an emerging collective behavior through communicating robots using an AIS overcomes some of the problems. The immune-network theory was used to suppress or encourage robot behavior (Lee *et al.* 1999; Singh and Thayer 2001) and for SANET approaches, such as collaborative mine detection (Opp and Sahin 2004).

Obviously, the scope of AISs spans a wide area of application domains. Broadly, these domains can be summarized as follows:

- fault and anomaly detection;

- data mining (machine learning, pattern recognition);

- agent-based systems;

- autonomous control and robotics;

- scheduling and other optimization problems;

- security of information systems.

As a particular example, the AIS approach to misbehavior detection in mobile ad hoc networks should be mentioned. Le Boudec and Sarafijanovic (2004) developed an AIS solution for application in networks relying on the Dynamic Source Routing (DSR) protocol. Two main reasons for misbehavior have been identified. First, node misbehavior can be the result of faulty software or hardware. Secondly, misbehavior can be the result of selfish nodes, which are trying to save energy, i.e. these nodes pretend to participate in DSR, but do not forward packets sent from other nodes.

The detection of both kinds of misbehaving nodes is essentially necessary for successfully operating an ad hoc network. The proposed AIS approach focuses on the detection of nodes that do not correctly execute the DSR protocol. It does not incorporate any actions to be performed after detection.

In order to get started, first, a mapping of the element of an AIS to specific parts of the investigated system is required, i.e representations of antigens and antibodies are needed.

In the proposed system, sequences of DSR protocol events represent the antigens, whereas patterns with the same format of antigens are used as antibodies.

In the developed system, an antigen is a data set consisting of protocol events recorded during a fixed time interval Δt with a maximum number of events N_s per data set. Each set is encoded using a symbolic representation of the protocol actions, i.e. A=`"RREQ sent"`, B=`"RREQ received"`, C=`"RREP sent"`, and so forth. Based on these symbols, genes are defined as specific sequences, e.g. the number of E in sequence or the number of E*A in sequence, where * represents one arbitrary label or no label at all.

Finally, the detection function is defined based on mappings of antigens consisting of such genes. More precisely, this mapping is explicitly done for capturing the correlation between protocol events. This mapping is done by simply counting the number of occurrences in the given time interval Δt. In normal protocol operation, several genes will correlate. The system can learn this correlation during a training phase. Then, in the case of misbehavior, these correlations will change and can be detected by the AIS.

18.4   Cellular signaling pathways

Molecular biology is the basis of all biological systems. It features a high specificity of information transfer. Interestingly, we find many similar structures in this domain of biology and in computer networking (Dressler and Krüger 2004; Krüger and Dressler 2005). Therefore, the investigation of structure and organization of intercellular communication seems to be valuable with regard to efficient networking strategies. The main focus of this section is to briefly introduce the information exchange in cellular environments, i.e. the so-called signaling pathways (Alberts *et al.* 1994; Pawson 1995). The most interesting aspect of the cellular information exchange is the specificity of the information handling. The studied mechanisms allow extraction of the following lessons to learn: efficient and specific response to a request, shortening of information pathways, and directing of messages to appropriate destinations.

18.4.1   Introduction to signaling pathways

Signaling in biological systems occurs at multiple levels and in many shapes. Basically, the term 'signaling' describes the interactions between single molecules (Weng *et al.* 1999). Briefly, cellular interactions can be viewed as processing in two steps. Initially, an extracellular molecule binds to a specific receptor on a target cell, converting the dormant receptor to an active state. Subsequently, the receptor stimulates intracellular biochemical pathways, leading to a cellular response (Pawson 1995). Thus, the following two cellular signaling techniques can be distinguished:

- Intracellular signaling – Intracellular signaling refers to the information processing capabilities of a single cell. Received information particles initiate complex signaling cascades that finally lead to the cellular response.

- Intercellular signaling – Communication among multiple cells is performed by intercellular signaling pathways. Essentially, the objective is to reach appropriate destinations and to induce a specific effect at this place.

Intracellular signaling

The intracellular information can be transferred by different mechanisms; one is the transfer via receptors sitting in the plasma membrane (on the cell surface). The cell-surface receptor becomes activated, for example, by a change in its sterical or chemical conformation (e.g. phosphorylation of defined amino acids). Simplified, one can regard this activation as the reception of a specific message. The activated receptor molecule is able to further activate intracellular molecules, resulting in a 'domino effect'. The effect of such a signal trans-duction pathway is mostly gene transcription. Other possibilities are the reorganization of intracellular structure such as the cell cytoskeleton or the internalization and externalization of molecules in and out of the cell. Gene transcription means that the cell responds to the incoming signal by the production of other factors which are then secreted (transported out of the cell), where it can induce signaling processes in the cell's direct environment.

This process is depicted in a simplified manner in Figure 18.8. For simplification, all processes of cellular signaling are shown in one cell. In reality, not all signaling cascades can take place in the same cell. That depends on whether a cell expresses the specific receptor. Furthermore, all the drawings are highly simplified. It needs to be mentioned that

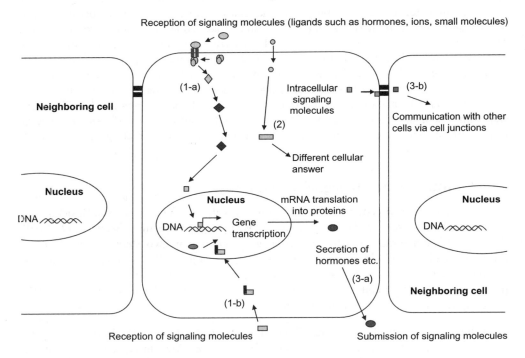

Figure 18.8 Intracellular signaling pathways. In the depicted cell, some examples of signal-ing cascades are shown. All cascades are initiated by the reception of a signaling molecule. After processing the information, a specific cellular answer is initiated. An effect could be, for example, the submission of a signaling molecule

this model does not show all known signaling cascades, but a number of common signaling pathways.

The following examples of signaling cascades are shown (the numbering corresponds to the processes depicted in Figure 18.8):

(1) Reception of signaling molecules via receptors

Cellular signaling cascades are often initiated by the reception of signaling molecules (ligands) via receptors.

(1-a) Receptors can be expressed on the cell surface. In consequence, ligands bind to cell surface receptors and initiate the activation of a cascade of intracellular molecules. Typical examples are several growth factors.

(1-b) Receptors can be expressed as intracellular receptors. In consequence, ligands have to enter the cell to bind the receptor. Examples are effects of steroide hormones such as cortisol.

Additional signaling molecules may affect the established signaling cascade towards the nucleus. The cellular answer is relying on the nucleus to initiate the desired process. In particular, a specific reaction is induced by gene transcription and the translation of mRNA into new proteins.

(2) Indirect stimulation of cellular processes

A signaling molecule can directly enter the cell and is processed in a biochemical reaction. The resulting product changes the behavior or state of the cell. For example, nitric oxide leads to smooth muscle contraction.

(3) Cellular answer, e.g. submission of signaling molecules

The cellular answer is a specific response according to the received signaling molecules and the current constitution of the cell. For example, signaling molecules can be created to send messages to other cells.

(3-a) In response to a received information particle, a new message can be created and submitted into the extracellular space, e.g. secretion of hormones.

(3-b) Additionally, messages can be forwarded to a neighboring cell via a paracellular pathway (via intracellular signaling molecules and a cell-junction), e.g. submission of signaling molecules.

This *specific response* is the key to information processing. It depends on the type of signal and the state of the cells (which receptors have been built and which of them are already occupied by particular proteins). Finally, a specific cellular response is induced: either the local state is manipulated and/or a new messaging protein is created.

Intercellular signaling

The intercellular information exchange works in analogue. Proteins, peptides and steroids are used as information particles (hormones) between cells. A signal is released into the bloodstream (the medium that carries it to distant cells) and induces an answer in these cells, which then pass on the information or can activate helper cells (e.g. the renin–angiotensin–aldosterone system (Guyton 1991) and the immune system). Obviously, this represents a kind of diffuse (probabilistic) communication scheme, depending on the concentration of the hormones. The interesting property of this transmission is that the information itself addresses the destination. During differentiation, a cell is programmed to express a subset of receptors in order to fulfill a specific function in the tissue. In consequence, hormones in the bloodstream affect only those cells expressing the correct receptor. This is the main reason for the specificity of cellular signal transduction. Of course, cells also express a variety of receptors which regulate the cellular metabolism, survival and death.

It is necessary to mention that the principles are not as simple as described here. Many of these signaling pathways are interfering and interacting. Different signaling molecules affect the same pathway. Inhibitory pathways interfere with the straightforward signal transduction. The stimulatory and inhibitory effects in particular enable efficient and specific response mechanisms. To sum up, the final effect is dependent up on the strongest signal.

Cellular signaling networks

As we have seen, cellular processes are regulated by interactions between various types of molecules, such as proteins, DNA and other small molecules. Among these, the interactions between proteins and the interactions between transcription factors and their target genes play a prominent role, controlling the activity of proteins and the expression levels of genes (Yeger-Lotem *et al*. 2004).

A key challenge for biology in the 21st century is to understand the structure and the dynamics of the complex intercellular web of interactions that contribute to the structure and function of a living cell (Barabási and Oltvai 2004).

Complex networks are studied across many fields of science. To uncover their structural design principles, network motifs have been defined as patterns of interconnections occurring in complex networks in numbers that are significantly higher than those in randomized networks (Milo *et al*. 2002).

The concept of network motifs is depicted in Figure 18.9. Please note that this is only a small sample of network motifs in integrated cellular networks (Milo *et al*. 2002; Yeger-Lotem *et al*. 2004). The three basic building blocks of complex networks are shown in this figure, together with application examples relevant to SANETs. Feed-forward motifs

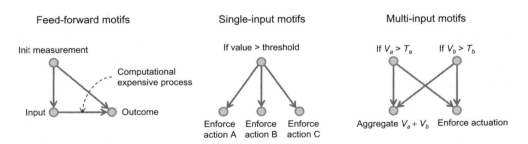

Figure 18.9 Typical network motifs in integrated cellular networks

represent network-inherent mechanisms for controlling (expensive) processes. This can also be seen as an amplification technique. Single-input motifs allow the initiation of multiple reactions on a single stimulus. Furthermore, multi-input motifs are depicted. The basic concept is twofold. First, inhibitory or controlling effects can be achieved, as two stimuli are required to continue the signaling cascade. Secondly, if the threshold, i.e. the multiple simultaneous stimuli, is exceeded, a number of parallel actions can be initiated at once.

18.4.2 Applicability in SANETs

The lessons to be learned from cellular signaling pathways are the efficient and, above all, the very specific response to a problem, the shortening of information pathways and the possibility of directing each problem to the adequate helper component. Therefore, the adaptation of mechanisms known from molecular and cell biology promises to enable more efficient information exchange.

The primary research objectives are new concepts for behavior patterns of network nodes, the improved efficiency and reliability of the entire communication system, and flexible self-organizing infrastructures. The two main concepts to be exploited in the context of SANETs are the following:

- signaling pathways based on specific signal cascades with stimulating and inhibitory functionality used for intracellular communication;

- diffuse (probabilistic) communication with specific encoding of the destination receptors for intercellular communication.

The same communication mechanisms can be emulated using the Rule-based Sensor Network (RSN) approach that we already discussed in Section 14.4. The underlying concepts and the necessary technical capabilities of this approach have been investigated for specific application in SANETs (Dressler 2006a, 2007).

Recapitulating the RSN approach, we see the following three basic concepts: data-centric operation, i.e. each message carries all necessary information to allow this specific handling, specific reaction on received data, i.e. a rule-based programming scheme is used to describe specific actions to be taken after the reception of particular information fragments, and simple local behavior control, i.e. simple state machines control each node, regardless of whether sensor or actor. Thus, there is no high-level control information considering the overall system available.

In order to understand the capabilities and advantages, we consider the following example (Dressler *et al.* 2007a). A huge number of sensors and actors should be deployed in a field. According to the sensor readings, the actors are demanded to initiate some specific tasks. In order to increase the significance of the measurements, at least two single measures need to be aggregated in order to prevent single measurement failures and other outliers.

Thus, both sensors and actors participate in this network by aggregating and forwarding measurement data according a simple rule set. Only actors will perform the first rule. This can be achieved by programming sensors and actors with slightly different programs, as shown below:

1. Actors only: if an aggregate is available, the local actuation task is initiated; according to the received data, the specific actuation might be data-dependent.

2. If multiple data messages have been received, aggregate them into a single aggregate; if the current node is a sensor, the aggregate is forwarded, whereas if the current node is an actor, the message is re-processed.

3. Forward non-aggregated messages with probability p to all neighboring nodes.

In RSN, the programming for sensor nodes would look like this:

```
# aggregate two or more data messages and forward the aggregate
if ALL ($type := data , :count > 1) then {
    !send(type:=aggregate, $value := @average of $value);
    !drop;
}
# probabilistic forwarding of single measures and aggregates
if :random < p then {
    !sendAll;
    !drop;
}
# clean up
!drop;
```

Consequently, the actor program would look like this:

```
# initiate actuation procedures
if $type := aggregate then {
    !actuate($value := @average of $value);
    !drop;
}
# aggregate two or more data messages and re-process the aggregate
if ALL ($type := data , :count > 1) then {
    !return(type:=aggregate, $value := @average of $value);
    !drop;
}
# probabilistic forwarding of single measures
if :random < p then {
    !sendAll;
    !drop;
}
# clean up
!drop;
```

In combination, the sensors and actors are both responsible for network-centric operation in the network. Both react specifically to received messages. In particular, sensor nodes have 'receptors' for normal data messages. If at least two of them have been received, a new message is created – an aggregate – and released to the local environment in the form of probabilistic message forwarding. A second receptor receives and probabilistically forwards aggregates.

Similarly, actors have receptors for data messages and aggregates. If more than a single data message has been received, a processing cascade (similar to the signaling cascade) is initiated. First, an aggregate is produced and, in a second step, this aggregate initiates the local actuation. A short cut is used if aggregates have been received. Single data messages are forwarded to enable neighboring nodes to perform the aggregation task.

18.5   Conclusion

In this chapter, we outlined and discussed the main principles of bio-inspired systems. After a short introduction on biological domains that are typically exploited as sources of inspiration for the development of bio-inspired systems, architectures and algorithms, we focused on the following three domains: swarm intelligence, the artificial immune system and cellular signaling pathways. For all these three domains, we introduced the main concepts and the most interesting properties that can be replicated in technical systems.

In addition, we studied a particular application example related to the context of Sensor and Actor Networks. In particular, we investigated ant-based routing techniques as proposed in the AntNet and AntHocNet approaches. We have been seen that similar techniques have been successfully applied to task-allocation problems. AIS approaches are most frequently cited in the context of pattern and anomaly detection. The discussed application example relates to fault and misbehavior detection in ad hoc routing. Finally, we have shown that RSN can be used to emulate the behavior of cellular signaling cascades, particularly the signaling networks.

In conclusion, it can be said that there are many interesting concepts, ideas and solutions available. Even though bio-inspired networking is a quite new research domain, there are already a number of powerful solutions described in the literature. And this encouraging course holds on.

Appendix IV

Bio-inspired Networking – Further Reading

In the following, a selection of some interesting textbooks, papers and articles is provided that is recommended for further studies of the concepts of self-organization. Additionally, a list of major journals and conferences is provided that are either directly focused on self-organization and related aspects or broadly interested in publishing special issues or organizing dedicated workshops in this domain. Obviously, this list cannot be complete or comprehensive. It is intended to provide a starting point for further research.

Textbooks

- A. Abraham, C. Grosan and V. Ramos, *Stigmergic Optimization*, vol. 31, Springer, 2006.

- B. Alberts, D. Bray, J. Lewis, M. Raff, K. Roberts and J. D. Watson, *Molecular Biology of the Cell*, 3rd edn, Garland Publishing, Inc., 1994.

- C. M. Bishop, *Neural Networks for Pattern Recognition*, Oxford University Press, 1996.

- E. Bonabeau, M. Dorigo and G. Theraulaz, *Swarm Intelligence: From Natural to Artificial Systems*, Oxford University Press, 1999.

- S. Camazine, J.-L. Deneubourg, N. R. Franks, J. Sneyd, G. Theraula and E. Bonabeau, *Self-Organization in Biological Systems*, Princeton University Press, 2003.

- L. N. de Castro and J. Timmis, *Artificial Immune Systems: A New Computational Intelligence Approach*, Springer, 2002.

- L. N. de Castro, *Fundamentals of Natural Computing: Basic Concepts, Algorithms, and Applications*, Chapman & Hall/CRC, 2006.

- F. Dressler and I. Carreras (eds), *Advances in Biologically Inspired Information Systems: Models, Methods, and Tools, Studies in Computational Intelligence (SCI)*, vol. 69, Springer, 2007.

- M. Eigen and P. Schuster, *The Hypercycle: A Principle of Natural Self Organization*, Springer, 1979.

- R. I. Freshney and M. G. Freshney, *Culture of Epithelial Cells*, 2nd edn, John Wiley & Sons Ltd, 2002.

- S. F. Gilbert, *Developmental Biology*, 7th edn, Sinauer Associates, 2003.

- D. E. Goldberg, *Genetic Algorithms in Search, Optimization and Machine Learning*, Kluwer Academic Publishers, 1989.

- C. A. Janeway, M. Walport and P. Travers, *Immunobiology: The Immune System in Health and Disease*, 5th edn, Garland Publishing, 2001.

- S. A. Kauffman, *The Origins of Order: Self-Organization and Selection in Evolution*, Oxford University Press, 1993.

- J. Kennedy and R. C. Eberhart, *Swarm Intelligence*, Morgan Kaufmann Publishers, 2001.

Papers and articles

- G. Di Caro and M. Dorgio, 'AntNet: Distributed Stigmergetic Control for Communication Networks', *Journal of Artificial Intelligence Research*, vol. 9, pp. 317–365, December 1998.

- G. Di Caro, F. Ducatelle and L. M. Gambardella, 'AntHocNet: An Adaptive Nature-Inspired Algorithm for Routing in Mobile Ad Hoc Networks', *European Transactions on Telecommunications, Special Issue on Self-organization in Mobile Networking*, vol. 16, pp. 443–455, 2005.

- M. Dorigo, V. Trianni, E. Sahin, Roderich Gro, T. H. Labella, G. Baldassarre, S. Nol, J.-L. Deneubourg, F. Mondada, D. Floreano and L. M. Gambardella, 'Evolving Self-Organizing Behaviors for a Swarm-bot', *Autonomous Robots*, vol. 17(2–3), pp. 223–245, 2004.

- F. Dressler, 'Bio-inspired Promoters and Inhibitors for Self-Organized Network Security Facilities', Proceedings of 1st IEEE/ACM International Conference on Bio-Inspired Models of Network, Information and Computing Systems (IEEE/ACM BIONETICS 2006), Cavalese, Italy, December 2006.

- F. Dressler, I. Dietrich, R. German and B. Krüger, 'Efficient Operation in Sensor and Actor Networks Inspired by Cellular Signaling Cascades', Proceedings of 1st ICST/ACM International Conference on Autonomic Computing and Communication Systems (Autonomics 2007), Rome, Italy, October 2007.

- S. A. Hofmeyr, 'Who Says Biology Need Be Destiny?', in CNET News.com, April 13 2004.

- J. O. Kephart, 'A Biologically Inspired Immune System for Computers', Proceedings of 4th International Workshop on Synthesis and Simulation of Living Systems, Cambridge, Massachusetts, USA, 1994, pp. 130–139.

- J. Kim and P. J. Bentley, 'The Artificial Immune System for Network Intrusion Detection: An Investigation of Clonal Selection with Negative Selection Operator,' Proceedings of Congres on Evolutionary Computation (CEC-2001), Seoul, Korea, July 2001, pp. 1244–1252.

- B. Krüger and F. Dressler, 'Molecular Processes as a Basis for Autonomous Networking', *IPSI Transactions on Advances Research: Issues in Computer Science and Engineering*, vol. 1(1), pp. 43–50, January 2005.

- T. H. Labella and F. Dressler, 'A Bio-Inspired Architecture for Division of Labour in SANETs', Proceedings of 1st IEEE/ACM International Conference on Bio-Inspired Models of Network, Information and Computing Systems (IEEE/ACM BIONETICS 2006), Cavalese, Italy, December 2006.

- J.-Y. Le Boudec and S. Sarafijanovic, 'An Artificial Immune System Approach to Misbehavior Detection in Mobile Ad-Hoc Networks', Proceedings of First International Workshop on Biologically Inspired Approaches to Advanced Information Technology (Bio-ADIT2004), Lausanne, Switzerland, January 2004, pp. 96–111.

- K. H. Low, W. K. Leow and M. H. Ang, 'Autonomic Mobile Sensor Network with Self-Coordinated Task Allocation and Execution', *IEEE Transactions on Systems, Man, and Cypernetics–Part C: Applications and Reviews*, vol. 36(3), pp. 315–327, March 2005.

- T. Pawson, 'Protein Modules and Signalling Networks', *Nature*, vol. 373(6515), pp. 573–580, February 1995.

Journals

- *Autonomous Robots* (Springer)

- *Communications Magazine* (IEEE)

- *Computer* (IEEE)

- *Information Sciences* (Elsevier)

- *Intelligent Systems* (IEEE)

- *International Journal of Intelligent Systems* (Wiley Interscience)

- *International Journal of Systems Science* (Taylor & Francis)

- *International Journal of Unconventional Computing* (Old City Publishing)

- *Journal of Intelligent Information Systems* (Kluwer Academic Publishers)

- *Journal of Selected Areas in Communications* (IEEE)

- *Nature* (Nature Publishing Group)

- *Science* (AAAS)

- *Transactions on Computational Biology and Bioinformatics* (IEEE/ACM)

- *Transactions on Computational Systems Biology* (Springer)

- *Transactions on Systems, Man, and Cybernetics* (IEEE)

Conferences

- Bio-ADIT – International Workshop on Biologically Inspired Approaches to Advanced Information Technology

- BIONETICS – IEEE/ACM International Conference on Bio-Inspired Models of Network, Information and Computing Systems

- CEC – IEEE Congress on Evolutionary Computation

- GECCO – Genetic and Evolutionary Computation Conference

- CIBCB - IEEE Symposium on Computational Intelligence in Bioinformatics and Computational Biology

- ECAL – European Conference on Advances in Artificial Life

- ICARIS – International Conferences on Artificial Immune Systems

- SMC – IEEE International Conference on Systems, Man, and Cybernetics

- SIS – IEEE Swarm Intelligence Symposium

- SSCI – IEEE Symposium Series on Computational Intelligence

- WCCI – IEEE World Congress on Computational Intelligence

Bibliography

Abraham A, Grosan C and Ramos V 2006 *Stigmergic Optimization* vol. 31 of *Studies in Computational Intelligence*. Springer.

Abramson N 1985 Development of the ALOHANET. *IEEE Transactions on Information Theory* **31**(2), 119–123.

Adamou M, Lee I and Shin I 2001 An Energy Efficient Real-Time Medium Access Control Protocol for Wireless Ad-Hoc Networks *22nd IEEE Real-Time Systems Symposium (RTSS 2001), WIP-Session*, London, UK.

Adleman LM 1994 Molecular Computation of Solutions to Combinatorial Problems. *Science* **226**(11), 1021–1024.

Agarwal S, Ahuja A, Singh JP and Shorey R 2000 Route-Lifetime Assessment Based Routing (RABR) Protocol for Mobile Ad-Hoc Networks *IEEE International Conference on Communications (IEEE ICC 2001*, vol. 3, pp. 1697–1701, New Orleans, LA, USA.

Akan OB and Akyildiz IF 2005 Event-to-Sink Reliable Transport in Wireless Sensor Networks. *IEEE/ACM Transactions on Networking (TON)* **13**(5), 1003–1016.

Akkaya K and Younis M 2004 Energy-Aware Routing of Time-Constrained Traffic in Wireless Sensor Networks. *Journal of Communication Systems, Special Issue on Service Differentiation and QoS in Ad Hoc Networks* **17**(6), 663–687.

Akkaya K and Younis M 2005 A Survey of Routing Protocols in Wireless Sensor Networks. *Elsevier Ad Hoc Networks* **3**(3), 325–349.

Akyildiz IF and Kasimoglu IH 2004 Wireless Sensor and Actor Networks: Research Challenges. *Elsevier Ad Hoc Networks* **2**, 351–367.

Akyildiz IF and Wang X 2005 A Survey on Wireless Mesh Networks. *IEEE Communications Magazine* **43**(9), S23–S30.

Akyildiz IF, Pompili D and Melodia T 2005a Underwater Acoustic Sensor Networks: Research Challenges. *Elsevier Ad Hoc Networks* **3**(3), 257–279.

Akyildiz IF, Su W, Sankarasubramaniam Y and Cayirci E 2002a A Survey on Sensor Networks. *IEEE Communications Magazine* **40**(8), 102–116.

Akyildiz IF, Su W, Sankarasubramaniam Y and Cayirci E 2002b Wireless Sensor Networks: A Survey. *Elsevier Computer Networks* **38**, 393–422.

Akyildiz IF, Wang X and Wang W 2005b Wireless Mesh Networks: A Survey. *Elsevier Computer Networks* **47**(4), 445–487.

Alberts B, Bray D, Lewis J, Raff M, Roberts K and Watson JD 1994 *Molecular Biology of the Cell* 3rd edn. Garland Publishing, Inc.

Allen JF 1983 Maintaining Knowledge about Temporal Intervals. *Communications of the ACM* **26**(11), 832–843.

Anantharaman V, Park SJ, Sundaresan K and Sivakumar R 2004 TCP Performance Over Mobile Ad Hoc Networks: A Quantitative Study. *Wireless Communications and Mobile Computing* **4**(2), 203–222.

Anastasi G, Conti M, Gregori E, Falchi A and Passarella A 2004 Performance Measurements of Mote Sensor Networks *7th ACM International Symposium on Modeling, Analysis and Simulation of Wireless and Mobile Systems (ACM MSWiM 2004)*, Venice, Italy.

Andersen D, Balakrishnan H, Kaashoek F and Morris R 2001 Resilient Overlay Networks *18th ACM Symposium on Operating Systems Principles (SOSP)*, pp. 131–145, Banff, Canada.

Arai T, Pagello E and Parker LE 2002 Editorial: Advances in Multi-Robot Systems. *IEEE Transactions on Robotics and Automation* **18**(5), 655–661.

Arvind K 1989 A New Probabilistic Algorithm for Clock Synchronization *Real Time Systems Symposium*, pp. 330–339, Santa Monica, CA, USA.

Ashby WR 1962 Principles of the Self-Organizing System In *Principles of Self-Organization* (ed. von Foerster H and Zopf GW) Pergamon Press pp. 255–278.

Asimov I 1968 *I, Robot*. Gordon Books.

Awerbuch B 1985 Complexity of Network Synchronization. *Journal of the ACM (JACM)* **32**(4), 804–823.

Baggio A 2005 Wireless Sensor Networks in Precision Agriculture *ACM Workshop on Real-World Wireless Sensor Networks (REALWSN 2005)*, Stockholm, Sweden.

Bai F and Helmy A 2004 A Survey of Mobility Modeling and Analysis in Wireless Adhoc Networks *Wireless Ad Hoc and Sensor Networks* Kluwer Academic Publishers.

Bai X, Kumary S, Xuany D, Yunz Z and Lai TH 2006 Deploying Wireless Sensors to Achieve Both Coverage and Connectivity *7th ACM International Symposium on Mobile Ad Hoc Networking and Computing (ACM Mobihoc 2006)*, pp. 131–142, Florence, Italy.

Banerjee S and Khuller S 2001 A Clustering Scheme for Hierarchical Control in Multi-Hop Wireless Networks *20th IEEE Conference on Computer Communications (IEEE INFOCOM 2001)*, Anchorage, Alaska, USA.

Barabási AL and Oltvai ZN 2004 Network Biology: Understanding the Cell's Functional Organization. *Nature Reviews Genetics* **5**(2), 101–113.

Barrett CL, Eidenberz SJ and Kroc L 2003 Parametric Probabilistic Sensor Network Routing *9th ACM International Conference on Mobile Computing and Networking (ACM MobiCom 2003)*, San Diego, CA, USA.

Barsukov Y 2004 Challenges and Solutions in Battery Fuel Gauging. Workbook, Texas Instruments.

Batalin MA and Sukhatme GS 2003a Coverage, Exploration and Deployment by a Mobile Robot and Communication Network *International Workshop on Information Processing in Sensor Networks*, pp. 376–391, Palo Alto, USA.

Batalin MA and Sukhatme GS 2003b Sensor Network-Based Multi-Robot Task Allocation *IEEE/RSJ International Conference on Intelligent Robots and Systems (IROS2003)*, pp. 1939–1944, Las Vegas, Nevada.

Batalin MA and Sukhatme GS 2003c Using a Sensor Network for Distributed Multi-Robot Task Allocation *IEEE International Conference on Robotics and Automation*, pp. 158–164, New Orleans, LA, USA.

Beckwith R, Teibel D and Bowen P 2004 Report from the Field: Results from an Agricultural Wireless Sensor *29th IEEE International Conference on Local Computer Networks (LCN)*, pp. 471–478.

Bellur B and Ogier RG 1999 A Reliable, Efficient Topology Broadcast Protocol for Dynamic Networks *18th IEEE Conference on Computer Communications (IEEE INFOCOM 1999)*, vol. 1, pp. 178–186, New York, NY.

Berkhin P 2002 Survey of Clustering Data Mining Techniques. Technical report, Accrue Software, Inc.

Bernardos C and Calderon M 2005 Survey of IP Address Autoconfiguration Mechanisms for MANETs draft-bernardos-manet-autoconf-survey-00.txt.

Bettstetter C 2001 Mobility Modeling in Wireless Networks: Categorization, Smooth Movement, and Border Effects. *ACM Mobile Computing and Communications Review* **5**(3), 55–67.

Bezdek JC 1981 *Pattern Recognition with Fuzzy Objective Function Algorithms*. Kluwer Academic Publishers.

Bharghavan V, Demers A, Shenker S and Zhang L 1994 MACAW: A Media Access Protocol for Wireless LAN's *ACM SIGCOMM'94*, pp. 212–225, London, UK.

Bhatt M, Chokshi R, Desai S, Panichpapiboon S, Wisitpongphan N and Tonguz OK 2003 Impact of Mobility on the Performance of Ad Hoc Wireless Networks *58th IEEE Vehicular Technology Conference (VTC2003-Fall)*, vol. 5, pp. 3025–3029, Orlando USA.

Birman KP, Hayden M, Ozkasap O, Xiao Z, Budiu M and Minsky Y 1999 Bimodal Multicast. *ACM Transactions on Computer Systems* **17**(2), 41–88.

Bishop CM 1996 *Neural Networks for Pattern Recognition*. Oxford University Press.

Blough DM and Santi P 2002 Investigating Upper Bounds on Network Lifetime Extension for Cell-Based Energy Conservation Techniques in Stationary Ad Hoc Networks *8th ACM International Conference on Mobile Computing and Networking (ACM MobiCom 2002)*, pp. 183–192, Atlanta, Georgia, USA.

Blum JJ, Eskandarian A and Hoffman LJ 2004 Challenges of Intervehicle Ad Hoc Networks. *IEEE Transactions on Intelligent Transportation Systems* **5**(4), 347–351.

Bokareva T, Bulusu N and Jha S 2004 A Performance Comparison of Data Dissemination Protocols for Wireless Sensor Networks *IEEE Globecom Wireless Ad Hoc and Sensor Networks Workshop*.

Bonabeau E, Dorigo M and Theraulaz G 1999 *Swarm Intelligence: From Natural to Artificial Systems*. Oxford University Press.

Boukerche A and Nikoletseas S 2004 Protocols for Data Propagation in Wireless Sensor Networks: A Survey In *Wireless Communications Systems and Networks* (ed. Guizani M) Kluwer Academic Publishers.

Boulis A, Ganeriwal S and Srivastava MB 2003a Aggregation in Sensor Networks: An Energy-Accuracy Trade-off *IEEE Workshop on Sensor Network Protocols and Applications (SNPA 2003)*, pp. 128–138.

Boulis A, Ganeriwal S and Srivastava MB 2003b Aggregation in Sensor Networks: An Energy-Accuracy Trade-off. *Elsevier Ad Hoc Networks* **1**, 317–331.

Braginsky D and Estrin D 2002 Rumor Routing Algorithm For Sensor Networks *First Workshop on Sensor Networks and Applications (WSNA)*, Atlanta, Georgia, USA.

Broadbent S and Hammersley J 1957 Percolation processes I. Crystals and Mazes. *Proceedings of the Cambridge Philosophical Society* **53**, 629–641.

Brooks RR and Iyengar SS 1997 *Multi-Sensor Fusion: Fundamentals and Applications With Software*. Prentice Hall PTR.

Brooks RR, Piretti M, Zhu M and Iyengar SS 2003a Distributed Adaptation Methods for Wireless Sensor Networks *IEEE Global Telecommunications Conference (IEEE GLOBECOM 2003)*, vol. 5, pp. 2967–2971, San Francisco, CA.

Brooks RR, Piretti M, Zhu M and Iyengar SS 2003b Emergent Routing Protocols in Wireless Ad Hoc Sensor Networks *SPIE Conference*, vol. 5205, pp. 155–163.

Bulushu N, Estrin D, Girod L and Heidemann J 2001 Scalable Coodination for Wireless Sensor Networks: Self-Configuring Localization Systems *6th International Symposium on Communication Theory and Applications (ISCTA'01)*, Ambleside, Lake District, UK.

Burrell J, Brooke T and Beckwith R 2004 Vineyard Computing: Sensor Networks in Agricultural Production. *Pervasive Computing* **3**(1), 38–45.

Buttyán L and Hubaux JP 2003 Stimulating Cooperation in Self-Organizing Mobile Ad Hoc Networks. *Mobile Networks and Applications* **8**(5), 579–592.

Cai Y, Hua KA and Phillips A 2005 Leveraging 1-hop Neighborhood Knowledge for Efficient Flooding in Wireless Ad Hoc Networks *International Performance, Computing, and Communication Conference (IPCCC'05)*, pp. 347–354, Phoenix, AZ.

Camazine S, Deneubourg JL, Franks NR, Sneyd J, Theraula G and Bonabeau E 2003 *Self-Organization in Biological Systems*. Princeton University Press.

Camp T, Boleng J and Davies V 2002 A Survey of Mobility Models for Ad Hoc Network Research. *Wireless Communications and Mobile Computing, Special Issue on Mobile Ad Hoc Networking: Research, Trends and Applications* **2**(5), 483–502.

Cardei M and Wu J 2004 Coverage in Wireless Sensor Networks In *Handbook of Sensor Networks* (ed. Ilyas M) CRC Press West Palm Beach, FL, USA.

Caruso A, Chessa S, De S and Urpi A 2005 GPS Free Coordinate Assignment and Routing in Wireless Sensor Networks *24th IEEE Conference on Computer Communications (IEEE INFOCOM 2005)*, Miami, FL, USA.

Cerpa A and Estrin D 2002 ASCENT: Adaptive Self-Configuring Sensor Networks Topologies *21st IEEE Conference on Computer Communications (IEEE INFOCOM 2002)*, New York, NY, USA.

Cerpa A and Estrin D 2004 ASCENT: Adaptive Self-Configuring Sensor Networks Topologies. *IEEE Transactions on Mobile Computing* **3**(3), 272–285.

Cerpa A, Busek N and Estrin D 2003 SCALE: A Tool for Simple Connectivity Assessment in Lossy Environments. Technical report, UCLA Center for Embedded Network Sensing (CENS).

Chakeres I and Perkins C 2007 Dynamic MANET On-Demand (DYMO) Routing draft-ietf-manet-dymo-07.txt.

Chakeres I and Royer E 2004 AODV Routing Protocol Implementation Design *International Workshop on Wireless Ad Hoc Networking (WWAN)*, Tokyo, Japan.

Chakeres ID and Macker JP 2006 Mobile Ad Hoc Networking and the IETF. *ACM SIGMOBILE Mobile Computing and Communications Review* **10**(1), 58–60.

Chandor A 1985 *The Penguin Dictionary of Computers* 3rd edn. Penguin.

Chandy KM and Lamport L 1985 Distributed Snapshots: Determining Global States of Distributed Systems. *ACM Transactions on Computer Systems* **3**(1), 63–75.

Chang JH and Tassiulas L 2004 Maximum Lifetime Routing in Wireless Sensor Networks. *IEEE/ACM Transactions on Networking (TON)* **12**(4), 609–619.

Chellappan S, Bai X, Ma B and Xuan D 2005 Sensor Networks Deployment Using Flip-Based Sensors *2nd IEEE International Conference on Mobile Ad-hoc and Sensor Systems (IEEE MASS 2005)*, Washington, DC.

Chen B, Jamieson K, Balakrishnan H and Morris R 2002 Span: An Energy-Efficient Coordination Algorithm for Topology Maintenance in Ad Hoc Wireless Networks. *ACM Wireless Networks Journal* **8**(5).

Chen F, Dressler F and Heindl A 2006 End-to-End Performance Characteristics in Energy-Aware Wireless Sensor Networks *Third ACM International Workshop on Performance Evaluation of Wireless Ad Hoc, Sensor, and Ubiquitous Networks (ACM PE-WASUN'06)*, pp. 41–47, Torremolinos, Malaga, Spain.

Chen YP, Liestman AL and Liu J 2004 Clustering Algorithms for Ad Hoc Wireless Networks In *Ad Hoc and Sensor Networks* (ed. Xiao Y and Pan Y) Nova Science Publisher.

Cheyer A and Martin D 2001 The Open Agent Architecture. *Journal of Autonomous Agents and Multi-Agent Systems* **4**(1), 143–148.

Chiasserini CF and Garetto M 2004 Modeling the Performance of Wireless Sensor Networks *23rd IEEE Conference on Computer Communications (IEEE INFOCOM 2004)*, Hongkong.

Chiasserini CF and Rao RR 2001 Improving Battery Performance by Using Traffic Shaping Techniques. *IEEE Journal on Selected Areas in Communications* **19**(7), 1385–1394.

Chiu CY, Chen GH and Wu EHK 2003 A Stability Aware Cluster Routing Protocol for Mobile Ad Hoc Networks. *Wireless Communications and Mobile Computing* **3**(4), 503–515.

Choksi A, Martin RP, BN and Pupala R 2002 Mobility Support for Diffusion-Based Ad-Hoc Sensor Networks. Technical report, Rutgers University, Department of Computer Science.

Chong CY and Kumar SP 2003 Sensor Networks: Evolution, Opportunities, and Challenges. *Proceedings of the IEEE* **91**(8), 1247–1256.

Chu M, Haussecker H and Zhao F 2002 Scalable Information-Driven Sensor Querying and Routing for Ad Hoc Heterogeneous Sensor Networks. *The International Journal of High Performance Computing Applications* **16**(3).

Clarke R 1993 Asimov's Laws of Robotics: Implications for Information Technology – Part 1. *IEEE Computer* **26**(12), 53–61.

Clarke R 1994 Asimov's Laws of Robotics: Implications for Information Technology – Part 2. *IEEE Computer* **27**(1), 57–66.

Clausen T and Jacquet P 2003 Optimized Link State Routing Protocol (OLSR) RFC 3626.

Clausen T, Dearlove C and Dean J 2006 MANET Neighborhood Discovery Protocol (NHDP) draft-ietf-manet-nhdp-00.txt.

Cramer C, Stanze O, Weniger K and Zitterbart M 2004 Demand-Driven Clustering in MANETs *International Workshop on Mobile Ad Hoc Networks and Interoperability Issues (MANETII04)*, Las Vegas, USA.

Cristian F 1989 A Probabilistic Approach to Distributed Clock Synchronization. *Distributed Computing* **3**, 146–158.

Culler D, Estrin D and Srivastava MB 2004 Overview of Sensor Networks. *Computer* **37**(8), 41–49.

Das SK, Banerjee N and Roy A 2006 Solving Optimization Problems in Wireless Networks Using Genetic Algorithms In *Handbook of Bioinspired Algorithms and Applications* (ed. Olariu S, Zomaya AY and Olariu O) CRC Press.

de Castro LN 2006 *Fundamentals of Natural Computing: Basic Concepts, Algorithms, and Applications*. Chapman & Hall/CRC.

de Castro LN and Timmis J 2002 *Artificial Immune Systems: A New Computational Intelligence Approach*. Springer.

De S, Qiao C and Das SK 2003 A Resource-Efficient QoS Routing Protocol for Mobile Ad Hoc Networks. *Wireless Communications and Mobile Computing* **3**(4), 465–486.

Deng J and Haas ZJ 1998 Dual Busy Tone Multiple Access (DBTMA): A New Medium Access Control for Packet Radio Networks *IEEE ICUPC'98*, Florence, Italy.

Dengler S, Awad A and Dressler F 2007 Sensor/Actuator Networks in Smart Homes for Supporting Elderly and Handicapped People *21st IEEE International Conference on Advanced Information Networking and Applications (AINA-07): First International Workshop on Smart Homes for Tele-Health (SmarTel'07)*, vol. II, pp. 863–868. IEEE, Niagara Falls, Canada.

Dewasurendra D and Mishra A 2004 Scalability of a Scheduling Scheme for Energy Aware Sensor Networks. *Wireless Communications and Mobile Computing* **4**(3), 289–303.

Di Caro G and Dorgio M 1998 AntNet: Distributed Stigmergetic Control for Communication Networks. *Journal of Artificial Intelligence Research* **9**, 317–365.

Di Caro G, Ducatelle F and Gambardella LM 2005 AntHocNet: An Adaptive Nature-Inspired Algorithm for Routing in Mobile Ad Hoc Networks. *European Transactions on Telecommunications, Special Issue on Self-organization in Mobile Networking* **16**, 443–455.

Dietrich I and Dressler F 2006 On the Lifetime of Wireless Sensor Networks. Technical report, University of Erlangen, Dept of Computer Science 7.

Dietrich I, Sommer C and Dressler F 2007 Simulating DYMO in OMNeT++. Technical report, University of Erlangen, Dept of Computer Science 7.

Dijkstra EW 1959 A Note on Two Problems in Connexion with Graphs. *Numerische Mathematik* **1**, 269–271.

Ding M, Chen D, Xing K and Cheng X 2005 Localized Fault-Tolerant Event Boundary Detection in Sensor Networks *24th IEEE Conference on Computer Communications (IEEE INFOCOM 2005)*, Miami, FL, USA.

Ding M, Cheng X and Xue G 2003 Aggregation Tree Construction in Sensor Networks *58th IEEE Vehicular Technology Conference (VTC2003-Fall)*, vol. 4, pp. 2168–2172.

Dorigo M, Caro GD and Gambardella LM 1999 Ant Algorithms for Discrete Optimization. *Artificial Life* **5**(2), 137–172.

Dorigo M, Maniezzo V and Colorni A 1996 The Ant System: Optimization by a Colony of Cooperating Agents. *IEEE Transactions on Systems, Man, and Cybernetics* **26**(1), 1–13.

Dorigo M, Trianni V, Sahin E, GroßR, Labella TH, Baldassarre G, Nol S, Deneubourg JL, Mondada F, Floreano D and Gambardella LM 2004 Evolving Self-Organizing Behaviors for a Swarm-bot. *Autonomous Robots* **17**(2-3), 223–245.

Dressler F 2005 Locality Driven Congestion Control in Self-Organizing Wireless Sensor Networks *3rd International Conference on Pervasive Computing (Pervasive 2005): International Workshop on Software Architectures for Self-Organization, and Software Techniques for Embedded and Pervasive Systems (SASO+STEPS 2005)*, Munich, Germany.

Dressler F 2006a Network-centric Actuation Control in Sensor/Actuator Networks based on Bio-Inspired Technologies *3rd IEEE International Conference on Mobile Ad Hoc and Sensor Systems (IEEE MASS 2006): 2nd International Workshop on Localized Communication and Topology Protocols for Ad Hoc Networks (LOCAN 2006)*, pp. 680–684, Vancouver, Canada.

Dressler F 2006b Self-Organization in Ad Hoc Networks: Overview and Classification. Technical report, University of Erlangen, Dept of Computer Science 7.

Dressler F 2006c Weighted Probabilistic Data Dissemination (WPDD). Technical report, University of Erlangen, Dept of Computer Science 7.

Dressler F 2007 Bio-inspired Network-centric Operation and Control for Sensor/Actuator Networks. *Transactions on Computational Systems Biology (TCSB)* **VIII**(LNBI 4780), 1–13.

Dressler F and Carreras I 2007 *Advances in Biologically Inspired Information Systems: Models, Methods, and Tools* vol. 69 of *Studies in Computational Intelligence (SCI)*. Springer.

Dressler F and Chen F 2007 Dynamic Address Allocation (DAA) for Self-organized Management and Control in Sensor Networks. *International Journal of Mobile Network Design and Innovation (IJMNDI)* **2**(2), 116–124.

Dressler F and Dietrich I 2006 Lifetime Analysis in Heterogeneous Sensor Networks *9th EUROMICRO Conference on Digital System Design – Architectures, Methods and Tools (DSD 2006)*, pp. 606–613, Dubrovnik, Croatia.

Dressler F and Krüger B 2004 Cell Biology as a Key to Computer Networking *German Conference on Bioinformatics 2004 (GCB'04), Poster Session*, Bielefeld, Germany.

Dressler F, Dietrich I and Krüger B 2007a Efficient Operation in Sensor and Actor Networks Inspired by Cellular Signaling Cascades *First ICST/ACM International Conference on Autonomic Computing and Communication Systems (Autonomics 2007)*, Rome, Italy.

Dressler F, Fuchs G, Truchat S, Yao Z, Lu Z and Marquardt H 2007b Profile-Matching Techniques for On-demand Software Management in Sensor Networks. *EURASIP Journal on Wireless Communications and Networking (JWCN), Special Issue on Mobile Multi-Hop Ad Hoc Networks: From Theory to Reality* **2007**, 10, Article ID 80619.

Droms R 1997 Dynamic Host Configuration Protocol RFC 2131.

Droms R 2004 Stateless Dynamic Host Configuration Protocol (DHCP) Service for IPv6 RFC 3736.

Du K, Wu J and Zhou D 2003 Chain-Based Protocols for Data Broadcasting and Gathering in the Sensor Networks *17th International Parallel and Distributed Processing Symposium (IPDPS)*, p. 260a.

Duarte-Melo EJ and Liu M 2002 Analysis of Energy Consumption and Lifetime of Heterogeneous Wireless Sensor Networks *IEEE Global Telecommunications Conference (IEEE GLOBECOM 2002)*, Taipei, Taiwan.

Durvy M and Thiran P 2005 Reaction-Diffusion Based Transmission Patterns for Ad Hoc Networks *24th IEEE Conference on Computer Communications (IEEE INFOCOM 2005)*, Miami, FL, USA.

Eigen M and Schuster P 1979 *The Hypercycle: A Principle of Natural Self Organization*. Springer.

Elson J and Estrin D 2001 Time Synchronization for Wireless Sensor Networks *2001 International Parallel and Distributed Processing Symposium (IPDPS)*, San Francisco, CA, USA.

Elson J and Römer K 2003 Wireless Sensor Networks: A New Regime for Time Synchronization. *ACM Computer Communication Review (CCR)* **33**(1), 149–154.

Elson J, Girod L and Estrin D 2002 Fine-Grained Network Time Synchronization using Reference Broadcasts *Fifth Symposium on Operating Systems Design and Implementation (OSDI 2002)*, Boston, MA.

Erdös P and Rényi A 1960 On the Evolution of Random Graphs. *Publications of the Mathematical Institute of the Hungarian Academy of Sciences* **5**, 17–61.

Estrin D, Culler D, Pister K and Sukhatme GS 2002 Connecting the Physical World with Pervasive Networks. *IEEE Pervasive Computing* **1**(1), 59–69.

Estrin D, Govindan R and Heidemann J 1999a Scalable Coordination in Sensor Networks. Technical report, USC/Information Sciences Institute.

Estrin D, Govindan R, Heidemann J and Kumar S 1999b Next Century Challenges: Scalable Coordination in Sensor Networks *5th ACM International Conference on Mobile Computing and Networking (ACM MobiCom 1999)*, pp. 263–270. ACM, Seattle, Washington, USA.

Fall K and Varadhan K 2007 The ns Manual. Technical report, The VINT Project.

Fernandess Y and Malkhi D 2002 K-Clustering in Wireless Ad Hoc Networks *2nd ACM Workshop on Principles of Mobile Computing*, pp. 31–37, Toulouse, France.

Foster I 2001 The Anatomy of the Grid: Enabling Scalable Virtual Organizations *1st IEEE/ACM International Symposium on Cluster Computing and the Grid (CCGRID'01)*, pp. 6–7.

Foster I and Kesselman C 2004 *The Grid 2: Blueprint for a New Computing Infrastructure* 2nd edn. Morgan Kaufmann.

Freshney RI and Freshney MG 2002 *Culture of Epithelial Cells* 2nd edn. John Wiley & Sons Ltd.

Fuchs G, Truchat S and Dressler F 2006 Distributed Software Management in Sensor Networks using Profiling Techniques *1st IEEE/ACM International Conference on Communication System Software and Middleware (IEEE/ACM COMSWARE 2006): 1st International Workshop on Software for Sensor Networks (SensorWare 2006)*, pp. 1–6, New Dehli, India.

Gage A, Murphy R, Valavanis K and Long M 2004 Affective Task Allocation for Distributed Multi-Robot Teams. Technical report, Center for Robot-Assisted Search and Rescue, University of South Florida.

Gale D 1960 *The Theory of Linear Economic Models*. McGraw-Hill Book Company.

Galmés S 2006 Lifetime Issues in Wireless Sensor Networks for Vineyard Monitoring *3rd IEEE International Conference on Mobile Ad Hoc and Sensor Systems (IEEE MASS 2006), Poster Session*, pp. 542–545, Vancouver, Canada.

Ganeriwal S, Elson J and Srivastava MB 2005 Time Synchronization In *Wireless Sensor Networks* (ed. Bulusu N and Jha S) Artech House Boston, London pp. 59–74.

Ganeriwal S, Kumar R and Srivastava MB 2003 Timing-sync Protocol for Sensor Networks *1st ACM Conference on Embedded Networked Sensor Systems (ACM SenSys 2003)*, Los Angeles, CA.

Garcia-Luna-Aceves JJ 2002 Flow-Oriented Protocols for Scalable Wireless Networks *5th ACM International Symposium on Modeling, Analysis and Simulation of Wireless and Mobile Systems (ACM MSWiM 2002)*, Atlanta, Georgia, USA.

Garcia-Molina H 1982 Elections in a Distributed Computing System. *IEEE Transactions on Computers* **31**(1), 48–59.

Gerhenson C and Heylighen F 2003 When Can we Call a System Self-organizing? *7th European Conference on Advances in Artificial Life (ECAL 2003)*, pp. 606–614, Dortmund, Germany.

Gerkey BP 2003 *On Multi-Robot Task Allocation* Ph.D. thesis University of Southern California.

Giridhar A and Kumar P 2005 Maximizing the Functional Lifetime of Sensor Networks *4th International Symposium on Information Processing in Sensor Networks (IPSN 2005)*, Los Angeles, California, USA.

Girod L, Stathopoulos T, Ramanathan N, Elson J, Estrin D, Osterweil E and Schoellhammer. T 2004 A System for Simulation, Emulation, and Deployment of Heterogeneous Sensor Networks *2nd ACM Conference on Embedded Networked Sensor Systems (ACM SenSys 2004)*.

Goldberg DE 1989 *Genetic Algorithms in Search, Optimization and Machine Learning*. Kluwer Academic Publishers.

Goldenfeld N and Kadanoff LP 1999 Simple Lessons from Complexity. *Science* **284**(5411), 87–89.

Grimmett G 1989 *Percolation*. Springer.

Gungor VC, Akan OB and Akyildiz IF 2007 A Real-Time and Reliable Transport Protocol for Wireless Sensor and Actor Networks. *IEEE/ACM Transactions on Networking (ToN)*.

Gupta P and Kumar P 2000 The Capacity of Wireless Networks. *IEEE Transactions on Information Theory* **46**(2), 388–404.

Guyton A 1991 Blood Pressure Control: Special Role of the Kidneys and Body Fluids. *Science* **252**(5014), 1813–1816.

Haas ZJ and Pearlman MR 2001 The Performance of Query Control Schemes for the Zone Routing Protocol. *IEEE/ACM Transactions on Networking (TON)* **9**, 427–438.

Haas ZJ, Halpern JY and Li L 2002a Gossip-Based Ad Hoc Routing *21st IEEE Conference on Computer Communications (IEEE INFOCOM 2002)*, pp. 1707–1716.

Haas ZJ, Pearlman MR and Samar P 2002b The Zone Routing Protocol (ZRP) for Ad Hoc Networks draft-ietf-manet-zone-zrp-04.txt.

Han CC, Kumar R, Shea R and Srivastava M 2005 Sensor Network Software Update Management: A Survey. *ACM International Journal on Network Management* **15**(4), 283–294.

Handziski V, Köpke A, Karl H, Frank C and Drytkiewicz W 2004 Improving the Energy Efficiency of Directed Diffusion Using Pervasive Clustering *1st European Workshop in Wireless Sensor Networks (EWSN)*, vol. LNCS 2920, pp. 172–187, Berlin, Germany.

Handziski V, Polastrey J, Hauer JH, Sharpy C, Wolisz A and Culler D 2005 Flexible Hardware Abstraction for Wireless Sensor Networks *2nd European Workshop on Wireless Sensor Networks (EWSN 2005)*, Istanbul, Turkey.

Hansmann U, Merk L, Nicklous MS and Stober T 2003 *Pervasive Computing: The Mobile World* Springer Professional Computing 2nd edn. Springer.

Hartigan JA 1975 *Clustering Algorithms*. John Wiley and Sons, Ltd.

Hartigan JA and Wong MA 1979 A K-Means Clustering Algorithm. *Applied Statistics* **28**(1), 100–108.

Hatler M and Chi C 2005 Wireless Sensor Networks: Growing Markets, Accelerating Demand. On world report, ON World.

Hedetniemi SM, Hedetniemi ST and Liestman A 1988 A Survey of Gossiping and Broadcasting in Communication Networks. *Networks* **18**(4), 319–349.

Hedrick C 1998 Routing Information Protocol RFC 1058.

Heinzelman WR, Chandrakasan A and Balakrishnan H 2000 Energy-Efficient Communication Proto-col for Wireless Microsensor Networks *33rd Hawaii International Conference on System Sciences.*

Heylighen F 1999 The Science of Self-Organization and Adaptivity. The Encyclopedia of Life Support Systems (EOLSS).

Hofmeyr SA and Forrest S 2000 Architecture for an Artificial Immune System. *Evolutionary Computation* **8**(4), 443–473.

Holland G and Vaidya N 1999 Analysis of TCP Performance over Mobile Ad Hoc Networks *5th ACM International Conference on Mobile Computing and Networking (ACM MobiCom 1999)*, pp. 219–230, Seattle, Wasington, USA.

Hollos D, Karl H and Wolisz A 2004 Regionalizing Global Optimization Algorithms to Improve the Operation of Large Ad Hoc Networks *IEEE Wireless Communications and Networking Conference*, Atlanta, Georgia, USA.

Hong X, Xu K and Gerla M 2002 Scalable Routing Protocols for Mobile Ad Hoc Networks. *IEEE Network* **16**, 11–21.

Howard A, Mataric MJ and Sukhatme GS 2002 An Incremental Self-Deployment Algorithm for Mobile-Sensor Networks. *Autonomous Robots* **13**(2), 113–126.

Hubaux JP, Boudec JYL, Giordano S, Hamdi M, Blazevic L, Buttyan L and Vojnovic M 2000 Towards Mobile Ad-Hoc WANs: Terminodes *IEEE Wireless Communications and Networking Conference (WCNC'2000)*, Chicago, USA.

Hubaux JP, Buttyán L and Capkun S 2001 The Quest for Security in Mobile Ad Hoc Networks *2nd ACM International Symposium on Mobile Ad Hoc Networking and Computing (ACM Mobihoc 2001)*, pp. 146–155. ACM, Long Beach, CA, USA.

Hui J and Culler D 2004 The Dynamic Behavior of a Data Dissemination Protocol for Network Programming at Scale *2nd ACM Conference on Embedded Networked Sensor Systems (ACM SenSys 2004)*, Baltimore, MD, USA.

Huitema C 1996 The Case for Packet Level FEC *IFIP Fifth International Workshop on Protocols for High Speed Networks*, pp. 109–120, Sophia Antipolis, France.

Hurler B, Hof HJ and Zitterbart M 2004 A General Architecture for Wireless Sensor Networks: First Steps *4th International Workshop on Smart Appliances and Wearable Computing*, pp. 442–444, Tokyo, Japan.

Ibriq J and Mahgoub I 2004 Cluster-Based Routing in Wireless Sensor Networks: Issues and Challenges *International Symposium on Performance Evaluation of Computer and Telecommunication Systems (SPECTS'04)*, pp. 759–766, San Jose, California, USA.

IEEE 1999 Wireless LAN Medium Access Control (MAC) and Physical Layer (PHY) Specification IEEE Standard 802.11-1999 edition.

IEEE 2006 Wireless Medium Access Control (MAC) and Physical Layer (PHY) Specifications for Low Rate Wireless Personal Area Networks (WPANs) IEEE Standard 802.15.4-2006.

Intanagonwiwat C, Govindan R and Estrin D 2000 Directed Diffusion: A Scalable and Robust Communication Paradigm for Sensor Networks *6th ACM International Conference on Mobile Computing and Networking (ACM MobiCom 2000)*, pp. 56–67, Boston, MA, USA.

ISO 1995 Open Distributed Processing Reference Model International Standard ISO/IEC IS 10746.

Iwata A, Chiang CC, Pei G, Gerla M and Chen TW 1999 Scalable Routing Strategies for Ad Hoc Wireless Networks. *IEEE Journal on Selected Areas in Communications, Special Issue on Ad-Hoc Networks* **17**(8), 1369–1379.

Jain R 1991 *The Art of Computer Systems Performance Analysis: Techniques for Experimental Design, Measurement, Simulation, and Modeling.* John Wiley and Sons, Inc.

Janeway CA, Walport M and Travers P 2001 *Immunobiology: The Immune System in Health and Disease* 5th edn. Garland Publishing.

Jeong J and Culler D 2004 Incremental Network Programming for Wireless Sensors *First IEEE International Conference on Sensor and Ad Hoc Communications and Networks (IEEE SECON)*.

Johnson D, Stack T, Fish R, Flickinger DM, Stoller L, Ricci R and Lepreau J 2006 Mobile Emulab: A Robotic Wireless and Sensor Network Testbed *25th IEEE Conference on Computer Communications (IEEE INFOCOM 2006)*, Barcelona, Spain.

Johnson DB 1994 Routing in Ad Hoc Networks of Mobile Hosts *Workshop on Mobile Computing Systems and Applications*, pp. 158–163. IEEE, Santa Cruz, CA.

Johnson DB and Maltz DA 1996 Dynamic Source Routing in Ad Hoc Wireless Networks In *Mobile Computing* (ed. Imielinski T and Korth HF) vol. 353 Kluwer Academic Publishers pp. 152–181.

Johnson DB, Hu Y and Maltz DA 2007 The Dynamic Source Routing Protocol (DSR) for Mobile Ad Hoc Networks for IPv4 RFC 4728.

Johnson DB, Maltz DA and Broch J 2001 DSR: The Dynamic Source Routing Protocol for Multi-Hop Wireless Ad Hoc Networks In *Ad Hoc Networking* (ed. Perkins CE) Addison-Wesley pp. 139–172.

Johnson SC 1967 Hierarchical Clustering Schemes. *Psychometrika* **32**(3), 241–254.

Juang P, Oki H, Wang Y, Martonosi M, Peh LS and Rubenstein D 2002 Energy-Efficient Computing for Wildlife Tracking: Design Tradeoffs and Early Experiences with ZebraNet. *ACM SIGOPS Operating Systems Review* **36**(5), 96–107.

Jung B and Sukhatme GS 2001 Cooperative Tracking using Mobile Robots and Environment-Embedded, Network Sensors *2001 International Symposium on Computational Intelligence in Robotics and Automation*, pp. 206–211, Banff, Alberta, Canada.

Jung ES and Vaidya N 2002 A Power Control MAC Protocol for Ad Hoc Networks *8th ACM International Conference on Mobile Computing and Networking (ACM MobiCom 2002)*.

Jung ES and Vaidya NH 2005 A Power Control MAC Protocol for Ad Hoc Networks. *ACM/Kluwer Wireless Networks (WINET)* **11**(1–2), 55–66.

Kahn JM, Katz R and Pister K 1999 Next Century Challenges: Mobile Networking for 'Smart Dust' *5th ACM International Conference on Mobile Computing and Networking (ACM MobiCom 1999)*, pp. 271–278, Seattle, Washington, USA.

Kahn JM, Katz R and Pister K 2000 Emerging Challenges: Mobile Networking for 'Smart Dust'. *Journal of Communications and Networking*.

Karl H and Willig A 2005 *Protocols and Architectures for Wireless Sensor Networks*. John Wiley & Sons Ltd.

Karn P 1990 MACA: A New Channel Access Method for Packet Radio *ARRL/CRRL Amateur Radio 9th Computer Networking Conference*, pp. 134–140, London, Ontario, Canada.

Karp B and Kung HT 2000 GPSR: Greedy Perimeter Stateless Routing for Wireless Networks *6th ACM International Conference on Mobile Computing and Networking (ACM MobiCom 2000)*, pp. 243–254, Boston, Massachusetts, USA.

Karp R, Elson J, Estrin D and Shenker S 2003 Optimal and Global Time Synchronization in Sensornets. Cens technical report, UCLA, Dept. of Computer Science.

Kashiwagi A, Urabe I, Kaneko K and Yomo T 2006 Adaptive Response of a Gene Network to Environmental Changes by Fitness-Induced Attractor Selection. *PLoS ONE* **1**(1), e49.

Kauffman SA 1993 *The Origins of Order: Self-Organization and Selection in Evolution*. Oxford University Press.

Keith EW 2000 *Core Jini* 2nd edn. Prentice Hall.

Kempe D, Kleinberg J and Demers A 2001 Spatial Gossip and Resource Location Protocols. *Journal of the ACM (JACM)* **51**(6), 943–967.

Kennedy J and Eberhart RC 2001 *Swarm Intelligence*. Morgan Kaufmann Publishers.

Kephart JO 1994 A Biologically Inspired Immune System for Computers *4th International Workshop on Synthesis and Simulation of Living Systems*, pp. 130–139. MIT Press, Cambridge, Massachusetts, USA.

Khan M and Misic J 2007 Security in IEEE 802.15.4 Cluster Based Networks In *Security in Wireless Mesh Networks* (ed. Zhang Y, Zheng J and Hu H) Auerbach Publications, CRC Press.

Kim J and Bentley PJ 2001 An Evaluation of Negative Selection in an Artificial Immune System for Network Intrusion Detection *Genetic and Evolutionary Computation Conference (GECCO-2001)*, pp. 1330–1337, San Francisco, CA.

Kim J and Bentley PJ 2002 Towards an Artificial Immune System for Network Intrusion Detection *IEEE Congress on Evolutionary Computation (CEC)*, pp. 1015–1020, Honolulu.

Kim Y, Lee JJ and Helmy A 2004 Modeling and Analyzing the Impact of Location Inconsistencies on Geographic Routing in Wireless Networks. *ACM SIGMOBILE Mobile Computing and Communications Review* **8**(1), 48–60.

Ko YB and Vaidya NH 1998 Location-Aided Routing (LAR) in Mobile Ad Hoc Networks *4th ACM International Conference on Mobile Computing and Networking (ACM MobiCom 1998)*, pp. 66–75, Dallas, Texas, United States.

Ko YB and Vaidya NH 2002 Flooding-Based Geocasting Protocols for Mobile Ad Hoc Networks. *Mobile Networks and Applications* **7**(6), 471–480.

Krieger MJB, Billeter JB and Keller L 2000 Ant-Like Task Allocation and Recruitment in Cooperative Robots. *Nature* **406**, 992–995.

Krishnamachari B, Estrin D and Wicker S 2002 The Impact of Data Aggregation in Wireless Sensor Networks *International Workshop Distributed Event Based System (DEBS)*.

Krüger B and Dressler F 2005 Molecular Processes as a Basis for Autonomous Networking. *IPSI Transactions on Advances Research: Issues in Computer Science and Engineering* **1**(1), 43–50.

Kumar V, Rus D and Singh S 2004 Robot and Sensor Networks for First Responders. *IEEE Pervasive Computing* **3**(4), 24–33.

Kwon TJ and Gerla M 2002 Efficient Flooding with Passive Clustering (PC) in Ad Hoc Networks. *ACM SIGCOMM Computer Communication Review*.

Kyasanur P, Choudhury RR and Gupta I 2006 Smart Gossip: An Adaptive Gossip-based Broadcasting Service for Sensor Networks *3rd IEEE International Conference on Mobile Ad Hoc and Sensor Systems (IEEE MASS 2006)*, pp. 91–100, Vancouver, Canada.

Labella TH and Dorigo M 2004 Efficiency and Task Allocation in Prey Retrieval *First International Workshop on Biologically Inspired Approaches to Advanced Information Technology (Bio-ADIT2004)*, pp. 32–47, Lausanne, Switzerland.

Labella TH and Dressler F 2006 A Bio-Inspired Architecture for Division of Labour in SANETs *1st IEEE/ACM International Conference on Bio-Inspired Models of Network, Information and Computing Systems (IEEE/ACM BIONETICS 2006)*, Cavalese, Italy.

Labella TH and Dressler F 2007 A Bio-Inspired Architecture for Division of Labour in SANETs In *Advances in Biologically Inspired Information Systems: Models, Methods, and Tools* (ed. Dressler F and Carreras I) vol. 69 of *Studies in Computational Intelligence (SCI)* Springer Berlin, Heidelberg, New York pp. 209–228.

Labella TH, Dorigo M and Deneubourg JL 2004 Self-Organised Task Allocation in a Group of Robots *7th International Symposium on Distributed Autonomous Robotic Systems (DARS04)*, Toulouse, France.

Lagoudakis MG, Berhault M, Koenig S, Keskinocak P and Kleywegt AJ 2004 Simple Auctions with Performance Guarantees for Multi-Robot Task Allocation *IEEE/RSJ International Conference on Intelligent Robots and Systems*, Sendai, Japan.

Lahiri K, Dey S, Panigrahi D and Raghunathan A 2002 Battery-Driven System Design: A New Frontier in Low Power Design *2002 Conference on Asia South Pacific Design Automation/VLSI Design*, p. 261.

Lamport L 1978 Time, Clocks, and the Ordering of Events in a Distributed System. *Communications of the ACM* **21**(4), 558–565.

Landsiedel O, Wehrle K and Götz S 2005 Accurate Prediction of Power Consumption in Sensor Networks *Second IEEE Workshop on Embedded Networked Sensors (EmNetS-II)*, Sydney, Australia.

Law AM and Kelton WD 2000 *Simulation, Modeling and Analysis* 3rd edn. McGraw-Hill Book Co.

Le Boudec JY and Sarafijanovic S 2004 An Artificial Immune System Approach to Misbehavior Detection in Mobile Ad-Hoc Networks *First International Workshop on Biologically Inspired Approaches to Advanced Information Technology (Bio-ADIT2004)*, pp. 96–111, Lausanne, Switzerland.

Lee DW, Jun HB and Sim KB 1999 Artificial Immune System for Realization of Cooperative Strategies and Group Behavior in Collective Autonomous Mobile Robots *4th International Symposium on Artificial Life and Robotics*, pp. 232–235.

Lee JJ, Krishnamachari B and Kuo CCJ 2004 Impact of Heterogeneous Deployment on Lifetime Sensing Coverage in Sensor Networks *IEEE Communications Society Conference on Sensor and Ad Hoc Communications and Networks (IEEE SECON 2004)*, pp. 367–376.

Leibnitz K, Wakamiya N and Murata M 2006 Biologically-Inspired Self-Adaptive Multi-Path Routing in Overlay Networks. *Communications of the ACM, Special Issue on Self-Managed Systems and Services* **49**(3), 63–67.

Levis P, Lee N, Welsh M and Culler D 2003 TOSSIM: Accurate and Scalable Simulation of Entire TinyOS Applications *ACM SensSys 2003*.

Levis P, Madden S, Gay D, Polastre J, Szewczyk R, Woo A, Brewer E and Culler D 2004 The Emergence of Networking Abstractions and Techniques in TinyOS *First USENIX/ACM Symposium on Networked Systems Design and Implementation (NSDI 2004)*, San Francisco, CA, USA.

Li Y, Ye W and Heidemann J 2004 Energy and Latency Control in Low Duty Cycle MAC Protocols. Technical report, USC Information Sciences Institute.

Liang B and Haas ZJ 1999 Predictive Distance-Based Mobility Management for PCS Networks *18th IEEE Conference on Computer Communications (IEEE INFOCOM 1999)*.

Liang YC and Smith AE 1999 An Ant System Approach to Redundancy Allocation *IEEE Congress on Evolutionary Computation (CEC)*, pp. 1478–1484, Washington, DC.

Lin CR and Gerla M 1997 Adaptive Clustering for Mobile Wireless Networks. *Selected Areas in Communications* **15**(7), 1265–1275.

Lindsey S and Raghavendra CS 2002 PEGASIS: Power Efficient Gathering in Sensor Information Systems *IEEE Aerospace Conference*, Big Sky, Montana, USA.

Lindsey S, Raghavendra C and Sivalingam K 2001 Data Gathering in Sensor Networks Using the Energy*Delay Metric *15th International Parallel and Distributed Processing Symposium (IPDPS), Workshop on Issues in Wireless Networks and Mobile Computing*, San Francisco, CA, USA.

Lindsey S, Raghavendra C and Sivalingam KM 2002 Data Gathering Algorithms in Sensor Networks Using Energy Metrics. *IEEE Transactions on Parallel and Distributed Systems* **13**(9), 924–935.

Liu B, Brass P, Dousse O, Nain P and Towsley D 2005 Mobility Improves Coverage of Sensor Networks *6th ACM International Symposium on Mobile Ad Hoc Networking and Computing (ACM Mobihoc 2005)*, pp. 300–308, Urbana-Champaign, IL.

Liu H, Wan P, Jia X, Liu X and Yao F 2006 Efficient Flooding Scheme Based on 1-hop Information in Mobile Ad Hoc Networks *25th IEEE Conference on Computer Communications (IEEE INFOCOM 2006)*, Barcelona, Spain.

Lorincz K, Malan DJ, Fulford-Jones TR, Nawoj A, Clavel A, Shnayder V, Mainland G, Welsh M and Moulton S 2004 Sensor Networks for Emergency Response: Challenges and Opportunities. *IEEE Pervasive Computing* **3**(4), 16–23.

Low KH, Leow WK and Ang MH 2005 Autonomic Mobile Sensor Network with Self-Coordinated Task Allocation and Execution. *IEEE Transactions on Systems, Man, and Cypernetics–Part C: Applications and Reviews* **36**(3), 315–327.

Lu C, Blum BM, Abdelzaher TF, Stankovic JA and He T 2002 RAP: A Real-Time Communication Architecture for Large-Scale Wireless Sensor Networks *8th IEEE Real-Time and Embedded Technology and Applications Symposium (RTAS'02)*.

Lu G, Sadagopan N, Krishnamachari B and Goel A 2005 Delay Efficient Sleep Scheduling in Wireless Sensor Networks *24th IEEE Conference on Computer Communications (IEEE INFOCOM 2005)*, Miami, FL, USA.

Luo J and Hubaux JP 2004 A Survey of Inter-Vehicle Communication. Technical report, School of Computer and Communication Sciences, EPFL.

Luo J, Eugster PT and Hubaux JP 2004 Pilot: Probabilistic Lightweight Group Communication System for Ad Hoc Networks. *IEEE Transactions on Mobile Computing* **3**(2), 164–179.

MacQueen J 1967 Some Methods for Classification and Analysis of Multivariate Observations *Berkeley Symposium on Mathematical Statistics and Probability*, pp. 281–297. University of California Press, Berkeley.

Madden SR, Franklin MJ, Hellerstein JM and Hong W 2005 TinyDB: An Acquisitional Query Processing System for Sensor Networks. *ACM Transactions on Database Systems (TODS)* **30**(1), 122–173.

Mainwaring A, Polastre J, Szewczyk R, Culler D and Anderson J 2002 Wireless Sensor Networks for Habitat Monitoring *First ACM Workshop on Wireless Sensor Networks and Applications*, Atlanta, GA, USA.

MANET WG 2006 Mobile Ad-hoc Networks (manet) Charter.

Margi C 2003 A Survey on Networking, Sensor Processing and System Aspects of Sensor Networks. Report, University of California, Santa Cruz.

Martin D, Cheyer A and Moran D 1999 The Open Agent Architecture: A Framework for Building Distributed Software Systems. *Applied Artificial Intelligence* **13**(1/2), 91–128.

Mataric MJ 1995a Designing and Understanding Adaptive Group Behavior. *Adaptive Behavior* **4**(1), 51–80.

Mataric MJ 1995b Issues and Approaches in the Design of Collective Autonomous Agents. *Robotics and Autonomous Systems* **16**(2), 321–331.

Mataric MJ, Sukhatme GS and Ostergaard EH 2003 Multi-Robot Task Allocation in Uncertain Environments. *Autonomous Robots* **14**, 255–263.

Mauve M and Widmer J 2001 A Survey on Position-Based Routing in Mobile Ad-Hoc Networks. *IEEE Network* **15**(6), 30–39.

Meguerdichian S, Koushanfar F, Potkonjak M and Srivastava MB 2001a Coverage Problems in Wireless Ad-hoc Sensor Networks *20th IEEE Conference on Computer Communications (IEEE INFOCOM 2001)*, pp. 1380–1387, Ankorange, Alaska, USA.

Meguerdichian S, Slijepcevic S, Karayan V and Potkonjak M 2001b Localized Algorithms in Wireless Ad-Hoc Networks: Location Discovery and Sensor Exposure *2nd ACM International Symposium on Mobile Ad Hoc Networking and Computing (ACM MobiHoc 2001)*, pp. 106–116, Long Beach, CA, USA.

Melodia T, Pompili D and Akyildiz IF 2006 A Communication Architecture for Mobile Wireless Sensor and Actor Networks *IEEE SECON 2006*, Reston, VA.

Melodia T, Pompili D, Gungor VC and Akyildiz IF 2005 A Distributed Coordination Framework for Wireless Sensor and Actor Networks *6th ACM International Symposium on Mobile Ad Hoc Networking and Computing (ACM Mobihoc 2005)*, pp. 99–110, Urbana-Champaign, Il, USA.

Melodia T, Pompili D, Gungor VC and Akyildiz IF 2007 Communication and Coordination in Wireless Sensor and Actor Networks. *IEEE Transactions on Mobile Computing*.

Mhatre V and Rosenberg C 2004 Design Guidelines for Wireless Sensor Networks: Communication, Clustering and Aggregation. *Elsevier Ad Hoc Networks* **2**(1), 45–63.

Mills DL 1990 On the Accuracy and Stability of Clocks Synchronized by the Network Time Protocol in the Internet System. *ACM Computer Communications Review* **20**(1), 65–75.

Mills DL 1991 Internet Tme Synchronization: The Network Time Protocol. *IEEE/ACM Transactions on Networking (TON)* **39**(10), 1482–1493.

Mills Dl 1992 Network Time Protocol(Version 3) Specification, Implementation and Analysis RFC 1305.

Milnor J 1985 On the Concept of Attractor. *Communications in Mathematical Physics* **99**(2), 177–195.

Milo R, Shen-Orr S, Itzkovitz S, Kashtan N, Chklovskii D and Alon U 2002 Network Motifs: Simple Building Blocks of Complex Networks. *Nature* **298**, 824–827.

Mindbranch 2004 U.S. Wireless Sensing Networks: R&D and Commercialization Activities. Market research report, Fuji-Keizai USA.

Moore GE 1965 Cramming More Components onto Integrated Circuits. *Electronics* **38**, 114–117.

Muraleedharan R and Osadciw LA 2003 Balancing The Performance of a Sensor Network Using an Ant System *37th Annual Conference on Information Sciences and Systems (CISS 2003)*, Baltimore, MD.

Murthy CSR and Manoj BS 2004 *Ad Hoc Wireless Networks*. Prentice Hall PTR.

Murthy S and Garcia-Luna-Aceves JJ 1996 An Efficient Routing Protocol for Wireless Networks. *ACM Mobile Networks and Applications, Special Issue on Routing in Mobile Communication Networks* pp. 83–197.

Narten T and Draves R 2001 Privacy Extensions for Stateless Address Autoconfiguration in IPv6 RFC 3041.

Navas JC and Imielinski T 1997 GeoCast: Geographic Addressing and Routing *3rd ACM International Conference on Mobile Computing and Networking (ACM MobiCom 1997)*, pp. 66–76, Budapest, Hungary.

Nesargi S and Prakash R 2002 MANETconf: Configuration of Hosts in a Mobile Ad Hoc Network *21st IEEE Conference on Computer Communications (IEEE INFOCOM 2002)*, vol. 2, pp. 1059–1068.

Neuman BC 1994 Scale in Distributed Systems In *Readings in Distributed Computing Systems* (ed. Casavant TL and Singhal M) IEEE Computer Society Press Los Alamitos, CA pp. 463–489.

Neumann Jv 1966 *Theory of Self-Reproducing Automata*. University of Illionois Press.

Ni SY, Tseng YC, Chen YS and Sheu JP 1999 The Broadcast Storm Problem in a Mobile Ad Hoc Network *5th ACM International Conference on Mobile Computing and Networking (ACM MobiCom 1999)*, pp. 151–162, Seattle, Washington.

Noury N, Herve T, Rialle V, Virone G and Mercier E 2001 Monitoring Behavior in Home Using a Smart Fall Sensor Network and Position Sensors *1st EMBS Special Topic Conference on Microtechnology in Medicine and Biology*, pp. 607–617. IEEE Computer Society Press.

Olariu S and Stojmenovic I 2006 Design Guidelines for Maximizing Lifetime and Avoiding Energy Holes in Sensor Networks with Uniform Distribution and Uniform Reporting *25th IEEE Conference on Computer Communications (IEEE INFOCOM 2006)*, Barcelona, Spain.

Opp WJ and Sahin F 2004 An Artificial Immune System Approach to Mobile Sensor Networks and Min Detection *IEEE International Conference on Systems, Man and Cybernetics*, vol. 1, pp. 947–952, The Hague, The Netherlands.

Ortiz CL, Rauenbusch TL, Hsu E and Vincent R 2003 Dynamic Resource-bounded Negotiation in Non-Additive Domains In *Distributed Sensor Networks: A Multiagent Perspective* (ed. Lesser V, Ortiz CL and Tambe M) Multiagent Systems, Artificial Societies, and Simulated Organizations Kluwer Acedemic Pubishers Boston pp. 61–107.

Parker LE 1994 ALLIANCE: An Architecture for Fault Tolerant, Cooperative Control of Heterogeneous Mobile Robots *IEEE/RSJ/GI International Conference on Intelligent Robots and Systems (IROS)*, pp. 776–783.

Parker LE 1998 ALLIANCE: An Architecture for Fault Tolerant Multirobot Cooperation. *IEEE Transactions on Robotics and Automation* **14**(2), 220–240.

Passing M and Dressler F 2006 Experimental Performance Evaluation of Cryptographic Algorithms on Sensor Nodes *3rd IEEE International Conference on Mobile Ad Hoc and Sensor Systems (IEEE MASS 2006): 2nd IEEE International Workshop on Wireless and Sensor Networks Security (WSNS'06)*, pp. 882–887, Vancouver, Canada.

Patchipulusu P 2001 *Dynamic Address Allocation Protocols for Mobile Ad Hoc Networks* Master's thesis Texas AM University.

Paul K, Bandyopadhyay S, Mukherjee A and Saha D 1999 Communication-Aware Mobile Hosts in Ad-Hoc Wireless Network *IEEE International Conference on Personal Wireless Communication (ICPWC 1999)*, pp. 83–87, Jaipur, India.

Pawson T 1995 Protein Modules and Signalling Networks. *Nature* **373**(6515), 573–580.

Pei G, Gerla M and Hong X 2000 LANMAR: Landmark Routing for Large Scale Wireless Ad Hoc Networks withGroup Mobility *1st ACM International Symposium on Mobile Ad Hoc Networking and Computing (ACM Mobihoc 2000)*, pp. 64–75, Boston, MA.

Perkins C and Royer E 1999 Ad Hoc On-Demand Distance Vector Routing *2nd IEEE Workshop on Mobile Computing Systems and Applications*, pp. 90–100, New Orleans, LA.

Perkins C, Belding-Royer E and Das S 2003 Ad Hoc On-Demand Distance Vector (AODV) Routing RFC 3561.

Perkins CE 2000 *Ad Hoc Networking* 1st edn. Addison-Wesley Professional.

Perkins CE and Bhagwat P 1994 Highly Dynamic Destination-Sequenced Distance-Vector Routing (DSDV) for Mobile Computers. *Computer Communications Review* pp. 234–244.

Perkins CE, Malinen JT, Wakikawa R, Belding-Royer EM and Sun Y 2001 IP Address Autoconfiguration for Ad Hoc Networks draft-ietf-manet-autoconf-01.txt.

Pleisch S, Balakrishnan M, Birman K and Renesse Rv 2006 MISTRAL: Efficient Flooding in Mobile Adhoc Networks *7th ACM International Symposium on Mobile Ad Hoc Networking and Computing (ACM MobiHoc 2006)*, Florence, Italy.

Prehofer C and Bettstetter C 2005 Self-Organization in Communication Networks: Principles and Design Paradigms. *IEEE Communications Magazine* **43**(7), 78–85.

Priyantha NB, Balakrishnan H, Demaine ED and Teller S 2005 Mobile-Assisted Localization in Wireless Sensor Networks *24th IEEE Conference on Computer Communications (IEEE INFOCOM 2005)*, Miami, FL, USA.

Qi H, Wang X, Iyengar SS and Chakrabarty K 2001 Multisensor Data Fusion in Distributed Sensor Networks Using Mobile Agents *International Conference on Information Fusion*, pp. 11–16.

Raghunathan V, Schurgers C, Park S and Srivastava MB 2002 Energy-Aware Wireless Microsensor Networks. *IEEE Signal Processing Magazine* **19**(2), 40–50.

Rahimi M, Shah H, Sukhatme GS, Heidemann J and Estrin D 2003 Studying the Feasibility of Energy Harvesting in a Mobile Sensor Network *IEEE International Conference on Robotics and Automation (ICRA'03)*, pp. 19–24.

Rajagopalan R and Varshney PK 2006 Data-Aggregation Techniques in Sensor Networks: A Survey. *IEEE Communication Surveys and Tutorials* **8**(4), 48–63.

Rajaraman R 2002 Topology Control and Routing in Ad Hoc Networks: A Survey. *ACM SIGACT News* **33**(2), 60–73.

Rauenbusch TW 2002 The Mediation Algorithm for Real Time Negotiation *First International Joint Conference on Autonomous Agents and Multiagent Systems (AAMAS'02), Poster Session*, pp. 1139–1140, Bologna, Italy.

Rauenbusch TW 2004 *Measuring Information Transmission for Team Decision Making* Ph.D. thesis Hardvard University.

Reichardt D, Miglietta M, Moretti L, Morsink P and Schulz W 2002 CarTALK 2000: Safe and Comfortable Driving Based Upon Inter-Vehicle-Communication *IEEE Intelligent Vehicle Symposium*, vol. 2, pp. 545–550, Versailles.

Resnick M 1997 *Turtles, Termites, and Traffic Jams: Explorations in Massively Parallel Microworlds*. MIT Press.

Römer K 2001 Time Synchronization in Ad Hoc Networks *2nd ACM International Symposium on Mobile Ad Hoc Networking and Computing (ACM Mobihoc 2001)*, pp. 173–182, Long Beach, USA.

Roundy S, Wright PK and Rabaey JM 2003 *Energy Scavenging for Wireless Sensor Networks*. Springer-Verlag.

Royer E and Perkins C 2000 An Implementation Study of the AODV Routing Protocol *IEEE Wireless Communications and Networking Conference*, Chicago, IL.

Sadagopan N, Krishnamachari B and Helmy A 2003 The ACQUIRE Mechanism for Efficient Querying in Sensor Networks *First International Workshop on Sensor Network Protocol and Applications*, Anchorage, Alaska.

Salhieh A and Schwiebert L 2002 Power Aware Metrics for Wireless Sensor Networks *14th IASTED Conference on Parallel and Distributed Computing and Systems (PDCS 2002)*, pp. 326–331, Boston, Massachusetts.

Sankarasubramaniam Y, Akan OB and Akyildiz IF 2003 ESRT: Event-to-Sink Reliable Transport in Wireless Sensor Networks *4th ACM International Symposium on Mobile Ad Hoc Networking and Computing (ACM Mobihoc 2003)*, pp. 177–188, Annapolis, Maryland, USA.

Sasson Y, Cavin D and Schiper A 2002 Probabilistic Broadcast for Flooding in Wireless Mobile Ad Hoc Networks *IEEE WCNC 2003*.

Schröder-Preikschat W, Kapitza R, Kleinöder J, Felser M, Karmeier K, Labella TH and Dressler F 2007 Robust and Efficient Software Management in Sensor Networks *2nd IEEE/ACM International Conference on Communication System Software and Middleware (IEEE/ACM COMSWARE 2007): 2nd IEEE/ACM International Workshop on Software for Sensor Networks (IEEE/ACM SensorWare 2007)*. IEEE, Bangalore, India.

Schulzrinne H, Wu X, Sidiroglou S and Berger S 2003 Ubiquitous Computing in Home Networks. *IEEE Communications Magazine* **41**(11), 128–135.

Scott DJ and Yasinsac A 2004 Dynamic Probabilistic Retransmission in Ad Hoc Networks *International Conference on Wireless Networks (ICWN '04)*, pp. 158–164, Las Vegas, Nevada, USA.

Shah GA, Bozyigit M, Akan OB and Baykal B 2006 Real-Time Coordination and Routing in Wireless Sensor and Actor Networks *6th International Conference on Next Generation Teletraffic and Wired/Wireless Advanced Networking (NEW2AN 2006)*, St Petersburg, Russia.

Shah RC, Roy S, Jain S and Brunette W 2003 Data MULEs: Modeling a Three-tier Architecture for Sparse Sensor Networks *First IEEE International Workshop on Sensor Network Protocols and Applications (SNPA'03)*, pp. 30–41.

Shah RC, Wiethölter S and Wolisz A 2005a When Does Opportunistic Routing Make Sense? *First International Workshop on Sensor Networks and Systems for Pervasive Computing (PerSeNS 2005)*, Kauai Island, USA.

Shah RC, Wiethölter S, Wolisz A and Rabaey JM 2005b Modelling and Analysis of Opportunistic Routing in Low Traffic Scenarios *3rd International Symposium on Modeling and Optimization in Mobile, Ad Hoc, and Wireless Networks (WiOpt'05)*, Trento, Italy.

Shen X, Chen J, Wang Z and Sun Y 2006 Grid Scan: A Simple and Effective Approach for Coverage Issue in Wireless Sensor Networks *IEEE International Conference on Communications (IEEE ICC 2006)*, Istanbul, Turkey.

Singh S and Kurose J 1994 Electing 'Good' leaders. *Journal of Parallel and Distributed Computing* **21**(2), 184–201.

Singh S, Woo M and Raghavendra CS 1998 Power-Aware Routing in Mobile Ad Hoc Networks *4th ACM International Conference on Mobile Computing and Networking (ACM MobiCom 1998)*, pp. 181–190, Dallas, Texas, United States.

Singh SPN and Thayer SM 2001 Immunology Directed Methods for Distributed Robotics: A Novel, Immunity-Based Architecture for Robust Control and Coordination *SPIE: Mobile Robots XVI*, vol. 4573, Newton, MA.

Solis I and Obraczka K 2006 In-Network Aggregation Trade-Offs for Data Collection in Wireless Sensor Networks. *International Journal of Sensor Networks (IJSNET)* **1**(3/4), 200–212.

Sommer C and Dressler F 2007 The DYMO Routing Protocol in VANET Scenarios *66th IEEE Vehicular Technology Conference (VTC2007-Fall)*. IEEE, Baltimore, Maryland, USA.

Steinmetz R and Wehrle K 2005 *Peer-to-Peer Systems and Applications* vol. LNCS 3485. Springer.

Su W and Akyildiz IF 2005 Time-Diffusion Synchronization Protocol for Wireless Sensor Networks. *IEEE/ACM Transactions on Networking (TON)* **13**(2), 384–397.

Sun Y and Belding-Royer EM 2003 Dynamic Address Configuration in Mobile Ad Hoc Networks. Technical report, Department of Computer Science, UCSB.

Sun Y and Belding-Royer EM 2004 A Study of Dynamic Addressing Techniques in Mobile Ad Hoc Networks. *Wireless Communications and Mobile Computing* **4**(3), 315–329.

Szewczyk R, Osterweil E, Polastre J, Hamilton M, Mainwaring A and Estrin D 2004 Habitat Monitoring with Sensor Networks. *Communications of the ACM, Special Issue on Wireless Sensor Networks* **47**(6), 34–40.

Talucci F and Gerla M 1997 MACA-BI (MACA By Invitation): A Wireless MAC Protocol for High Speed Ad Hoc Networking *IEEE 6th International Conference on Universal Personal Communications (ICUPC'97)*, vol. 2, pp. 913–917, San Diego, CA, USA.

Tanenbaum AS, Gamage C and Crispo B 2006 Taking Sensor Networks from the Lab to the Jungle. *IEEE Computer* **39**(8), 98–100.

Tannenbaum AS and van Steen M 2002 *Distributed Systems: Principles and Paradigms*. Prentice-Hall, Inc.

Thakoor S, Chahl J, Hine B and Zornetzer S 2004 BEES: Exploring Mars with Bioinspired Technologies. *IEEE Computer* **37**(9), 38–47.

Thakoor S, Chahl J, Srinivasan MV, Young L, Werblin F, Hine B and Zornetzer S 2002 Bioinspired Engineering of Exploration Systems for NASA and DoD. *Artificial Life Journal* **8**(4), 357–369.

Theraulaz G and Bonbeau E 1999 A Brief History of Stigmergy. *Artificial Life Journal* **5**(2), 97–116.

Toh CK 2001 Maximum Battery Life Routing to Support Ubiquitous Mobile Computing in Wireless Ad Hoc Networks. *IEEE Communications Magazine* **39**(6), 138–147.

Toner S and O'Mahony D 2003 Self-Organising Node Address Management in Ad Hoc Networks *8th IFIP International Conference on Personal Wireless Communications (PWC 2003)*, vol. LNCS 2775, pp. 476–483. Springer, Venice, Italy.

Truchat S, Fuchs G, Meyer S and Dressler F 2006 An Adaptive Model for Reconfigurable Autonomous Services Using Profiling. *International Journal of Pervasive Computing and Communications (JPCC), Special Issue on Pervasive Management* **2**(3), 247–259.

Turing AM 1952 The Chemical Basis for Morphogenesis. *Philosophical Transactions of the Royal Society of London. Series B, Biological Sciences* **237**(641), 37–72.

Vaidya NH 2002 Weak Duplicate Address Detection in Mobile Ad Hoc Networks *3rd ACM International Symposium on Mobile Ad Hoc Networking and Computing (ACM Mobihoc 2002)*, pp. 206–216, Lausanne, Switzerland.

Vaidyanathan K, Sur S, Narravula S and Sinha P 2004 Data Aggregation Techniques in Sensor Networks. Technical report, Ohio State University.

von Foerster H 1960 On Self-Organizing Systems and their Environments In *Self-Organizing Systems* (ed. Yovitts MC and Cameron S) Pergamon Press pp. 31–50.

von Foerster H and Zopf GW 1962 *Principles of Self-Organization*. Pergamon Press.

Waldo J 2000 *The Jini Specification* 2nd edn. Prentice Hall.

Wang G, Cao G and Porta TL 2004 Movement-Assisted Sensor Deployment *23rd IEEE Conference on Computer Communications (IEEE INFOCOM 2004)*, Hong Kong.

Wang L and Xiao Y 2006 A Survey of Energy-Efficient Scheduling Mechanisms in Sensor Networks. *Mobile Networks and Applications* **11**(5), 723–740.

Wang M and Suda T 2001 The Bio-Networking Architecture: A Biologically Inspired Approach to the Design of Scalable, Adaptive, and Survivable/Available Network Applications *1st IEEE Symposium on Applications and the Internet (SAINT)*, San Diego, CA, USA.

Wang W, Srinivasan V and Chua KC 2005 Using Mobile Relays to Prolong the Lifetime of Wireless Sensor Networks *11th ACM International Conference on Mobile Computing and Networking (ACM MobiCom 2005)*, pp. 270–283, Cologne, Germany.

Warneke B, Last M, Liebowitz B and Pister K 2001 Smart Dust: Communicating with a Cubic-Millimeter Computer. *Computer* **34**(1), 44–51.

Warneke BA, Scott MD, Leibowitz BS, Zhou L, Bellew CL, Chediak JA, Kahn JM, Boser BE and Pister KS 2002 An Autonomous 16mm^3 Solar-Powered Node for Distributed Wireless Sensor Networks *IEEE International Conference on Sensors 2002*, Orlando, FL.

Webb B 2000 What Does Robotics Offer Animal Behaviour? *Animal Behavior* **60**(5), 545–558.

Weiser M 1991 The Computer for the 21st Century. *Scientific American* **265**(3), 94–104.

Weiser M 1993a Some Computer Science Issues in Ubiquitous Computing. *Communications of the ACM* **36**(7), 75–84.

Weiser M 1993b Ubiquitous Computing. *IEEE Computer* **26**(10), 71–72.

Weng G, Bhalla US and Iyengar R 1999 Complexity in Biological Signaling Systems. *Science* **284**(5411), 92–96.

Weniger K 2003 Passive Duplicate Address Detection in Mobile Ad Hoc Networks *IEEE Wireless Communications and Networking Conference (WCNC)*, New Orleans, USA.

Weniger K 2005 PACMAN: Passive Autoconfiguration for Mobile Ad Hoc Networks. *IEEE Journal on Selected Areas in Communications (JSAC)* **23**(3), 507–519.

Weniger K and Zitterbart M 2002 IPv6 Autoconfiguration in Large Scale Mobile Ad-Hoc Networks *European Wireless 2002*, pp. 142–148, Florence, Italy.

Weniger K and Zitterbart M 2004 Address Autoconfiguration in Mobile Ad Hoc Networks: Current Approaches and Future Directions. *IEEE Network Magazine, Special Issue on Ad Hoc Networking: Data Communications and Topology Control* **18**(4), 6–11.

Whitesides GM and Ismagilov RF 1999 Complexity in Chemistry. *Science* **284**(5411), 89–92.

Williams B and Camp T 2002 Comparison of Broadcasting Techniques for Mobile Ad Hoc Networks *3rd ACM international Symposium on Mobile Ad Hoc Networking and Computing (ACM Mobihoc 2002)*, pp. 194–205, Lausanne, Switzerland.

Winfree AT 1972 Spiral Waves of Chemical Activity. *Science* **175**(4022), 634–636.

Wischhof L, Ebner A and Rohling H 2005 Information Dissemination in Self-Organizing Intervehicle Networks. *IEEE Transactions on Intelligent Transportation Systems* **6**(1), 90–101.

Wischhof L, Ebner A, Rohling H, Lott M and Halfmann R 2003 SOTIS: A Self-Organizing Traffic Information System *57th IEEE Vehicular Technology Conference (VTC2003-Spring)*, Jeju, South Korea.

Wisniewski S 2004 *Wireless and Cellular Networks*. Prentice Hall.

Wolfram S 2002 *A New Kind of Science*. Wolfram Media.

Wu J and Yan S 2005 SMART: A Scan-Based Movement-Assisted Sensor Deployment Method in Wireless Sensor Networks *24th IEEE Conference on Computer Communications (IEEE INFOCOM 2005)*, Miami, FL, USA.

Xing G, Wang X, Zhang Y, Lu C, Pless R and Gill C 2005 Integrated Coverage and Connectivity Configuration for Energy Conservation in Sensor Networks. *ACM Transactions on Sensor Networks (TOSN)* **1**(1), 36–72.

Xu Y, Heidemann J and Estrin D 2001 Geography-Informed Energy Conservation for Ad Hoc Routing *7th ACM International Conference on Mobile Computing and Networking (ACM MobiCom 2001)*, Rome, Italy.

Yao Y and Gehrke J 2002 The Cougar Approach to In-Network Query Processing in Sensor Networks. *ACM SIGMOD Record* **31**(3), 9–18.

Yao Z and Dressler F 2007 Dynamic Address Allocation for Management and Control in Wireless Sensor Networks *40th Hawaii International Conference on System Sciences (HICSS-40)*, p. 292b. IEEE, Waikoloa, Big Island, Hawaii.

Yao Z, Lu Z, Marquardt H, Fuchs G, Truchat S and Dressler F 2006 On-Demand Software Management in Sensor Networks Using Profiling Techniques *ACM Second International Workshop on Multi-Hop Ad Hoc Networks: From Theory to Reality 2006 (ACM REALMAN 2006), Demo Session*, pp. 113–115, Florence, Italy.

Yates FE, Garfinkel A, Walter DO and Yates GB 1987 *Self-Organizing Systems: The Emergence of Order*. Plenum Press.

Ye W and Heidemann J 2003 Medium Access Control in Wireless Sensor Networks. Technical report, USC Information Sciences Institute.

Ye W, Heidemann J and Estrin D 2002 An Energy-Efficient MAC Protocol for Wireless Sensor Networks *21st International Annual Joint Conference of the IEEE Computer and Communications Societies (INFOCOM)*, vol. 3, pp. 1567–1576, New York, NY, USA.

Ye W, Heidemann J and Estrin D 2004 Medium Access Control with Coordinated Adaptive Sleeping for Wireless Sensor Networks. *IEEE/ACM Transactions on Networking (TON)* **12**(3), 493–506.

Yeger-Lotem E, Sattath S, Kashtan N, Itzkovitz S, Milo R, Pinter RY, Alon U and Margalit H 2004 Network Motifs in Integrated Cellular Networks of Transcription regulation and Protein–Protein Interaction. *PNAS* **101**(16), 5934–5939.

Younis M, Akkaya K and Kunjithapatham A 2003 Optimization of Task Allocation in a Cluster-Based Sensor Network *8th IEEE International Symposium on Computers and Communications*, pp. 329–340, Kemer-Antalya, Turkey.

Younis O and Fahmy S 2004 HEED: A Hybrid, Energy-Efficient, Distributed Clustering Approach for Ad-hoc Sensor Networks. *IEEE Transactions on Mobile Computing* **3**(4), 366–379.

Yu Y, Govindan R and Estrin D 2001 Geographical and Energy Aware Routing: A Recursive Data Dissemination Protocol for Wireless Sensor Networks. Technical report, UCLA Computer Science Department.

Zambonelli F, Gleizes MP, Mamei M and Tolksdorf R 2004 Spray Computers: Frontiers of Self-Organization for Pervasive Computing *13th IEEE International Workshops on Enabling Technologies: Infrastructure for Collaborative Enterprises (WETICE'04)*, pp. 403–408.

Zhang Q and Agrawal DP 2005 Dynamic Probabilistic Broadcasting in MANETs. *Journal of Parallel and Distributed Computing* **65**(2), 220–233.

Zhou Z, Das S and Gupta H 2005 Fault Tolerant Connected Sensor Cover with Variable Sensing and Transmission Ranges *Second Annual IEEE Communications Society Conference on Sensor and Ad Hoc Communications and Networks (IEEE SECON 2005)*, pp. 594–604.

Zimmer C 1999 Complex Systems: Life After Chaos. *Science* **284**(5411), 83–86.

Zou Y and Chakrabarty K 2003 Sensor Deployment and Target Localization Based on Virtual Forces *22nd IEEE Conference on Computer Communications (IEEE INFOCOM 2003)*, pp. 1293–1303, San Franciso, CA, USA.

Zykov VS, Mikhailov AS and Müller SC 1998 Wave Instabilities in Excitable Media with Fast Inhibitor Diffusion. *Physical Review Letters* **81**(13), 2811–2814.

Index